AF307914

ISNM 105:
International Series of Numerical Mathematics
Internationale Schriftenreihe zur Numerischen Mathematik
Série Internationale d'Analyse Numérique
Vol. 105

Edited by
K.-H. Hoffmann, München; H. D. Mittelmann, Tempe;
J. Todd, Pasadena

Springer Basel AG

Numerical Methods in Approximation Theory, Vol. 9

Edited by

D. Braess
L. L. Schumaker

Springer Basel AG

Editors

Prof. Dr. Dietrich Braess
Fakultät und Institut
für Mathematik
Ruhr-Universität Bochum
Universitätsstr. 150, Geb. NA
D–W–4630 Bochum 1
Germany

Prof. Dr. Larry L. Schumaker
Stevenson Professor
Dept. of Mathematics
Vanderbilt University
1326 Stevenson Center
Nashville, TN 37240
USA

A CIP catalogue record for this book is available from the Library of Congress,
Washington D.C., USA

Deutsche Bibliothek Cataloging-in-Publication Data

Numerical methods of approximation theory. – Basel ; Boston ; Berlin : Birkhäuser.
  Bis Vol. 8 mit dem Parallelt.: Numerische Methoden der Approximationstheorie.
  Bis Bd. 4 u.d.T.: Numerische Methoden der Approximationstheorie
NE: Tagung über Numerische Methoden der Approximationstheorie;
  Numerische Methoden der Approximationstheorie
Vol. 9. Ed. by D. Braess ; L. L. Schumaker. – 1992
  (International series of numerical mathematics ; Vol. 105)
  ISBN 978-3-0348-9702-0   ISBN 978-3-0348-8619-2 (eBook)
  DOI 10.1007/978-3-0348-8619-2
NE: Braess, Dietrich [Hrsg.]; GT

© 1992 Springer Basel AG
Originally published by Birkhäuser Verlag Basel in 1992
Softcover reprint of the hardcover 1st edition 1992
Printed from the authors' camera-ready manuscripts on acid-free paper
ISBN 978-3-0348-9702-0

Lothar Collatz 6. 7. 1910 – 26. 9. 1990

CONTENTS

Preface ix

Contributors xiii

Blending Approximations with Sine Functions
   G. Baszenski, F.-J. Delvos, and S. Jester 1

Quasi-interpolation in the Absence of Polynomial Reproduction
   R. K. Beatson and W. A. Light 21

Estimating the Condition Number for Multivariate Interpolation
Problems
   P. Binev and K. Jetter 41

Wavelets on a Bounded Interval
   Charles K. Chui and Ewald Quak 53

Quasi-Kernel Polynomials and Convergence Results
for Quasi-Minimal Residual Iterations
   Roland W. Freund 77

Rate of Approximation of Weighted Derivatives
by Linear Combinations of SMD Operators
   Margareta Heilmann 97

Approximation by Multivariate Splines:
an Application of Boolean Methods
   Rong-Qing Jia 117

$L^{m,\ell,s}$–Splines in $\mathbb{R}^d$
   A. Le Méhauté and A. Bouhamidi 135

Constructive Multivariate Approximation via Sigmoidal Functions
with Applications to Neural Networks
   Burkhard Lenze 155

Spline-Wavelets of Minimal Support
  T. Lyche and K. Mørken 177

Necessary Conditions for Local Best Chebyshev Approximations
by Splines with Free Knots
  Bernd Mulansky 195

C^1 Interpolation on Higher-Dimensional Analogs
of the 4-Direction Mesh
  A. Neff and J. Peters 207

Tabulation of Thin Plate Splines on a Very Fine
Two-Dimensional Grid
  M. J. D. Powell 221

The L_2-Approximation Orders of Principal Shift-Invariant Spaces
Generated by a Radial Basis Function
  Amos Ron 245

A Multi–Parameter Method for Nonlinear Least–Squares
Approximation
  R. Schaback 269

Analog VLSI Networks
  W. Schempp 285

Converse Theorems for Approximation on Discrete Sets II
  G. Schmeisser 301

A Dual Method for Smoothing Histograms using Nonnegative C^1-Splines
  Jochen W. Schmidt 317

Segment Approximation By Using Linear Functionals
  Manfred Sommer 331

Construction of Monotone Extensions to Boundary Functions
  Cornelis Traas 347

PREFACE

This book is the official proceedings of a conference on *Numerical Methods in Approximation Theory* which was held at the Mathematisches Forschungs-institut in Oberwolfach during the week of November 24–30, 1991. It contains refereed and edited papers by 20 of the 49 participants.

The book is dedicated to the memory of Prof. Lothar Collatz who maintained a long and active interest in numerical approximation. It is the ninth in a series of volumes published by Birkhäuser resulting from conferences on the subject held at Oberwolfach, and co-organized by Prof. Collatz.

We now briefly describe the contents of the book. The paper of BASZEN-SKI, DELVOS and JESTER deals with blending using sine double series expansions of functions defined on the unit square. In addition to giving explicit error estimates for partial sums and for interpolating sine polynomials, they also show that Boolean sums yield almost the same asymptotic error estimates as the conventional tensor-product approach, but with a reduced number of terms.

The paper of BEATSON and LIGHT discusses approximation by quasi-interpolants which are sums of scaled translates of a one-parameter family of functions. They do not require reproduction of low degree polynomials, but nevertheless are able to give error bounds and analyze quasi-interpolation based on Gaussians and exponentials.

BINEV and JETTER deal with multivariate interpolation using shifts of a single basis function. They treat both gridded data and scattered data. As examples, they consider box splines and certain radial basis functions.

The subject of the paper by CHUI and QUAK is the currently very hot topic of wavelets. The aim of their paper is to present two different approaches to multiresolution analysis on a bounded interval. In particular, they explicitly construct certain (non-orthogonal) wavelets in this case.

The contribution of FREUND is a continuation of his earlier work on quasi-kernel polynomials and their connection with the so-called quasi-minimal residual algorithm (QMR) for solving general nonsingular non-Hermitian linear systems. In particular, he derives bounds on the norms of such polynomials, and uses them to obtain convergence theorems for the QMR method and certain of its variants.

The Durrmeyer modification of the Szász-Mirakjan operators is the subject of the paper by HEILMANN. The rate of simultaneous approximation by linear combinations is related to the Ditzian-Totik modulus of smoothness, and both direct and inverse theorems are established.

JIA investigates multivariate smooth splines on nonuniform rectangular grids, and develops a general theory of Boolean methods in such a way that it can be applied to noncommutative operators. Based on this theory, an explicit quasi-interpolant is constructed so that it gives rise to an efficient scheme of approximation by multivariate smooth splines. This scheme is shown to achieve the optimal rate of approximation.

LE MÉHAUTÉ and BOUHAMIDI introduce a space of splines which are a natural generalization of the thin plate splines of Duchon, and investigate their properties. As an application, they characterize thin plate splines under tension.

In the paper of LENZE, sigmoidal functions are used to generate approximation operators for multivariate functions of bounded variation. He starts with Lebesgue-Stieltjes type convolution operators, and then uses numerical quadrature to pass to point-evaluation operators, and to give local and global approximation results for them. He also treats an interesting application to neural networks with one hidden layer consisting of so-called sigma-pi units.

The paper of LYCHE and MØRKEN is intimately connected with wavelets. In particular, they show how to determine a basis for the orthogonal complement of certain spline spaces in related larger ones.

Chebyshev approximation of real continuous functions from the class of polynomial splines of degree n with at most k free knots is discussed in the paper of MULANSKY. The analysis is based on introducing the notion of an extended tangent cone which also contains discontinuous splines. An improved necessary condition (which can also be formulated as an alternant criterion) for local best approximations is derived.

NEFF and PETERS give sharp necessary and sufficient conditions on data at the vertices of a certain 4-direction mesh which allow interpolation of data by m-variate, C^1 piecewise polynomials of degree $m+1$. They also show that for degree $m+2$ and higher, values and normals at the vertices can be stably interpolated, and exhibit a unit-norm C^2 Lagrange function for each vertex.

In his paper, POWELL presents an algorithm for evaluating a thin plate spline at all lattice points of a very fine square grid (with as many as 100 million points). He shows that the amount of work required by the algorithm is bounded by a small constant multiple of the number of grid points plus a constant multiple of $n\epsilon^{-1/3}|\log h|$, where ϵ is a given tolerance on the calculated values of the spline, and where h is the mesh size of the fine grid. In addition, he gives numerical results, and analyzes the errors of the subtabulation procedures.

RON considers approximating from the L_2-closure of a set of finite linear combinations of the shifts of a radial basis function, and gives a thorough analysis of the least-squares approximation orders from such spaces. The results apply to polyharmonic splines, multiquadrics, the Gaussian kernel and other functions, and include the derivation of spectral orders. For stationary refinements it is shown that the saturation class is trivial, i.e., no non-zero function in the underlying Sobolev space can be approximated to a better rate.

The paper of SCHABACK deals with the discrete nonlinear least-squares approximation problem of minimizing $\sum_{j=1}^{m} f_j^2(x)$ for m smooth functions $f_j : \mathbb{R}^n \to \mathbb{R}$. A numerical method is proposed which first minimizes each f_j separately, and then applies a penalty strategy to gradually force the different minimizers to coalesce. An application to discrete rational approximation is discussed, and numerical examples are given.

The retina forms a multilayered precortical structure which collects and preprocesses the information that reaches the visual cortex. To simulate neural network computation by the analog VLSI implementation technology, an analog model of the first stages of retinal processing has been constructed on a single silicon chip by CMOS VLSI circuitry and applied to machine vision. The purpose of the paper of SCHEMPP is to study the exactly solvable hexagonal resistive networks which model the horizontal cell layer of the retina. An overview of free-space multilayer architectures of hybrid optoelectronic interconnection networks is also given.

Converse theorems describe regularity properties of a function in terms of the speed of convergence of its best approximations on an interval. The paper of SCHMEISSER deals with such results for approximation in the L^p norm ($p \in [1, +\infty)$) by polynomials, trigonometric polynomials and entire functions of exponential type. He extends some of the known results to approximation on a sequence of discrete sets instead of an interval. The main tools are upper estimates of an L^p norm in terms of an l^p norm for the three classes of approximating functions under consideration.

SCHMIDT investigates smoothing histograms under positivity constraints. The approach is to minimize a certain objective function over a class of quadratic C^1-splines. This leads to finite dimensional programming problems with a partially separable structure which can be solved efficiently via dualization.

Properties of linear functionals which annihilate finite dimensional Haar or weak Chebyshev subspaces of $C[a, b]$ are studied in the paper of SOMMER. It is shown that such functionals can be used to compute optimal sets of knots for piecewise approximation of certain functions in $C[a, b]$ with respect to best L_∞- and L_1-approximation.

Given monotonically increasing, smooth, univariate functions along the edges of the unit square, TRAAS studies the problem of constructing an extension F to the whole square which is monotone and of class C^1. He presents a nonlinear method which defines F in terms of a set of level lines, each of which is represented as a cubic Bézier curve. As the level changes, the corresponding control points shift along trajectories which contain appropriate kinks.

In conclusion, the editors would like to thank Professor Martin Barner, the director of the Forschungsinstitut, for making this conference possible, and Birkhäuser-Verlag for agreeing to publish the proceedings in its ISNM series. We would also like to acknowledge the assistance of Gerda Schumaker and Sonya Stanley who assisted in preparing the camera-ready manuscript.

Dietrich Braess, Bochum, Germany
Larry L. Schumaker, Nashville, TN
August 1, 1992

CONTRIBUTORS

GÜNTER BASZENSKI, *Fachhochschule Dortmund, Fachbereich Nachrichten-technik, Sonnenstraße 96–100, 4600 Dortmund 1, Germany* [guenter.baszenski@ruba.rz.ruhr-uni-bochum.dbp.de]

RICK BEATSON, *Dept. of Mathematics, Univ. of Canterbury, Canterbury, New Zealand* [rkb@math.canterbury.ac.nz]

PETER BINEV, *Dept. of Mathematics, University of Sofia, Sofia 1156, Bulgaria*

A. BOUHAMIDI, *Département de Mathématiques et d'Informatique, Faculté des Sciences. Université de Nantes, 2 rue de la Houssinière, 44072 Nantes Cedex 03, France*

CHARLES K. CHUI, *Center for Approximation Theory, Texas A&M University, College Station, Texas 77843-3368* [cchui@tamu.edu]

FRANZ-JÜRGEN DELVOS, *Lehrstuhl für Mathematik I, Universität Siegen, Hölderlinstraße 3, 5900 Siegen, Germany*

ROLAND W. FREUND, *AT&T Bell Laboratories, 600 Mountain Avenue, Room 2C-420, P.O. Box 636, Murray Hill, NJ 07974-0636* [freund@research.att.com]

MARGARETA HEILMANN, *University of Dortmund, Vogelpothsweg 87, Dortmund, Germany* [uma019@ddohrz11.bitnet]

STEPHANIE JESTER, *Fakultät für Wirtschaftswissenschaften, Ruhr-Universität Bochum, Universitätsstraße 150, 4630 Bochum 1, Germany*

KURT JETTER, *Dept. of Mathematics, University of Duisburg, 4100 Duisburg, Germany* [hn277je@unidui.uni-duisburg.de]

RONG-QING JIA, *Department of Mathematics, University of Alberta, Edmonton, Canada T6G 2G1* [jia@xihu.math.ualberta.ca]

ALAIN LE MÉHAUTÉ, *Département de Mathématiques et d'Informatique, Faculté des Sciences. Université de Nantes, 2 rue de la Houssinière, 44072 Nantes Cedex 03, France* [alm@cicb.fr]

BURKHARD LENZE, *FernUniversität Hagen, Fb. Mathematik, D-5800 Hagen, Germany* [ma112@dhafeu11.bitnet]

WILL LIGHT, *Dept. of Mathematics, University of Leicester, Leicester, England* [pwl@leicester.ac.uk]

TOM LYCHE, *Institutt for Informatikk, University of Oslo, P.O.Box 1080, Blindern, 0316 Oslo 3, Norway* [tom@ifi.uio.no]

KNUT MØRKEN, *Institutt for Informatikk, University of Oslo, P.O.Box 1080, Blindern, 0316 Oslo 3, Norway* [knutm@ifi.uio.no]

BERND MULANSKY, *Institut für Numerische Mathematik, Technische Universität, Mommsenstrasse 13, D-8027 Dresden, Germany* [mulansky@urzdfn.mathematik.tu-dresden.dbp.de]

ANDY NEFF, *IBM – TJ Watson Research Center, PO Box 218, Yorktown Heights, NY 10598* [caneff@watson.ibm.com]

JÖRG PETERS, *Department of the Mathematical Sciences, Rensselaer Polytechnic Institute, Troy, NY 12180* [jorg@cs.rpi.edu]

MIKE POWELL, *Department of Applied Mathematics, and Theoretical Physics, University of Cambridge, Silver Street, Cambridge CB3 9EW, England* [mjdp@amtp.cam.ac.uk]

EWALD QUAK, *Center for Approximation Theory, Texas A&M University, College Station, Texas 77843-3368* [egq4460@tamvenus.bitnet]

AMOS RON, *Computer Sciences Department, University of Wisconsin, 1210 W. Dayton St., Madison, WI 53706* [amos@cs.wisc.edu]

ROBERT SCHABACK, *Georg–August–Universität Göttingen, Lotzestrasse 16–18, D–3400 Göttingen, Germany* [schaback@namu01.gwdg.de]

WALTER SCHEMPP, *Lehrstuhl für Mathematik I, Hölderlinstrasse 3, D-5900 Siegen, Germany* [schempp@hrz.uni-siegen.dbp.de]

GERHARD SCHMEISSER, *Mathematisches Inst., Univ. Erlangen-Nürnberg, Bismarckstrasse $1\frac{1}{2}$, D-85 Erlangen, Germany* [mi@cnve.rrze.uni-erlangen.dbp.de]

JOCHEN W. SCHMIDT, *Institut für Numerische Mathematik, Technische Universität, D-O-8027 Dresden, Germany* [numath@urzdfn.mathematik.tu-dresden.dbp.de]

MANFRED SOMMER, *Katholische Universität Eichstätt, Mathematisch-Geographische Fakultät, Ostenstrasse 26-28, 8078 Eichstätt, Germany* [sommer@urz.ku-eichstaett.dbp.de]

CORNELIS TRAAS, *Department of Mathematics, University of Twente, 7500 AE Enschede, The Netherlands* [traas@math.utwente.nl]

Numerical Methods of Approximation Theory, Vol. 9
Dietrich Braess and Larry L. Schumaker (eds.), pp. 1–19.
International Series of Numerical Mathematics, Vol. 105
Copyright © 1992 by Birkhäuser Verlag, Basel
ISBN 3-7643-2746-4.

Blending Approximations with Sine Functions

G. Baszenski, F.-J. Delvos, and S. Jester

Dedicated to the memory of Lothar Collatz

Abstract. We consider sine double series expansions of functions defined on the unit square. We derive error estimates for partial sums and for interpolating sine polynomials assuming that the asymptotic growth of the series coefficients is known. The constants in the error estimates are explicitly computed. We show that Boolean sums yield almost the same asymptotic error estimates as the conventional tensor product approach but with a reduced number of terms. An example is given which shows that the asymptotic error (when the number of points increases) is exactly of the stated order.

§1. Korobov Spaces

Let $f(s,t) \in L_2(J)$ where $J = [0,1]^2$ is the unit square. Then the *sine double series expansion* of $f(s,t)$ is

$$f(s,t) \sim 4 \sum_{k\in\mathbb{N}} \sum_{l\in\mathbb{N}} \widehat{f}_{k,l} \sin k\pi s \, \sin l\pi t \qquad (1.1)$$

with coefficients given by

$$\widehat{f}_{k,l} = \iint_J f(\sigma,\tau) \sin k\pi\sigma \, \sin l\pi\tau \, d\tau \, d\sigma. \qquad (1.2)$$

If a rate of decay of the magnitude of these coefficients $\widehat{f}_{k,l}$ is known, then asymptotic error estimates for partial sums of the expansion (1.1) can conveniently be given.

We therefore introduce the so-called *Korobov spaces*

$$E^{(\alpha,\alpha)}(J) = \{f \in L_2(J) : |\widehat{f}_{k,l}| = \mathcal{O}((kl)^{-\alpha}), \quad k,l \to \infty\}$$

which depend on a positive real parameter α.

Smooth functions are members of a Korobov space with large index α if their function values and normal derivatives of even order at the boundary are zero:

Theorem 1.1. *If for $p \in \mathbb{N}_0$ we have $f \in C^{(2p+1,2p+1)}(J)$ and*

$$\begin{aligned}
f^{(2k,0)}(0,t) &= f^{(2k,0)}(1,t) = 0 \quad k = 0,\ldots,p-1, \quad 0 \leq t \leq 1 \\
f^{(0,2l)}(s,0) &= f^{(0,2l)}(s,1) = 0 \quad l = 0,\ldots,p-1, \quad 0 \leq s \leq 1,
\end{aligned} \tag{1.3}$$

then

$$|\widehat{f}_{k,l}| \leq \frac{V(f^{(2p,2p)})}{(kl\pi^2)^{2p+1}} \quad \text{for all } k,l \in \mathbb{N}$$

with

$$\begin{aligned}
V(g) := \sum_{\sigma=0}^{1} \sum_{\tau=0}^{1} |g(\sigma,\tau)| + \sum_{\sigma=0}^{1} \int_0^1 |g^{(0,1)}(\sigma,\tau)|\, d\tau \\
+ \sum_{\tau=0}^{1} \int_0^1 |g^{(1,0)}(\sigma,\tau)|\, d\sigma + \int_0^1 \int_0^1 |g^{(1,1)}(\sigma,\tau)|\, d\sigma d\tau,
\end{aligned}$$

and therefore $f \in E^{(2p+1,2p+1)}(J)$.

Proof: Integrating $2p+1$ times by parts with respect to each σ and τ, we obtain for all $k, l \in \mathbb{N}$:

$$\widehat{f}_{k,l} = \iint_J f(\sigma, \tau)\, \sin k\pi\sigma\, \sin l\pi\tau\, d\tau\, d\sigma$$

$$\vdots$$

$$
\begin{aligned}
= \frac{1}{(kl\pi^2)^{2p+1}} \bigg(& f^{(2p,2p)}(0,0) - (-1)^k f^{(2p,2p)}(1,0) \\
& - (-1)^l f^{(2p,2p)}(0,1) + (-1)^{k+l} f^{(2p,2p)}(1,1) \\
& + \int_0^1 f^{(2p+1,2p)}(\sigma,0)\cos k\pi\sigma\, d\sigma \\
& - (-1)^l \int_0^1 f^{(2p+1,2p)}(\sigma,1)\cos k\pi\sigma\, d\sigma \\
& + \int_0^1 f^{(2p,2p+1)}(0,\tau)\cos l\pi\tau\, d\tau \\
& - (-1)^k \int_0^1 f^{(2p,2p+1)}(1,\tau)\cos l\pi\tau\, d\tau \\
& + \int_0^1 \int_0^1 f^{(2p+1,2p+1)}(\sigma,\tau)\cos k\pi\sigma\,\cos l\pi\tau\, d\tau\, d\sigma \bigg).
\end{aligned}
$$

Taking absolute values on both sides and using the triangle inequality, we obtain the stated result (details of the computation are carried out in [5] for univariate functions). ∎

A smooth function which does not satisfy the boundary conditions (1.3) above can be boundary-corrected by an interpolating function [1,2,4]. This is the essence of the so-called Krylov-Lanczos method. The following example shows that Theorem 1.1 characterizes proper subsets of the Korobov spaces:

Example 1.2. *For each $p \in \mathbb{N}_0$ the function*

$$f(s,t) = \sum_{k=1}^\infty \sum_{l=1}^\infty \frac{(-1)^{k+l}}{(2k-1)^{2p+1}(2l-1)^{2p+1}} \sin(2k-1)\pi s\, \sin(2l-1)\pi t$$

satisfies $f(s,t) \in E^{(2p+1,2p+1)}(J)$ but $f(s,t) \notin C^{(2p+1,2p+1)}(J)$.

Proof: For $p = 0$ the function $f(s,t)$ is

$$f(s,t) = g(s)\, g(t) \tag{1.4}$$

with

$$g(t) = \ln\left|\tan(\tfrac{\pi}{4} - \tfrac{\pi}{2}t)\right| = 2\sum_{k=1}^{\infty} \frac{(-1)^{k+1}}{2k-1}\sin(2k-1)\pi t. \tag{1.5}$$

$g(t)$ is integrable and the series converges pointwise to $g(t)$, even at the poles $t = \nu + 1/2$, $\nu \in \mathbb{Z}$, where both the function and the series have the value $-\infty$ (see [15, p. 93f] for details). Therefore

$$f(s,t) = \sum_{k=1}^{\infty}\sum_{l=1}^{\infty} \frac{(-1)^{k+l}}{(2k-1)(2l-1)}\sin(2k-1)\pi s \,\sin(2l-1)\pi t$$

has poles at $s = 1/2$ and at $t = 1/2$ which implies $f(s,t) \notin C^{(1,1)}(J)$.

Integrating $2p$ times with respect to both variables s and t we obtain the stated result for arbitrary $p \in \mathbb{N}_0$. ∎

Similar effects are known in classical Fourier theory where the transition to conjugate Fourier series leaves the magnitude of the Fourier coefficients unchanged but singularities can be introduced.

§2. The Blending Projector and its Associated Remainder Projector

We assume that a chain of projectors which operate on univariate functions is given:

$$P_1 \le P_2 \le \cdots \le P_n.$$

By definition, $P_q \le P_r$ means $P_q P_r = P_r P_q = P_q$. In particular, all projectors commute.

In the following, we shall define parametric extensions of these projectors with respect to the first and second variable of bivariate functions which commute as well and form two chains:

$$P_1' \le P_2' \le \cdots \le P_n', \qquad P_1'' \le P_2'' \le \cdots \le P_n''.$$

Their Boolean sum is defined as

$$B_n = \bigoplus_{r=1}^{n} P_r' P_{n+1-r}'',$$

(where $A \oplus B = A + B - AB$ by definition) and an ordinary sum representation is [7]

$$B_n = \sum_{r=1}^{n} P_r' P_{n+1-r}'' - \sum_{r=1}^{n-1} P_r' P_{n-r}''. \tag{2.1}$$

This is a linear combination of tensor products. The remainder projector for B_n has the sum representation [7]

$$
\begin{aligned}
B_n^{\mathbf{c}} &= P_n''^{\mathbf{c}} + \sum_{r=1}^{n-1} P_r'^{\mathbf{c}} P_{n-r}''^{\mathbf{c}} + P_n'^{\mathbf{c}} - \sum_{r=1}^{n} P_r'^{\mathbf{c}} P_{n+1-r}''^{\mathbf{c}} \\
&= P_n''^{\mathbf{c}} + \sum_{r=1}^{n-1} P_r'^{\mathbf{c}} \left(P_{n+1-r}'' - P_{n-r}'' \right) + P_n'^{\mathbf{c}} P_1''
\end{aligned}
\tag{2.2}
$$

(where $A^{\mathbf{c}} = I - A$ denotes the remainder projector of A).

§3. Pseudo-Hyperbolic Partial Sum Error Estimates

We choose P_q as a partial sum with 2^q terms of sine series expansions:

$$
P_q f(t) = F_{2^q} f(t) = 2 \sum_{k=1}^{2^q} \widehat{f}_k \sin k\pi t, \qquad \text{with } \widehat{f}_k = \int_0^1 f(\tau) \sin k\pi\tau \, d\tau.
$$

Parametric extensions are defined as

$$
F_{2^q}' f(s,t) = 2 \sum_{k=1}^{2^q} \widehat{f}_k(t) \sin k\pi s, \qquad \text{where } \widehat{f}_k(t) = \int_0^1 f(\sigma,t) \sin k\pi\sigma \, d\sigma,
$$

which is equal to the double series

$$
F_{2^q}' f(s,t) = 4 \sum_{k=1}^{2^q} \sum_{l=1}^{\infty} \widehat{f}_{k,l} \sin k\pi s \sin l\pi t
$$

with coefficients defined by (1.2) if the sine double series (1.1) converges to $f(s,t)$. Similarly, we take the other parametric extensions as

$$
F_{2^r}'' f(s,t) = 2 \sum_{k=1}^{2^r} \widehat{\widehat{f}}_l(s) \sin l\pi t \qquad \text{where } \widehat{\widehat{f}}_l(s) = \int_0^1 f(s,\tau) \sin l\pi\tau \, d\tau
$$

or

$$
F_{2^r}'' f(s,t) = 4 \sum_{k=1}^{\infty} \sum_{l=1}^{2^r} \widehat{f}_{k,l} \sin k\pi s \sin l\pi t
$$

if convergence of the expansion (1.1) to $f(s,t)$ holds. Under this assumption a product of these projectors has the form

$$
F_{2^q}' F_{2^r}'' f(s,t) = 4 \sum_{k=1}^{2^q} \sum_{l=1}^{2^r} \widehat{f}_{k,l} \sin k\pi s \sin l\pi t.
\tag{3.1}
$$

The associated remainder projectors can also be expanded into double series. To show this we need the following estimates:

Lemma 3.1. *Assume that* $\alpha \in \mathbb{R}$, $\alpha > 1$ *and* $L_1, L_2 \in \mathbb{N}_0$, $L_1 < L_2$. *Then*

a) $\displaystyle \sum_{\lambda=L_1+1}^{L_2} \lambda^{-\alpha} \leq \begin{cases} \dfrac{L_1^{-\alpha+1} - L_2^{-\alpha+1}}{\alpha - 1} & \text{if } L_1 > 0 \\[2ex] \dfrac{\alpha - L_2^{-\alpha+1}}{\alpha - 1} & \text{if } L_1 = 0. \end{cases}$

b) $\displaystyle \sum_{\lambda=L_1+1}^{\infty} \lambda^{-\alpha} \leq \begin{cases} \dfrac{L_1^{-\alpha+1}}{\alpha - 1} & \text{if } L_1 > 0 \\[2ex] \dfrac{\alpha}{\alpha - 1} & \text{if } L_1 = 0. \end{cases}$

Proof: a) follows from the integral estimates

$$\sum_{\lambda=L_1+1}^{L_2} \lambda^{-\alpha} \leq \begin{cases} \displaystyle\int_{L_1}^{L_2} x^{-\alpha}\, dx & \text{if } L_1 > 0 \\[2ex] 1 + \displaystyle\int_{1}^{L_2} x^{-\alpha}\, dx & \text{if } L_1 = 0 \end{cases}$$

and b) is obtained by taking the limit $L_2 \to \infty$ in a). $\blacksquare$

Proposition 3.2. *If* $f(s,t) \in E^{(\alpha,\alpha)}(J)$ *with* $\alpha > 1$, *then the sine double series expansion* (1.1) *of* $f(s,t)$ *converges uniformly and for* $M_1, M_2 \in \mathbb{N}$, *the remainder projectors of* F'_{M_1}, F''_{M_2} *satisfy*

a) $\displaystyle F'_{M_1}{}^{\mathbf{c}} f(s,t) = 4 \sum_{k=M_1+1}^{\infty} \sum_{l=1}^{\infty} \widehat{f}_{k,l} \sin k\pi s \, \sin l\pi t$

b) $\displaystyle F''_{M_2}{}^{\mathbf{c}} f(s,t) = 4 \sum_{k=1}^{\infty} \sum_{l=M_2+1}^{\infty} \widehat{f}_{k,l} \sin k\pi s \, \sin l\pi t$

c) $\displaystyle F'_{M_1}{}^{\mathbf{c}} F''_{M_2}{}^{\mathbf{c}} f(s,t) = 4 \sum_{k=M_1+1}^{\infty} \sum_{l=M_2+1}^{\infty} \widehat{f}_{k,l} \sin k\pi s \, \sin l\pi t$

d) $\displaystyle F'_{M_1}{}^{\mathbf{c}} F''_{M_2} f(s,t) = 4 \sum_{k=M_1+1}^{\infty} \sum_{l=1}^{M_2} \widehat{f}_{k,l} \sin k\pi s \, \sin l\pi t.$

Proof: Let $C > 0$ with $\left|\widehat{f}_{k,l}\right| \leq C(kl)^{-\alpha} (k,l \in \mathbb{N})$. With the previous lemma absolute convergence of the double series expansion of $f(s,t)$ follows directly from

$$4\sum_{k=1}^{\infty} \sum_{l=1}^{\infty} \left|\widehat{f}_{k,l} \sin k\pi s \, \sin l\pi t\right| \leq 4C \sum_{k=1}^{\infty} \sum_{l=1}^{\infty} (kl)^{-\alpha} \leq 4C \left(\frac{\alpha}{\alpha - 1}\right)^2 < \infty. \quad \blacksquare$$

These expansions give the following error estimates

Lemma 3.3. For $f(s,t) \in E^{(\alpha,\alpha)}(J)$, where $\alpha > 1$ and $M_1, M_2, M_3 \in \mathbb{N}$ with $M_2 \le M_3$, we have the estimates:

a) $\left\| F'_{M_1}{}^c f(s,t) \right\|_\infty \le \dfrac{4C\alpha}{(\alpha-1)^2} M_1^{-\alpha+1} = \mathcal{O}(M_1^{-\alpha+1}) \quad \text{as } M_1 \to \infty$

b) $\left\| F''_{M_2}{}^c f(s,t) \right\|_\infty \le \dfrac{4C\alpha}{(\alpha-1)^2} M_2^{-\alpha+1} = \mathcal{O}(M_2^{-\alpha+1}) \quad \text{as } M_2 \to \infty$

c) $\left\| F'_{M_1}{}^c F''_{M_2}{}^c f(s,t) \right\|_\infty \le \dfrac{4C}{(\alpha-1)^2}(M_1 M_2)^{-\alpha+1}$

$$= \mathcal{O}\big((M_1 M_2)^{-\alpha+1}\big) \quad \text{as } M_1, M_2 \to \infty$$

d) $\left\| F'_{M_1}{}^c F''_{M_2} f(s,t) \right\|_\infty \le \dfrac{4C\alpha}{(\alpha-1)^2} M_1^{-\alpha+1} = \mathcal{O}(M_1^{-\alpha+1}) \quad \text{as } M_1 \to \infty$

e) $\left\| F'_{M_1}{}^c (F''_{M_3} - F''_{M_2}) f(s,t) \right\|_\infty \le \dfrac{4C}{(\alpha-1)^2} M_1^{-\alpha+1}(M_2^{-\alpha+1} - M_3^{-\alpha+1})$

$$= \mathcal{O}\big((M_1 M_2)^{-\alpha+1}\big) \quad \text{as } M_1, M_2 \to \infty.$$

Proof: We show only a) and e) since the other inequalities follow with the same techniques. Let $\left| \widehat{f}_{k,l} \right| \le C(kl)^{-\alpha}$ for all $k,l \in \mathbb{N}$. Using Lemma 3.1 we obtain from the expansions in Proposition 3.2:

$$\left| F'_{M_1}{}^c f(s,t) \right| = \left| 4 \sum_{k=M_1+1}^{\infty} \sum_{l=1}^{\infty} \widehat{f}_{k,l} \sin k\pi s \, \sin l\pi t \right| \le 4 \sum_{k=M_1+1}^{\infty} \sum_{l=1}^{\infty} C(kl)^{-\alpha}$$

$$= 4C \sum_{k=M_1+1}^{\infty} k^{-\alpha} \sum_{l=1}^{\infty} l^{-\alpha} \le \frac{4C\alpha}{(\alpha-1)^2} M_1^{-\alpha+1}.$$

Inequality e) follows from

$$\left| F'_{M_1}{}^c (F''_{M_3} - F''_{M_2}) f(s,t) \right| \le 4 \sum_{k=M_1+1}^{\infty} \sum_{k=M_2+1}^{M_3} C(kl)^{-\alpha}$$

$$\le 4C \sum_{k=M_1+1}^{\infty} k^{-\alpha} \sum_{l=M_2+1}^{M_3} l^{-\alpha} \le \frac{4C}{(\alpha-1)^2} M_1^{-\alpha+1}(M_2^{-\alpha+1} - M_3^{-\alpha+1}). \quad \blacksquare$$

The Boolean sum (2.1) becomes by (3.1)

$$B_n f(s,t) = 4 \sum_{r=1}^{n} \sum_{k=1}^{2^r} \sum_{l=1}^{2^{n+1-r}} \widehat{f}_{k,l} \sin k\pi s \sin l\pi t$$

$$- 4 \sum_{r=1}^{n-1} \sum_{k=1}^{2^r} \sum_{l=1}^{2^{n-r}} \widehat{f}_{k,l} \sin k\pi s \sin l\pi t$$

$$= 4 \sum_{r=1}^{n-1} \sum_{k=1}^{2^r} \sum_{l=2^{n-r}+1}^{2^{n+1-r}} \widehat{f}_{k,l} \sin k\pi s \sin l\pi t$$

$$+ 4 \sum_{k=1}^{2^n} \sum_{l=1}^{2} \widehat{f}_{k,l} \sin k\pi s \sin l\pi t \tag{3.2}$$

$$= 4 \sum_{(k,l)\in\mathcal{I}_n} \widehat{f}_{k,l} \sin k\pi s \sin l\pi t$$

with

$$\mathcal{I}_n = \bigcup_{r=1}^{n} \{(k,l) \in \mathbb{N}^2 : 1 \le k \le 2^r,\ 1 \le l \le 2^{n+1-r}\}. \tag{3.3}$$

Note that this is a certain partial sum of the sine double series expansion (1.1). The Boolean sum remainder formula (2.2) yields in connection with the expansions of the remainder projectors derived in Proposition 3.2 the following representation for the error of the partial sum (3.2):

$$B_n{}^{\mathbf{c}} f(s,t) = F_{2^n}''{}^{\mathbf{c}} f(s,t) + \sum_{r=1}^{n-1} F_{2^r}'{}^{\mathbf{c}} \left(F_{2^{n+1-r}}'' - F_{2^{n-r}}'' \right) f(s,t)$$

$$+ F_{2^n}'{}^{\mathbf{c}} F_2'' f(s,t)$$

$$= 4 \sum_{k=1}^{\infty} \sum_{l=2^n+1}^{\infty} \widehat{f}_{k,l} \sin k\pi s \sin l\pi t$$

$$+ 4 \sum_{r=1}^{n-1} \sum_{k=2^r+1}^{\infty} \sum_{l=2^{n-r}+1}^{2^{n+1-r}} \widehat{f}_{k,l} \sin k\pi s \sin l\pi t \tag{3.4}$$

$$+ \ 4 \sum_{k=2^n+1}^{\infty} \sum_{l=1}^{2} \widehat{f}_{k,l} \sin k\pi s \sin l\pi t$$

$$= 4 \sum_{(k,l)\in\mathbb{N}^2\setminus\mathcal{I}_n} \widehat{f}_{k,l} \sin k\pi s \sin l\pi t,$$

with $\mathcal{I}_n$ as in (3.3). Error estimates for the partial sum can now easily be established as follows:

Theorem 3.4. *Let* $f \in E^{(\alpha,\alpha)}(J)$ *and* $\alpha > 1$. *Then we have for* $n \in \mathbb{N}_0$ *and* $N = 2^n$:

$$\left\| B_n{}^{\mathbf{c}} f(s,t) \right\|_\infty \leq \frac{4C(2\alpha + n - 1)}{(\alpha - 1)^2} \cdot N^{-\alpha+1} = \mathcal{O}\left(\frac{\log_2 N}{N^{\alpha-1}} \right) \quad \text{as } N \to \infty.$$

A tensor product operator with comparable remainder is $F'_N F''_N$, *where*

$$\left\| (F'_N F''_N)^{\mathbf{c}} f(s,t) \right\|_\infty \leq \frac{8C\alpha}{(\alpha - 1)^2} \cdot N^{-\alpha+1} = \mathcal{O}\left(\frac{1}{N^{\alpha-1}} \right) \quad \text{as } N \to \infty.$$

Proof: If $\left| \widehat{f}_{k,l} \right| \leq C(kl)^{-\alpha}$ for $k, l \in \mathbb{N}$, then we can conclude by estimates of the previous lemma and by the Boolean sum remainder formula (3.4)

$$\left\| B_n{}^{\mathbf{c}} f(s,t) \right\|_\infty \leq \left\| F''_{2^n}{}^{\mathbf{c}} f(s,t) \right\|_\infty + \sum_{r=1}^{n-1} \left\| F'_{2^r}{}^{\mathbf{c}} (F''_{2^{n+1-r}} - F''_{2^{n-r}}) f(s,t) \right\|_\infty$$
$$+ \left\| F'_{2^n}{}^{\mathbf{c}} F''_2 f(s,t) \right\|_\infty$$
$$\leq \frac{4C}{(\alpha - 1)^2} \cdot N^{-\alpha+1}(\alpha + n - 1 + \alpha).$$

The tensor product is a Boolean sum with only one term. By the remainder formula the error estimate in this case becomes

$$\left\| (F'_N F''_N)^{\mathbf{c}} f(s,t) \right\|_\infty \leq \left\| F''_N{}^{\mathbf{c}} f(s,t) \right\|_\infty + \left\| F'_N{}^{\mathbf{c}} F''_N f(s,t) \right\|_\infty$$
$$\leq \frac{8C\alpha}{(\alpha - 1)^2} \cdot N^{-\alpha+1}. \quad \blacksquare$$

This comparison shows that the asymptotic error of the Boolean sum is greater than the corresponding error of a tensor product summation by a factor of only $n = \log_2 N$. The number of terms in the partial sum of the expansion is $N \log_2 N$ in the Boolean sum but N^2 for the tensor product.

The following example shows that the $\log_2 N$ factor in the Boolean sum error estimate is necessary:

Example 3.5. *For $\alpha > 1$ and*

$$f(s,t) = \sum_{k=1}^{\infty} \sum_{l=1}^{\infty} \frac{(-1)^{k+l}}{(kl)^{\alpha}} \sin(2k+1)\pi s \sin(2l+1)\pi t,$$

we have $f(s,t) \in E^{(\alpha,\alpha)}(J)$ and

$$\left\| B_n^{c} f(s,t) \right\|_{\infty} \geq \frac{C \log_2 N}{N^{\alpha-1}}$$

for all $N = 2^n \in \mathbb{N}$, where C is a positive value independent of N.

Proof: For $k, l \in \mathbb{N}$ we have

$$\left(\frac{2k+1}{k} \frac{2l+1}{l} \right)^{\alpha} \leq 9^{\alpha},$$

which is equivalent to

$$k^{-\alpha} l^{-\alpha} \leq 9^{\alpha} (2k+1)^{-\alpha} (2l+1)^{-\alpha}$$

or $|\widehat{f}_{k,l}| \leq C(kl)^{-\alpha}$ for all $k, l \in \mathbb{N}$ with positive C. Thus $f(s,t) \in E^{(\alpha,\alpha)}(J)$.
To show the lower error bound, we need the integral estimate

$$\sum_{\lambda=L_1}^{L_2-1} \lambda^{-\alpha} \geq \int_{L_1}^{L_2} x^{-\alpha}\, dx = \frac{L_1^{-\alpha+1} - L_2^{-\alpha+1}}{\alpha-1}. \tag{3.5}$$

We show that the proposed error is already attained as pointwise error at $(s,t) = (\frac{1}{2}, \frac{1}{2})$. For this choice of (s,t) the term with frequency $(2k+1, 2l+1)$ of the double series equals

$$\frac{(-1)^{k+l}}{(kl)^{\alpha}} \sin(2k+1)\frac{\pi}{2} \sin(2l+1)\frac{\pi}{2} = (kl)^{-\alpha}.$$

Since only terms with odd indices $k, l \geq 3$ are nonzero, from (3.4), (3.5) we then obtain for the error

$$B_n^{\mathrm{c}} f \left(\tfrac{1}{2}, \tfrac{1}{2}\right) = \sum_{2k+1=3}^{\infty} \sum_{2l+1=2^n+1}^{\infty} (kl)^{-\alpha} + \sum_{r=1}^{n-1} \sum_{2k+1=2^r+1}^{\infty} \sum_{2l+1=2^{n-r}+1}^{2^{n+1-r}} (kl)^{-\alpha}$$

$$+ \sum_{2k+1=2^n+1}^{\infty} \sum_{2l+1=3}^{2} (kl)^{-\alpha}$$

$$= \sum_{k=1}^{\infty} \sum_{l=2^{n-1}}^{\infty} (kl)^{-\alpha} + \sum_{r=1}^{n-1} \sum_{k=2^{r-1}}^{\infty} \sum_{l=2^{n-r-1}}^{2^{n-r}-1} (kl)^{-\alpha}$$

$$\geq \frac{1}{(\alpha-1)^2} \left((2^{n-1})^{-\alpha+1} + \right.$$

$$\left. + \sum_{r=1}^{n-1} (2^{r-1})^{-\alpha+1} \left((2^{n-r-1})^{-\alpha+1} - (2^{n-r})^{-\alpha+1} \right) \right)$$

$$= \frac{(2^n)^{-\alpha+1}}{(\alpha-1)^2} \left(2^{\alpha-1} + (n-1)\, 4^{\alpha-1} - (n-1)\, 2^{\alpha-1} \right)$$

$$\geq C n \, (2^n)^{-\alpha+1}$$

with a positive constant C if n is taken large enough. ∎

§4. Blending Interpolation with Sine Functions

The second blending construction in this article is based on trigonometric interpolation projectors with sine functions. We need pointwise convergence of the expansion (1.1) in order to have pointwise well-defined function values, and therefore assume in this section that $f(s,t) \in E^{(\alpha,\alpha)}(J)$ where $\alpha > 1$. Projectors operating on univariate functions are defined for $M \in \mathbb{N}$ and for $f(t) \in E^{\alpha}[0,1]$ with $\alpha > 1$ as follows:

$$T_M f(t) = 2 \sum_{k=1}^{M-1} f_{k,M}^* \, \sin k\pi t$$

with

$$f_{k,M}^* = \frac{1}{M} \sum_{j=1}^{M-1} f(t_{j,M}) \, \sin k\pi t_{j,M} = \widehat{f}_k + \sum_{\mu=1}^{\infty} \left(\widehat{f}_{2\mu M+k} - \widehat{f}_{2\mu M-k} \right)$$

where $t_{j,M} = j/M$ $(j = 0, \ldots, M)$.

T_M interpolates at the knots $t_{j,M}$, $j = 0, \ldots, M$ (note that $f(0) = f(1) = 0$ for $f(t) \in E^\alpha[0,1]$ with $\alpha > 1$). These interpolation properties and the series representation for $f^*_{k,M}$ in terms of $\widehat{f}_l$ above follow from the pointwise convergence of the sine series to $f(t)$, and from the following discrete orthogonality relation [4]:

Lemma 4.1. For $k, l, M \in \mathbb{N}$ with $1 \le k \le M - 1$ and $t_{j,M} = j/M$, $j = 0, \ldots, M$, we have

$$\sum_{j=1}^{M-1} \sin k\pi t_{j,M} \sin l\pi t_{j,M} = \begin{cases} M/2, & \text{if } l = 2\mu M + k, \ \mu \in \mathbb{N}_0 \\ -M/2, & \text{if } l = 2\mu M - k, \ \mu \in \mathbb{N} \\ 0, & \text{otherwise.} \end{cases}$$

For $M_1, M_2 \in \mathbb{N}$ and for $f(s,t) \in E^{(\alpha,\alpha)}(J)$, $\alpha > 1$, parametric extensions are defined as

$$T'_{M_1} f(s,t) = 2 \sum_{k=1}^{M_1-1} f^*_{k,M_1}(t) \ \sin k\pi s \tag{4.1}$$

with

$$\begin{aligned}
f^*_{k,M_1}(t) &= \frac{1}{M_1} \sum_{j=1}^{M_1-1} f(t_{j,M_1}, t) \ \sin k\pi t_{j,M_1} \\
&= 2 \sum_{l=1}^{\infty} \left(\widehat{f}_{k,l} + \sum_{\mu=1}^{\infty} \left(\widehat{f}_{2\mu M_1+k,l} - \widehat{f}_{2\mu M_1-k,l} \right) \right) \sin l\pi t
\end{aligned} \tag{4.2}$$

and

$$T''_{M_2} f(s,t) = 2 \sum_{l=1}^{M_2-1} f^{**}_{l,M_2}(s) \ \sin l\pi t \tag{4.3}$$

with

$$\begin{aligned}
f^{**}_{l,M_2}(s) &= \frac{1}{M_2} \sum_{j=1}^{M_2-1} f(s, t_{j,M_2}) \ \sin l\pi t_{j,M_2} \\
&= 2 \sum_{k=1}^{\infty} \left(\widehat{f}_{k,l} + \sum_{\nu=1}^{\infty} \left(\widehat{f}_{k,2\nu M_2+l} - \widehat{f}_{k,2\nu M_2-l} \right) \right) \sin k\pi s.
\end{aligned} \tag{4.4}$$

These projectors yield interpolants along lines in J:

$$T'_{M_1} f(t_{j,M_1}, t) = f(t_{j,M_1}, t) \quad (j = 0, \ldots, M_1,\ 0 \le t \le 1)$$
$$T''_{M_2} f(s, t_{j,M_2}) = f(s, t_{j,M_2}) \quad (j = 0, \ldots, M_2,\ 0 \le s \le 1).$$

Note that if $f(s,t) \in E^{(\alpha,\alpha)}(J)$, then $f^*_{k,M_1}(t) \in E^{\alpha}[0,1]$, and therefore $T'_{M_1} f(s,t) \in E^{(\alpha,\alpha)}(J)$. In the same way, we have $T''_{M_2} f(s,t) \in E^{(\alpha,\alpha)}(J)$. Products of these parametric extensions give tensor product interpolants

$$T'_{M_1} T''_{M_2} f(s,t) = 4 \sum_{k=1}^{M_1-1} \sum_{l=1}^{M_2-1} f^*_{k,l;M_1,M_2} \sin k\pi s \sin l\pi t$$

with

$$f^*_{k,l;M_1,M_2} = \frac{1}{M_1 M_2} \sum_{j=1}^{M_1-1} \sum_{h=1}^{M_2-1} f(t_{j,M_1}, t_{h,M_2}) \sin k\pi t_{j,M_1} \sin l\pi t_{h,M_2}$$

$$= \widehat{f}_{k,l} + \sum_{\mu=1}^{\infty} \left(\widehat{f}_{2\mu M_1+k,l} - \widehat{f}_{2\mu M_1-k,l} \right)$$

$$+ \sum_{\nu=1}^{\infty} \left(\widehat{f}_{k,2\nu M_2+l} - \widehat{f}_{k,2\nu M_2-l} \right)$$

$$+ \sum_{\mu=1}^{\infty} \sum_{\nu=1}^{\infty} \left(\widehat{f}_{2\mu M_1+k,2\nu M_2+l} - \widehat{f}_{2\mu M_1-k,2\nu M_2+l} \right.$$

$$\left. - \widehat{f}_{2\mu M_1+k,2\nu M_2-l} + \widehat{f}_{2\mu M_1-k,2\nu M_2-l} \right)$$

which interpolate at the rectangular grid of points

$$(t_{j,M_1}, t_{h,M_2}), \quad j = 0, \ldots, M_1, \quad h = 0, \ldots, M_2.$$

Relations (4.1) – (4.4) immediately give double series expansions of the projectors T'_{M_1}, T''_{M_2} and of the associated remainder projectors $T'_{M_1}{}^{\mathbf{c}}$, $T''_{M_2}{}^{\mathbf{c}}$. The following expansions will be needed later.

Proposition 4.2. *For $f(s,t) \in E^{(\alpha,\alpha)}(J)$, $\alpha > 1$, and for $M_1, M_2 \in \mathbb{N}$, we have the sine double series expansions (which converge uniformly):*

a) $T''_{M_2} f(s,t) = 4 \sum_{k=1}^{\infty} \sum_{l=1}^{\infty} a_{k,l} \sin k\pi s \sin l\pi t$

with $a_{k,l} = \begin{cases} \widehat{f}_{k,l} + \sum_{\nu=1}^{\infty} \left(\widehat{f}_{k,2\nu M_2+l} - \widehat{f}_{k,2\nu M_2-l} \right) & \text{for } k \in \mathbb{N},\, l < M_2 \\ 0 & \text{for } k \in \mathbb{N},\, l \geq M_2 \end{cases}$

b) $T''_{2M_2} f(s,t) - T''_{M_2} f(s,t) = 4 \sum_{k=1}^{\infty} \sum_{l=1}^{\infty} b_{k,l} \sin k\pi s \sin l\pi t$ with

$b_{k,l} = \begin{cases} \sum_{\nu=1}^{\infty} \left(\widehat{f}_{k,(4\nu-2)M_2-l} - \widehat{f}_{k,(4\nu-2)M_2+l} \right) & \text{for } k \in \mathbb{N},\, l < M_2 \\ \widehat{f}_{k,l} + \sum_{\nu=1}^{\infty} \left(\widehat{f}_{k,4\nu M_2+l} - \widehat{f}_{k,4\nu M_2-l} \right) & \text{for } k \in \mathbb{N},\, M_2 \leq l < 2M_2 \\ 0 & \text{for } k \in \mathbb{N},\, l \geq 2M_2 \end{cases}$

c) $T'_{M_1}{}^{\mathbf{c}} f(s,t) = 4 \sum_{k=1}^{\infty} \sum_{l=1}^{\infty} c_{k,l} \sin k\pi s \sin l\pi t$

with $c_{k,l} = \begin{cases} \sum_{\mu=1}^{\infty} \left(\widehat{f}_{2\mu M_1-k,l} - \widehat{f}_{2\mu M_1+k,l} \right) & \text{for } k < M_1,\, l \in \mathbb{N} \\ \widehat{f}_{k,l} & \text{for } k \geq M_1,\, l \in \mathbb{N} \end{cases}$

d) $T''_{M_2}{}^{\mathbf{c}} f(s,t) = 4 \sum_{k=1}^{\infty} \sum_{l=1}^{\infty} d_{k,l} \sin k\pi s \sin l\pi t$

with $d_{k,l} = \begin{cases} \sum_{\nu=1}^{\infty} \left(\widehat{f}_{k,2\nu M_2-l} - \widehat{f}_{k,2\nu M_2+l} \right) & \text{for } k \in \mathbb{N},\, l < M_2 \\ \widehat{f}_{k,l} & \text{for } k \in \mathbb{N},\, l \geq M_2. \end{cases}$

We construct chains of interpolation projectors by doubling the number of knots from one projector to the next, starting with a spacing of $1/2$:

$$T'_2 \leq T'_4 \leq \cdots \leq T'_{2^n}, \quad T''_2 \leq T''_4 \leq \cdots \leq T''_{2^n}.$$

The Boolean sum interpolation projector obtained from these chains is given by

$$B_n f(s,t) = \bigoplus_{r=1}^{n} T'_{2^r} T''_{2^{n+1-r}} f(s,t)$$

$$= \sum_{r=1}^{n} T'_{2^r} T''_{2^{n+1-r}} f(s,t) - \sum_{r=1}^{n-1} T'_{2^r} T''_{2^{n-r}} f(s,t).$$

Its range and interpolation knot set are:

$$\operatorname{Im} B_n = \sum_{r=1}^{n} \operatorname{span} \left\{ \sin k\pi s \, \sin l\pi t : 1 \leq k < 2^r, \ 1 \leq l < 2^{n+1-r} \right\}$$

$$\operatorname{prec} B_n = \bigcup_{r=1}^{n} \left\{ \left(\frac{j}{2^r}, \frac{h}{2^{n+1-r}} \right) : 1 \leq j < 2^r, \ 1 \leq h < 2^{n+1-r} \right\}.$$

In order to get error estimates we first derive appraisals for the individual product terms contained in the Boolean sum remainder formula:

Proposition 4.3. *For $f(s,t) \in E^{(\alpha,\alpha)}(J)$, where $\alpha > 1$ and $r, n, M \in \mathbb{N}$ with $r < n$, $N = 2^n$ we have the estimates*

a) $\left\| T_N''{}^{\mathbf{c}} f(s,t) \right\|_\infty \leq \dfrac{8C\alpha}{(\alpha-1)^2} \left(1 + \dfrac{\alpha-1}{2N} \right) N^{-\alpha+1}$
$$= \mathcal{O}(N^{-\alpha+1}) \quad \text{as } N \to \infty$$

b) $\left\| T_N'{}^{\mathbf{c}} T_M'' f(s,t) \right\|_\infty \leq \dfrac{8C\alpha}{(\alpha-1)^2} \left(1 + \dfrac{\alpha-1}{2N} \right) N^{-\alpha+1}$
$$= \mathcal{O}(N^{-\alpha+1}) \quad \text{as } N \to \infty$$

c) $\left\| T_{2^r}'{}^{\mathbf{c}} (T_{2^{n+1-r}}'' - T_{2^{n-r}}'') f(s,t) \right\|_\infty$
$$\leq \frac{16C}{(\alpha-1)^2} \left(1 + (\alpha-1)\left(\frac{1}{4} + \frac{1}{N} \right) + \frac{(\alpha-1)^2}{4N} \right) N^{-\alpha+1}$$
$$= \mathcal{O}(N^{-\alpha+1}) \quad \text{as } N \to \infty.$$

Proof: Let $|\widehat{f}_{k,l}| \leq C(kl)^{-\alpha}$. From the series expansions in Proposition 4.2 we obtain

a) $\left| T_N''{}^{\mathbf{c}} f(s,t) \right| \leq 4 \displaystyle\sum_{k=1}^{\infty} \sum_{l=1}^{\infty} |d_{k,l}|$

$$\leq 4 \sum_{k=1}^{\infty} \sum_{l=1}^{N-1} \sum_{\nu=1}^{\infty} \left(|\widehat{f}_{k,2\nu N - l}| + |\widehat{f}_{k,2\nu N + l}| \right) + 4 \sum_{k=1}^{\infty} \sum_{l=N}^{\infty} |\widehat{f}_{k,l}|$$

$$\leq 4 \sum_{k=1}^{\infty} \sum_{l=N+1}^{\infty} C(kl)^{-\alpha} + 4 \sum_{k=1}^{\infty} \sum_{l=N}^{\infty} C(kl)^{-\alpha}$$

$$\leq \frac{8C\alpha}{(\alpha-1)} \left(\frac{1}{\alpha-1} + \frac{1}{2N} \right) N^{-\alpha+1}$$

by the integral estimates in Lemma 3.1 (note that a sum of the form $\sum_{l=1}^{N-1}\sum_{\nu=1}^{\infty}(a_{2\nu N-l}+a_{2\nu N+l})$ contains only elements with index at least $N+1$, and no element occurs more than one time).

b) Inserting the series representation of $T_M'' f(s,t)$ into $T_N'^{\,\mathbf{c}}$, we can conclude by Proposition 4.2 that the double series coefficients of $T_N'^{\,\mathbf{c}} T_M'' f(s,t)$ are given by:

$$
\varphi_{k,l} = \begin{cases}
\begin{aligned}
&\sum_{\mu=1}^{\infty}\Big(\widehat{f}_{2\mu N-k,l} \\
&\quad + \sum_{\nu=1}^{\infty}(\widehat{f}_{2\mu N-k,2\nu M+l}-\widehat{f}_{2\mu N-k,2\nu M-l}) \\
&\quad - \widehat{f}_{2\mu N+k,l} \\
&\quad - \sum_{\nu=1}^{\infty}(\widehat{f}_{2\mu N+k,2\nu M+l}-\widehat{f}_{2\mu N+k,2\nu M-l})\Big)
\end{aligned} & \text{if } l < M,\, k < N \\[2pt]
\widehat{f}_{k,l}+\sum_{\nu=1}^{\infty}(\widehat{f}_{k,2\nu M+l}-\widehat{f}_{k,2\nu M-l}) & \text{if } l < M,\, k \geq N \\[2pt]
0 & \text{if } l \geq M.
\end{cases}
$$

It follows that

$$
\left|T_N'^{\,\mathbf{c}} T_2'' f(s,t)\right| \leq 4\sum_{k=1}^{\infty}\sum_{l=1}^{\infty}|\varphi_{k,l}|
$$

$$
\leq 4\sum_{k=1}^{N-1}\sum_{l=1}^{M-1}\sum_{\mu=1}^{\infty}\Big(|\widehat{f}_{2\mu N-k,l}| + \sum_{\nu=1}^{\infty}\big(|\widehat{f}_{2\mu N-k,2\nu M+l}|+|\widehat{f}_{2\mu N-k,2\nu M-l}|\big)
$$

$$
+ |\widehat{f}_{2\mu N+k,l}| + \sum_{\nu=1}^{\infty}\big(|\widehat{f}_{2\mu N+k,2\nu M+l}| + |\widehat{f}_{2\mu N+k,2\nu M-l}|\big)\Big)
$$

$$
+ 4\sum_{k=N}^{\infty}\sum_{l=1}^{M-1}\Big(|\widehat{f}_{k,l}| + \sum_{\nu=1}^{\infty}\big(|\widehat{f}_{k,2\nu M+l}| + |\widehat{f}_{k,2\nu M-l}|\big)\Big)
$$

$$
\leq 4\sum_{k=N+1}^{\infty}\sum_{l=1}^{M-1}\Big(|\widehat{f}_{k,l}| + \sum_{\nu=1}^{\infty}\big(|\widehat{f}_{k,2\nu M+l}| + |\widehat{f}_{k,2\nu M-l}|\big)\Big)
$$

$$
+ 4\sum_{k=N}^{\infty}\sum_{l=1}^{M-1}\Big(|\widehat{f}_{k,l}| + \sum_{\nu=1}^{\infty}\big(|\widehat{f}_{k,2\nu M+l}| + |\widehat{f}_{k,2\nu M-l}|\big)\Big)
$$

$$
\leq 8\sideset{}{'}\sum_{k=N}^{\infty}\sum_{l=1}^{\infty} C(kl)^{-\alpha}
$$

$$
\leq \frac{8C\alpha}{(\alpha-1)}\left(\frac{1}{\alpha-1}+\frac{1}{2N}\right)N^{-\alpha+1}.
$$

(Here $\sum'^{\infty}_{k=N} a_k$ is an abbreviation for $\frac{1}{2}a_N + \sum^{\infty}_{k=N+1} a_k$).

c) is shown in the same way. From Proposition 4.2 we get by inserting the series coefficients of $(T''_{2^{n+1}-r} - T''_{2^n-r})f(s,t)$ into the formula for $T'_{2^r}{}^{\mathbf{c}}$ a double series representation. Rearranging sums as before and using the triangle inequality for the absolute value, we arrive at

$$\left| T'_{2^r}{}^{\mathbf{c}}(T''_{2^{n+1}-r} - T''_{2^n-r})f(s,t) \right| \le 16 \sum^{\infty}_{k=2^r}{}' \sum^{\infty}_{l=2^{n-r}}{}' \left| \widehat{f}_{k,l} \right|,$$

which gives the stated error bound when $\left| \widehat{f}_{k,l} \right|$ is replaced by the majorant $C(kl)^{-\alpha}$ and the integral estimates from Lemma 3.1 are used. ∎

Theorem 4.4. *Let $f \in E^{(\alpha,\alpha)}(J)$ where $\alpha > 1$. For $n \in \mathbb{N}$ and $N = 2^n$ the Boolean sum interpolation error is*

$$\left\| B_n{}^{\mathbf{c}} f(s,t) \right\|_{\infty} \le \frac{16C}{(\alpha-1)^2} \left(n + \frac{\alpha-1}{4}(n+3) + \frac{(\alpha-1)^2}{16}(n-1) \right) \cdot N^{-\alpha+1}$$

$$= \mathcal{O}\left(\frac{\log_2 N}{N^{\alpha-1}} \right) \quad \text{as } N \to \infty.$$

The tensor product interpolant $T'_N T''_N f(s,t)$ with $N \times N$ knots yields the error estimate

$$\left\| (T'_N T''_N)^{\mathbf{c}} f(s,t) \right\|_{\infty} \le \frac{16C\alpha}{(\alpha-1)^2} \left(1 + \frac{\alpha-1}{4} \right) \cdot N^{-\alpha+1}$$

$$= \mathcal{O}\left(\frac{1}{N^{\alpha-1}} \right) \quad \text{as } N \to \infty.$$

Proof: This follows directly from the remainder formulas

$$B_n{}^{\mathbf{c}} f(s,t) = T''_{2^n}{}^{\mathbf{c}} f(s,t) + \sum^{n-1}_{r=1} T'_{2^r}{}^{\mathbf{c}} (T''_{2^{n+1}-r} - T''_{2^n-r})f(s,t)$$

$$+ T'_{2^n}{}^{\mathbf{c}} T''_2 f(s,t)$$

$$(T'_N T''_N)^{\mathbf{c}} f(s,t) = T''_N{}^{\mathbf{c}} f(s,t) + T'_N{}^{\mathbf{c}} T''_N f(s,t)$$

and from Proposition 4.3. ∎

References

1. Baszenski, G., *Zur Konvergenzbeschleunigung von Orthogonal-Doppelreihen*, dissertation, Universität Siegen 1983.

2. Baszenski, G., and F.-J. Delvos, Accelerating the convergence of bivariate Fourier expansions, in *Approximation Theory IV* C. K. Chui, L. L. Schumaker, J. D. Ward (eds.), Academic Press, New York 1983, 335–340.

3. Baszenski, G., and F.-J. Delvos, Boolean methods in Fourier approximation, in *Topics in Multivariate Approximation*, C. K. Chui, L. L. Schumaker, F. I. Utreras (eds.), Academic Press, New York, 1987, 1–12.

4. Baszenski, G., and F.-J. Delvos, A variant of the Krylov-Lanczos method for bivariate trigonometric interpolation, in *Constructive Theory of Fuctions '87* Bl. Sendov et al. (eds.), Publishing House of the Bulgarian Academy of Sciences, Sofia, 1988, 33–39.

5. Baszenski, G., and F.-J. Delvos, Error Estimates for Sine Series Expansions. Math. Nachr. **139** (1988), 155–166.

6. Baszenski, G., and F.-J. Delvos, A Discrete Fourier Transform Scheme for Boolean Sums of Trigonometric Operators, in *Multivariate Approximation Theory IV* C. K. Chui, W. Schempp, K. Zeller (eds.), ISNM 90, Birkhäuser, Basel, 1989, 15–24.

7. Delvos, F.-J., and H. Posdorf, Nth order Blending, in *Constructive Theory of Functions of Several Variables*, W. Schempp, K. Zeller (eds.), Springer-Verlag, Berlin, 1977, 53–64.

8. Gordon, W. J., Blending function methods of bivariate and multivariate interpolation and approximation. SIAM J. Numer. Anal. **8** (1971), 158–177.

9. Gordon, W. J., and E. W. Cheney, Bivariate and multivariate interpolation with noncommutative projectors, in *Linear Spaces and Approximation*, P.L. Butzer, B. Sz.-Nagy (eds.), Birkhäuser, Basel, 1977, 381–387.

10. Hlawka, E., F. Firneis, and P. Zinterhofer, *Zahlentheoretische Methoden in der numerischen Mathematik*. R. Oldenbourg Verlag, Wien 1981.

11. Hua, Lookeng, and Yuan Wang, *Applications of Number Theory to Numerical Analysis*. Springer-Verlag/Science Press, Beijing 1981.

12. Jester, S., *Trigonometrische Interpolation von Funktionen mit Sinusdoppelreihenentwicklung*, Diplomarbeit, Ruhr-Universität Bochum, 1991.

13. Lanczos, C., *Discourse on Fourier Series*, Hafner, New York 1966.

14. Tasche, M., Zur Konvergenzbeschleunigung von Fourier-Reihen. Math. Nachr. **90** (1979), 123–134.

15. Tolstov, G. P., *Fourier Series*, Dover, New York 1976.
16. Vetterli, M., and H. J. Nussbaumer, Algorithmes de transformations de Fourier et en cosinus mono et bidimensionnels. Ann. Télécommun. **40** (1985), 466–476.

Günter Baszenski
Fachhochschule Dortmund
Fachbereich Nachrichtentechnik
Sonnenstraße 96–100
4600 Dortmund 1
Germany
guenter.baszenski@ruba.rz.ruhr-uni-bochum.dbp.de

Franz-Jürgen Delvos
Lehrstuhl für Mathematik I
Universität Siegen
Hölderlinstraße 3
5900 Siegen
Germany

Stephanie Jester
Fakultät für Wirtschaftswissenschaften
Ruhr-Universität Bochum
Universitätsstraße 150
4630 Bochum 1
Germany

Numerical Methods of Approximation Theory, Vol. 9
Dietrich Braess and Larry L. Schumaker (eds.), pp. 21–39.
International Series of Numerical Mathematics, Vol. 105

Quasi-interpolation in the Absence of Polynomial Reproduction

R. K. Beatson and W. A. Light

Dedicated to the memory of Lothar Collatz

Abstract. This paper discusses approximation by quasi-interpolants K_h which are constructed using the usual scaling and translation operations on a one-parameter family of functions $\{\psi_\lambda\}_{\lambda>0}$. A particular feature of the analysis is the fact that we do not require K_h to reproduce low degree polynomials. In the first part of the paper, estimates are obtained for $|(f - K_h f)(x)|$ as $h \to 0$, when ψ_λ decays polynomially at infinity. These estimates are then used to analyse quasi-interpolation to functions in $W_\infty^k(\mathbb{R}^n)$ based on Gaussians and Exponentials.

§1. Introduction

One of the common ways of generating approximations defined over the whole of $\mathbb{R}^n$ is to use the technique of quasi-interpolation. One begins with a function ψ, which is usually a finite linear combination of translates of some basic function ϕ. For example, ϕ might be a truncated power function and ψ a B-spline, or ϕ might be a B-spline and ψ a combination of B-splines. The choice of the specific linear combination which generates ψ from ϕ is usually determined by the desire to have ψ decay rapidly at infinity. Then for suitable functions $f \in C(\mathbb{R}^n)$, one defines quasi-interpolants Lf and $K_h f$ by

$$(Lf)(x) = \sum_{\nu \in \mathbb{Z}^n} f(\nu)\psi(x - \nu), \quad x \in \mathbb{R}^n,$$

and

$$(K_h f)(x) = (S_{1/h} L S_h f)(x) = \sum_{\nu \in \mathbb{Z}^n} f(\nu h) \psi\left(\frac{x}{h} - \nu\right), \qquad x \in \mathbb{R}^n, \, h > 0.$$

One usually tries to further arrange the choice of ψ so that L, and consequently K_h, acts as the identity on π_{k-1}, the space of all polynomials of total degree at most $k - 1$ on $\mathbb{R}^n$. This leads in a standard way to an error estimate of the form

$$\|f - K_h f\|_\infty = \mathcal{O}(h^k) \quad \text{as } h \to 0.$$

Quasi-interpolation has been used in a variety of approximation settings including univariate splines, multivariate splines, and radial basis functions. Recent work in the radial basis function context is well-documented in Powell [3].

There are two problems with the above approach. Firstly, it requires some polynomial reproduction capabilities of the quasi-interpolant. For this reason, it does not apply to the Gaussian ϕ given by $\phi(x) = e^{-c\|x\|_2^2}$, $x \in \mathbb{R}^n$. No finite combination of translates of this ϕ yields a quasi-interpolant which reproduces π_0. Secondly, this approach requires that the function f be available over the whole of $\mathbb{R}^n$. In applications this is typically not true; f is usually only available at certain points within some compact domain in $\mathbb{R}^n$.

In this paper we address the first of the problems outlined above. We begin in Section 2 with an elementary abstract treatment of approximation of functions $f \in W_\infty^k(\mathbb{R}^n)$ by quasi-interpolation when ϕ, and therefore ψ, depends on a parameter c, and polynomial reproduction may be absent. In Section 3 this theory is applied to the Gaussian $\phi(x) = e^{-c\|x\|_2^2}$ and the exponential $\phi(x) = e^{-c\|x\|_1}$. The reader should note that approximation by one-parameter families has been previously discussed in Buhmann and Dyn [2], and in de Boor and Ron [1]. In [1], the distance of a given function with certain smoothness properties from the space of dilates of integer translates of a one-parameter family of functions is estimated very accurately. This important work applies in particular to the Gaussian. Their estimates are obtained for a class of functions which are characterised by conditions on the Fourier transform. In contrast, our estimates are for the error in quasi-interpolation and for functions in the Sobolev space $W_\infty^k(\mathbb{R}^n)$.

§2. Error Estimates

In this section we discuss the approximation order which can be achieved by a one-parameter family of functions in $C(\mathbb{R}^n)$. For $k \in \mathbb{N}$, we define

$$E_k = \{f \in C(\mathbb{R}^n) : \text{ there exists } \epsilon = \epsilon(f) > 0 \text{ such that}$$
$$\sup\{|f(x)|(1 + \|x\|)^{n+k+\epsilon} : x \in \mathbb{R}^n\} < \infty\}$$

Let $\{\psi_\lambda\}_{\lambda>0}$ be a one-parameter family of functions in E_k. Using this family, we define the quasi-interpolation operators L_λ by

$$(L_\lambda f)(x) = \sum_{\nu \in \mathbb{Z}^n} f(\nu)\psi_\lambda(x - \nu), \quad x \in \mathbb{R}^n.$$

Whenever $f \in C(\mathbb{R}^n)$ satisfies the growth condition $|f(x)| = \mathcal{O}(\|x\|^k)$ as $\|x\| \to \infty$, the above sum is absolutely convergent for each $x \in \mathbb{R}^n$, and $L_\lambda f$ is well-defined.

In the remainder of the paper it will be convenient to have the following notation available: If $f \in L^1(\mathbb{R}^n)$ then its Fourier transform $\widehat{f}$ will be defined by $\widehat{f}(t) = \int_{\mathbf{R}^n} f(x)e^{-ixt}dx$. α will be a multi-index in $\mathbb{Z}^n$, so $\alpha = (\alpha_1, \ldots, \alpha_n) \in \mathbb{Z}^n$. The normalised monomial $x^\alpha/\alpha!$ will be denoted by $V_\alpha(x)$, where $\alpha! = \alpha_1! \ldots \alpha_n!$. By $V_\alpha(D)$ we will mean $(1/\alpha!)D^\alpha$, although some authors define $V_\alpha(D)$ to be $(1/\alpha!)(-iD)^\alpha$. The shift operator T_x is defined by $(T_x g)(y) = g(y - x)$ for $x, y \in \mathbb{R}^n$ and $g \in C(\mathbb{R}^n)$. The dilation operator S_h is defined by $(S_h g)(x) = g(xh)$. The reflection operator B is defined by $(Bf)(x) = f(-x)$, $x \in \mathbb{R}^n$ and the functions e_x for $x \in \mathbb{R}^n$ by $e_x(y) = e^{ixy}$, $y \in \mathbb{R}^n$. When all the derivatives of order j are almost everywhere bounded, we define the seminorm $|f|_{j,\infty}$ as $\sum_{|\alpha|=j} \|D^\alpha f\|_\infty$. The Sobolev space $W_\infty^k(\mathbb{R}^n)$ consists of all functions f for which $\|f\|_{k,\infty} = \sum_{j=0}^k |f|_{j,\infty}$ is finite.

Theorem 2.1. *Let $k \in \mathbb{N}$ and let $\{\psi_\lambda\}_{\lambda>0} \subset E_k$. For each $\lambda > 0$, define*

$$A(\lambda, \alpha) = \sup_{x \in [0,1]^n} |\delta_{0\alpha} - (L_\lambda T_x V_\alpha)(x)|, \quad \alpha \in \mathbb{Z}^n, \alpha \geq 0, |\alpha| \leq k,$$

where $\delta_{0\alpha}$ is the usual Kronecker delta. Also define

$$B(\lambda, \alpha) = \sup_{x \in [0,1]^n} \sum_{\nu \in \mathbb{Z}^n} |\psi_\lambda(x - \nu)V_\alpha(\nu - x)|, \quad \alpha \in \mathbb{Z}^n, \alpha \geq 0, |\alpha| \leq k.$$

If $f \in C^k(\mathbb{R}^n)$ with the seminorm $|f|_{k,\infty}$ finite, then

$$|(f - L_\lambda f)(x)| \leq \sum_{|\alpha|<k} A(\lambda,\alpha)|(D^\alpha f)(x)| + \sum_{|\alpha|=k} B(\lambda,\alpha)\|D^\alpha f\|_\infty, \quad x \in \mathbb{R}^n.$$

Proof: Firstly, we will show that the quantities $A(\lambda,\alpha)$ and $B(\lambda,\alpha)$ of the statement are well-defined. We note that

$$(L_\lambda T_x V_\alpha)(x) = \sum_{\nu \in \mathbb{Z}^n} V_\alpha(\nu - x)\psi_\lambda(x - \nu).$$

Hence $A(\lambda,\alpha) \leq \delta_{0\alpha} + B(\lambda,\alpha)$ whenever $B(\lambda,\alpha)$ is finite. Now since $\psi_\lambda \in E_k$, there are positive constants C_λ and ϵ_λ with

$$|\psi_\lambda(x)| \leq C_\lambda(1 + \|x\|_\infty)^{-n-k-\epsilon_\lambda}, \quad \text{for } x \in \mathbb{R}^n.$$

Thus for $|\alpha| \leq k$,

$$B(\lambda,\alpha) \leq \sup_{x \in [0,1]^n} \sum_{\nu \in \mathbb{Z}^n} C_\lambda(1 + \|x - \nu\|_\infty)^{-n-k-\epsilon_\lambda}(1 + \|x - \nu\|_\infty)^{|\alpha|}$$

$$\leq C_\lambda \sup_{x \in [0,1]^n} \sum_{\nu \in \mathbb{Z}^n} (1 + \|x - \nu\|_\infty)^{-n-\epsilon_\lambda} < \infty,$$

as required.

Next note that for $\mu \in \mathbb{Z}^n$, $|\alpha| \leq k$, and $x \in \mathbb{R}^n$,

$$(L_\lambda T_x V_\alpha)(x) = \sum_{\nu \in \mathbb{Z}^n} \psi_\lambda(x - \nu)V_\alpha(\nu - x)$$

$$= \sum_{\nu \in \mathbb{Z}^n} \psi_\lambda(x + \mu - \nu)V_\alpha(\nu - (x + \mu))$$

$$= (L_\lambda T_{x+\mu} V_\alpha)(x + \mu).$$

Hence the function $x \mapsto (L_\lambda T_x V_\alpha)(x)$ is 1-periodic in each component of x. Thus

$$\sup_{x \in \mathbb{R}^n} |\delta_{0\alpha} - (L_\lambda T_x V_\alpha)(x)| = A(\lambda,\alpha), \quad |\alpha| \leq k.$$

The same periodicity argument shows that

$$\sup_{x \in \mathbb{R}^n} \sum_{\nu \in \mathbb{Z}^n} |\psi_\lambda(x - \nu)V_\alpha(\nu - x)| = B(\lambda,\alpha), \quad |\alpha| \leq k.$$

We now apply Taylor's theorem to obtain the error estimate. Fix $x \in \mathbb{R}^n$ and $f \in C^k(\mathbb{R}^n)$ with $|f|_{k,\infty}$ finite. For $y \in \mathbb{R}^n$, write

$$f(y) = \sum_{|\alpha|<k} (D^\alpha f)(x) V_\alpha(y-x) + \sum_{|\alpha|=k} (D^\alpha f)(\xi_y) V_\alpha(y-x)$$
$$= p(y) + r_x(y),$$

where $p \in \pi_{k-1}$ is the Taylor polynomial of degree $k-1$ to f at x, r_x is the Taylor remainder, and ξ_y lies on the line segment joining x to y. Finally,

$$|f(x) - (L_\lambda f)(x)| \leq |f(x) - (L_\lambda p)(x)| + |(L_\lambda p)(x) - (L_\lambda f)(x)|$$
$$= |f(x) - (L_\lambda p)(x)| + |(L_\lambda r_x)(x)|$$
$$= \left| f(x) - \sum_{|\alpha|<k} (D^\alpha f)(x) \sum_{\nu \in \mathbb{Z}^n} \psi_\lambda(x-\nu) V_\alpha(\nu-x) \right|$$
$$+ \left| \sum_{\nu \in \mathbb{Z}^n} \psi_\lambda(x-\nu) \sum_{|\alpha|=k} (D^\alpha f)(\xi_\nu) V_\alpha(\nu-x) \right|$$
$$\leq \sum_{|\alpha|<k} |(D^\alpha f)(x)| |\delta_{0\alpha} - (L_\lambda T_x V_\alpha)(x)|$$
$$+ \sum_{|\alpha|=k} \|D^\alpha f\|_\infty \sum_{\nu \in \mathbb{Z}^n} |\psi_\lambda(x-\nu) V_\alpha(\nu-x)|$$
$$= \sum_{|\alpha|<k} |(D^\alpha f)(x)| A(\lambda,\alpha) + \sum_{|\alpha|=k} \|D^\alpha f\|_\infty B(\lambda,\alpha). \qquad \blacksquare$$

Corollary 2.2. *Assume the hypotheses of Theorem 2.1. Then for $\lambda = \lambda(h)$, $f \in W_\infty^k(\mathbb{R}^n) \cap C^k(\mathbb{R}^n)$ and with $K_h = S_{1/h} L_\lambda S_h$,*

$$\|f - K_h f\|_\infty \leq \sum_{|\alpha|<k} A(\lambda,\alpha) h^{|\alpha|} \|D^\alpha f\|_\infty + \sum_{|\alpha|=k} B(\lambda,\alpha) h^k \|D^\alpha f\|_\infty.$$

Proof: We have, using Theorem 2.1,

$$\|f - K_h f\|_\infty = \|S_{1/h} S_h f - S_{1/h} L_\lambda S_h f\|_\infty$$
$$= \|S_h f - L_\lambda S_h f\|_\infty$$
$$\leq \sum_{|\alpha|<k} A(\lambda,\alpha) \|D^\alpha S_h f\|_\infty + \sum_{|\alpha|=k} B(\lambda,\alpha) \|D^\alpha S_h f\|_\infty$$
$$= \sum_{|\alpha|<k} A(\lambda,\alpha) h^{|\alpha|} \|D^\alpha f\|_\infty + \sum_{|\alpha|=k} B(\lambda,\alpha) h^k \|D^\alpha f\|_\infty. \qquad \blacksquare$$

Of course, the bound in Corollary 2.2 is not immediately informative, since it involves not only h^k, but a hidden dependence on h in the values of $A(\lambda, \alpha)$ and $B(\lambda, \alpha)$. We now indicate how the Poisson summation formula (Stein and Weiss, [4]) can be used to estimate $A(\lambda, \alpha)$. For all $x \in \mathbb{R}^n$,

$$
\begin{aligned}
(L_\lambda T_x V_\alpha)(x) &= \sum_{\nu \in \mathbb{Z}^n} (T_x V_\alpha)(\nu) \psi_\lambda(x - \nu) \\
&= \sum_{\nu \in \mathbb{Z}^n} (T_x V_\alpha)(\nu)(B\psi_\lambda)(\nu - x) \\
&= \sum_{\nu \in \mathbb{Z}^n} (T_x(V_\alpha B\psi_\lambda))(\nu).
\end{aligned}
$$

Now the Poisson summation formula, if applicable, would yield, for all $x \in \mathbb{R}^n$,

$$
\begin{aligned}
(L_\lambda T_x V_\alpha)(x) &= \sum_{\nu \in \mathbb{Z}^n} \big(T_x(V_\alpha B\psi_\lambda)\big)\widehat{\ }(2\pi\nu) \\
&= \sum_{\nu \in \mathbb{Z}^n} \big(e_{-x} V_\alpha(iD)(B\widehat{\psi}_\lambda)\big)(2\pi\nu) \\
&= \sum_{\nu \in \mathbb{Z}^n} e^{-2\pi i x\nu}(-1)^{|\alpha|}[V_\alpha(iD)\widehat{\psi}_\lambda](-2\pi\nu) \\
&= \sum_{\nu \in \mathbb{Z}^n} e^{2\pi i x\nu}[V_\alpha(-iD)\widehat{\psi}_\lambda](2\pi\nu),
\end{aligned}
$$

so that

$$
A(\lambda, \alpha) = \sup_{x \in [0,1]^n} \Big| \delta_{0\alpha} - \sum_{\nu \in \mathbb{Z}^n} e^{2\pi i x\nu}[V_\alpha(-iD)\widehat{\psi}_\lambda](2\pi\nu)\Big|.
$$

A sufficient condition for this application of the Poisson summation formula to be valid is the absolute convergence of the sums

$$
\sum_{\nu \in \mathbb{Z}^n} V_\alpha(\nu - x)\psi_\lambda(x - \nu) \quad \text{and} \quad \sum_{\nu \in \mathbb{Z}^n} e^{2\pi i x\nu}[V_\alpha(-iD)\widehat{\psi}_\lambda](2\pi\nu).
$$

Note that in both the above sums, the value of x has little to do with the absolute convergence, since the first sum is 1-periodic in x, while in the second sum the parameter x appears only as the argument of a function with modulus one.

In the applications we have in mind, the multivariate results can be lifted from univariate results via tensor product arguments. We outline a general form of this procedure. Let $\theta_h \in C(\mathbb{R})$ satisfy $\sum_{\nu \in \mathbb{Z}} |\theta_h(x - \nu)| \leq A$ for $x \in \mathbb{R}$, $0 < h$. Let $g \in W_\infty^k(\mathbb{R}) \cap C^k(\mathbb{R})$ and define a quasi-interpolant by

$$(K_h g)(t) = \sum_{\nu \in \mathbb{Z}} g(h\nu)\theta_h\left(\frac{t}{h} - \nu\right), \quad t \in \mathbb{R}, \, h > 0.$$

Suppose now there exists a constant $C > 0$ and a function $\rho : (0, 1) \to \mathbb{R}$ such that for all $g \in W_\infty^k(\mathbb{R}) \cap C^k(\mathbb{R})$,

$$\|g - K_h g\|_\infty \leq C\rho(h)\|g\|_{k,\infty}, \quad 0 < h < 1.$$

For each $h > 0$ define $\psi_h \in C(\mathbb{R}^n)$ by

$$\psi_h(x) = \Pi_{i=1}^n \theta_h(x_i) \quad \text{where } x = (x_1, \ldots, x_n) \in \mathbb{R}^n.$$

The functions ψ_h are now used to construct another quasi-interpolant. Fix $f \in W_\infty^k(\mathbb{R}^n) \cap C^k(\mathbb{R}^n)$. For each $x \in \mathbb{R}^n$ and $1 \leq j \leq n$, define

$$(K_h^j f)(x) = \sum_{\nu_j \in \mathbb{Z}} f(x_1, \ldots, x_{j-1}, h\nu_j, x_{j+1}, \ldots, x_n)\theta_h\left(\frac{x_j}{h} - \nu_j\right).$$

For $x \in \mathbb{R}^n$ and $1 \leq j \leq n$, set $x^j = (x_1, \ldots, x_{j-1}, x_{j+1}, \ldots, x_n)$ and define $g(x^j, \cdot) \in C(\mathbb{R})$ by $g(x^j, t) = f(x_1, \ldots, x_{j-1}, t, x_{j+1}, \ldots, x_n)$. Then we can write

$$(K_h^j f)(x) = \sum_{\nu_j \in \mathbb{Z}} g(x^j, h\nu_j)\theta_h\left(\frac{x_j}{h} - \nu_j\right).$$

Furthermore, for $x \in \mathbb{R}^n$,

$$|f(x) - (K_h^j f)(x)| \leq C\rho(h)\|g(x^j, \cdot)\|_{k,\infty} \leq C\rho(h)\|f\|_{k,\infty}.$$

Now set $Q_h = \Pi_{j=1}^n K_h^j = K_h \otimes K_h \otimes \cdots \otimes K_h$. Set $q_1 = f$ and $q_j = K_h^{j-1} q_{j-1}$, $j = 2, 3, \ldots, n+1$. Then a straightforward induction argument gives

$$|f(x) - (Q_h f)(x)| = |f(x) - \left[\left(\Pi_{j=1}^n K_h^j f\right)\right](x)|$$

$$\leq \sum_{j=1}^n |q_j(x) - (K_h^j q_j)(x)|$$

$$\leq C'\rho(h)\|f\|_{k,\infty},$$

where C' is a suitable constant. Thus, in the tensor product case, the univariate and n-variate error estimates are of the same order of magnitude.

§3. Applications

In this section we show how the analysis of the previous section yields order of convergence estimates for quasi-interpolation by Gaussians and exponentials. We begin with the Gaussian

$$\phi_\lambda(x) = \frac{\lambda}{\sqrt{2\pi}} e^{-\lambda^2 x^2/2}, \quad x \in \mathbb{R}.$$

It is well-known that

$$\widehat{\phi}_\lambda(t) = e^{-t^2/(2\lambda^2)}, \quad t \in \mathbb{R}.$$

Lemma 3.1. *Let ϕ_λ, $\lambda > 0$, be defined as above. Then, for $j = 0, 1, 2, \ldots$, the function f_λ defined by $f_\lambda(x) = x^j \phi_\lambda(x)$ satisfies*

$$\|f_\lambda\|_\infty = \frac{j^{j/2}}{\sqrt{2\pi}} \lambda^{1-j} e^{-j/2}.$$

Lemma 3.2. *Let $f \in C(\mathbb{R})$ be defined by $f(t) = e^{-t^2/(2\lambda^2)}$, $t \in \mathbb{R}$. Then for all $k \in \mathbb{N}$, $f^{(k)} = p_k f$, where*

$$p_k(t) = \left(\frac{-1}{\lambda}\right)^k q_k\left(\frac{t}{\lambda}\right), \quad t \in \mathbb{R}.$$

Here each q_k is a monic polynomial of exact degree k, even or odd as k is even or odd, and whose coefficients are independent of λ. Moreover, if k is even, $q_k(0) \neq 0$, and if k is odd, $q_k'(0) \neq 0$.

Proof: Note that

$$f'(t) = \frac{-t}{\lambda^2} e^{-t^2/(2\lambda^2)} = \frac{-t}{\lambda^2} f(t),$$

so that $p_1 = -t/\lambda^2$ and $q_1(t) = t$. Assume the required result holds for $k = 1, 2, \ldots, m$. Then

$$f^{(m+1)}(t) = p_m'(t)f(t) + p_m(t)f'(t)$$
$$= \left\{ \left(\frac{-1}{\lambda}\right)^m \frac{1}{\lambda} q_m'\left(\frac{t}{\lambda}\right) + \left(\frac{-1}{\lambda}\right)^{m+1} \left(\frac{t}{\lambda}\right) q_m\left(\frac{t}{\lambda}\right) \right\} f(t)$$
$$= \left(\frac{-1}{\lambda}\right)^{m+1} \left\{ \frac{t}{\lambda} q_m\left(\frac{t}{\lambda}\right) - q_m'\left(\frac{t}{\lambda}\right) \right\} f(t).$$

Setting

$$q_{m+1}(t) = t q_m(t) - q_m'(t)$$

completes the inductive step. ∎

Lemma 3.3. *Let* $j = 0, 1, 2, \ldots$, *and let* θ *be a positive real number satisfying* $\theta \leq 1/2$ *if* $j = 0$ *and* $\ln(\theta) \leq -j$ *if* $j \geq 1$. *Then*

$$\theta < \sum_{k=1}^{\infty} k^j \theta^{k^2} < 2\theta.$$

Proof: For $j = 0$,

$$\theta < \sum_{k=1}^{\infty} \theta^{k^2} < \sum_{k=1}^{\infty} \theta^k = \frac{\theta}{1-\theta} < 2\theta.$$

For $j \geq 1$ observe that for $k \geq 2$,

$$\begin{aligned}
\ln(k^j \theta^{k^2}) &= k^2 \ln \theta + j \ln k \\
&\leq k^2 \ln \theta - \ln \theta \ln k \\
&\leq k^2 \ln \theta - k \ln \theta \\
&\leq k \ln \theta.
\end{aligned}$$

The above inequality is strict if $k > 2$ and so

$$\theta < \sum_{k=1}^{\infty} k^j \theta^{k^2} < \sum_{k=1}^{\infty} \theta^k = \frac{\theta}{1-\theta} < 2\theta. \quad \blacksquare$$

With these lemmas in place, we are able to proceed with the analysis of the order of approximation. Note that the arguments involving the Poisson summation formula which were presented towards the end of Section 2 are valid here. (The required sums are absolutely convergent.) Thus when ϕ_λ is the Gaussian, and $\psi_\lambda = \phi_\lambda$,

$$A(\lambda, \alpha) = \sup_{x \in [0,1]} \left| \delta_{0\alpha} - \sum_{\nu \in \mathbb{Z}} e^{2\pi i x \nu} \left[V_\alpha(-iD)\widehat{\phi}_\lambda \right] (2\pi\nu) \right|.$$

We begin with $\alpha = 0$ and use Lemma 3.3 to obtain

$$\left| 1 - \sum_{\nu \in \mathbb{Z}} e^{2\pi i x \nu} e^{-(2\pi\nu)^2/(2\lambda^2)} \right| \leq 2 \sum_{\nu=1}^{\infty} e^{-2\pi^2 \nu^2/\lambda^2} < 4 e^{-2\pi^2/\lambda^2}, \tag{3.0}$$

providing λ is sufficiently small to ensure that $e^{-2\pi^2/\lambda^2} \leq 1/2$. For $\alpha > 0$ we have

$$\alpha! \left| \sum_{\nu \in \mathbb{Z}} e^{2\pi i x \nu} \left[V_\alpha(-iD)\widehat{\phi}_\lambda \right] (2\pi\nu) \right|$$

$$\leq \alpha! \left| \left[V_\alpha(-iD)\widehat{\phi}_\lambda \right](0) \right| + 2\alpha! \sum_{\nu=1}^{\infty} \left| \left[V_\alpha(-iD)\widehat{\phi}_\lambda \right](2\pi\nu) \right|$$

$$= \left| \widehat{\phi}_\lambda^{(\alpha)}(0) \right| + 2 \sum_{\nu=1}^{\infty} \left| \widehat{\phi}_\lambda^{(\alpha)}(2\pi\nu) \right|.$$

Applying Lemma 3.2, $\widehat{\phi}_\lambda^{(\alpha)}(0) = 0$ if α is odd and $\widehat{\phi}_\lambda^{(\alpha)}(0) = c_\alpha \lambda^{-\alpha}$ if α is even, where the parameter c_α is independent of λ. From Lemma 3.2 again,

$$\widehat{\phi}_\lambda^{(\alpha)}(2\pi\nu) = \left(\frac{-1}{\lambda} \right)^\alpha \sum_{\ell=0}^{\alpha} a_\ell \left(\frac{2\pi\nu}{\lambda} \right)^\ell \widehat{\phi}_\lambda(2\pi\nu).$$

Now further restrict λ so that $e^{-2\pi^2/\lambda^2} \leq \min\{1/2, e^{-k-1}\}$. Using Lemma 3.3 and adjusting c_α upwards, we see that for $0 \leq \alpha \leq k+1$,

$$\sum_{\nu=1}^{\infty} \left| \widehat{\phi}_\lambda^{(\alpha)}(2\pi\nu) \right| \leq \left(\frac{1}{\lambda} \right)^\alpha \sum_{\ell=0}^{\alpha} \frac{a_\ell}{\lambda^\ell} \sum_{\nu=1}^{\infty} (2\pi\nu)^\ell e^{-2\pi^2\nu^2/\lambda^2}$$

$$\leq 2\lambda^{-\alpha} \sum_{\ell=0}^{\alpha} a_\ell \left(\frac{2\pi}{\lambda} \right)^\ell e^{-2\pi^2/\lambda^2}$$

$$\leq c_\alpha \lambda^{-2\alpha} e^{-2\pi^2/\lambda^2}. \tag{3.1}$$

It now follows that for $0 \leq \alpha \leq k+1$,

$$\left| \sum_{\nu \in \mathbb{Z}} e^{2\pi i x \nu} \left[V_\alpha(-iD)\widehat{\phi}_\lambda \right] (2\pi\nu) \right| \leq \frac{c_\alpha \lambda^{-\alpha}}{\alpha!} (\epsilon_\alpha + \lambda^{-\alpha} e^{-2\pi^2/\lambda^2}), \tag{3.2}$$

where $\epsilon_\alpha = 1$ if α is even and $\epsilon_\alpha = 0$ if α is odd. That is, when ϕ_λ is the Gaussian, $\psi_\lambda = \phi_\lambda$, $0 < \alpha \leq k$, and λ is sufficiently small, $A(\lambda, \alpha)$ is bounded above by the right hand side of (3.2).

We now want to make a choice $\lambda = \lambda(h)$ such that we may draw the conclusion from Corollary 2.2 that for $f \in W_\infty^k(\mathbb{R})$, $\|f - K_h f\|_\infty$ is $\mathcal{O}(h^k)$ or something similar, perhaps involving a lower power of h, or some $|\ln h|$

terms. Unfortunately, for rates of convergence of $\mathcal{O}(h^\beta)$ where $\beta \geq 2$, there is now a conflict. Considering the estimate (3.0) of $A(\lambda, 0)$ we see that we must choose λ such that $2\pi^2/\lambda^2 \geq \beta|\ln h|$. Such a choice of λ then makes our estimate (3.2) of $A(\lambda, 2)$ increase without bound as $h \to 0$. Since ϕ_λ is positive, (3.2) also provides a bound for $B(\lambda, 2)$. Hence, our estimates cannot yield an error bound of the form $\mathcal{O}(h^\beta)$ for any $\beta \geq 2$. The way around this problem is to form a linear combination ψ_λ of translates of the function ϕ_λ so that $\widehat{\psi}_\lambda^{(\alpha)}(0) = \delta_{0\alpha}$, $\alpha < k$. Accordingly, we write $\psi_\lambda(x) = \sum_{j \in J} \mu_j \phi_\lambda(x - j)$ where J is a finite subset of $\mathbb{Z}$. Then

$$\widehat{\psi}_\lambda(t) = \sum_{j \in J} \mu_j e^{-ijt} \widehat{\phi}_\lambda(t), \quad t \in \mathbb{R}.$$

Lemma 3.4. *Let* $m = 2\beta$ *be a nonnegative even integer and* $g \in C(\mathbb{R})$ *be an even function satisfying* $g(0) = 1$, *and having* $m + 1$ *continuous derivatives in a neighbourhood of the origin. Let*

$$b_\alpha = D^\alpha \left(\frac{1}{g}\right)(0), \quad 0 \leq \alpha \leq m + 1. \tag{3.3}$$

Then there exist coefficients μ_j, $-\beta \leq j \leq \beta$, *such that*

(i) $\mu_{-j} = \mu_j$.

(ii) if

$$v(t) = \sum_{j=-\beta}^{\beta} \mu_j e^{-ijt} = 2 \sum_{j=0}^{\beta}{}' \mu_j \cos jt$$

then

$$v^{(\alpha)}(0) = b_\alpha, \quad 0 \leq \alpha \leq m + 1. \tag{3.4}$$

(iii) if $y = gv$ *then*

$$y^{(\alpha)}(0) = \delta_{0\alpha}, \quad 0 \leq \alpha \leq m + 1. \tag{3.5}$$

(iv) there exists a constant C *depending on* m *alone and not on* g, *such that*

$$\|\mu\|_\infty \leq C\|b\|_\infty.$$

Remark: If g has only m continuous derivatives in a neighbourhood of the origin then Lemma 3.4 continues to hold with m replacing $m + 1$ in each of (3.3), (3.4) and (3.5).

Proof: Since $g(0) = 1$ and g has $m + 1$ continuous derivatives in a neighbourhood of the origin, the same is true of $1/g$. Thus we can write

$$\frac{1}{g(t)} = p(t) + \mathcal{O}(t^{m+1}),$$

for small t, where p is the Taylor polynomial of degree m corresponding to $1/g$ at $t = 0$. Our aim is to choose $\mu_{-\beta}, \ldots, \mu_\beta$ so that $v(t) = p(t) + \mathcal{O}(t^{m+1})$ for small t. The definition of each b_α implies $p(t) = \sum_{\ell=0}^{m} \frac{b_\ell}{\ell!} t^\ell$. Hence, we require for small t

$$\sum_{\ell=0}^{m} \frac{(-it)^\ell}{\ell!} \sum_{j=-\beta}^{\beta} \mu_j j^\ell = \sum_{\ell=0}^{m} \frac{b_\ell}{\ell!} t^\ell.$$

Equating coefficients of t^ℓ gives

$$\sum_{j=-\beta}^{\beta} \mu_j j^\ell = (-i)^\ell b_\ell, \quad 0 \le \ell \le m. \tag{3.6}$$

This system of equations has a unique solution, since the associated matrix is the transpose of a Vandermonde matrix. By symmetry considerations and the fact that $b_1 = b_3 = \cdots = b_{2\beta-1} = 0$, we see that $\mu_{-j} = \mu_j$, $-\beta \le j \le \beta$. This establishes (i) and, using also the evenness of $1/g$ and v, (ii). From (ii)

$$v(t) = \frac{1}{g(t)} + \mathcal{O}(t^{m+1}) \quad \text{as } t \to 0,$$

and hence

$$y(t) = g(t)v(t) = 1 + \mathcal{O}(t^{m+1}) \quad \text{as } t \to 0.$$

This and the evenness of y establishes (iii). Finally, writing (3.6) in the form $A\mu = b$, where $\mu = (\mu_{-\beta}, \ldots, \mu_\beta)^T$ and $b = (b_0, \ldots, b_m)^T$, we can observe that A depends only on m, and in particular is independent of g. The inequality $\|\mu\|_\infty \le \|A^{-1}\|_\infty \|b\|_\infty$ then establishes (iv). $\blacksquare$

We now return to the problem of approximating a function $f \in W_\infty^k(\mathbb{R}) \cap C^k(\mathbb{R})$ by quasi-interpolation. If k is even, set $m = k - 2$; otherwise, set $m = k-1$. Suppose the functions $\{\phi_\lambda\}_{\lambda>0}$ in E_k are even and satisfy $\widehat{\phi}_\lambda(0) = 1$

and $\widehat{\phi}_\lambda \in C^{k-1}(\mathbb{R})$ for $\lambda > 0$. With $\beta = m/2$ and $g = \widehat{\phi}_\lambda$, choose $\mu_{-\beta}, \ldots, \mu_\beta$ in accordance with Lemma 3.4. Set

$$\psi_\lambda(x) = \mu_0 \phi_\lambda(x) + \sum_{j=1}^{\beta} \mu_j \left\{ \phi_\lambda(x-j) + \phi_\lambda(x+j) \right\}, \quad x \in \mathbb{R}. \qquad (3.7)$$

The quasi-interpolant will then be

$$(L_\lambda f)(x) = \sum_{\nu \in \mathbb{Z}} f(h\nu) \psi_\lambda \left(\frac{x}{h} - \nu \right), \quad x \in \mathbb{R}.$$

The form of ψ_λ ensures that $\widehat{\psi}_\lambda = \widehat{\phi}_\lambda v$ where

$$v(t) = \sum_{j=-\beta}^{\beta} \mu_j e^{-ijt} = 2 \sum_{j=0}^{\beta}{}' \mu_j \cos jt, \quad t \in \mathbb{R}.$$

Also from Lemma 3.4, $\widehat{\psi}_\lambda^{(\alpha)}(0) = \delta_{0\alpha}$, $\alpha = 0, 1, \ldots, k-1$. For $\nu \in \mathbb{Z}$ with $\nu \neq 0$, the Leibnitz rule together with the 2π–periodicity of v gives for $\alpha = 0, \ldots, k-1$

$$\widehat{\psi}_\lambda^{(\alpha)}(2\pi\nu) = \sum_{\ell=0}^{\alpha} \binom{\alpha}{\ell} v^{(l)}(2\pi\nu) \widehat{\phi}_\lambda^{(\alpha-\ell)}(2\pi\nu)$$

$$= \sum_{\ell=0}^{\alpha} \binom{\alpha}{\ell} v^{(l)}(0) \widehat{\phi}_\lambda^{(\alpha-\ell)}(2\pi\nu)$$

$$= \sum_{\ell=0}^{\alpha} \binom{\alpha}{\ell} b_\ell \widehat{\phi}_\lambda^{(\alpha-\ell)}(2\pi\nu), \qquad (3.8)$$

where, as in Lemma 3.4, $b_\alpha = D^\alpha(1/\widehat{\phi}_\lambda)(0)$. We now need to compute bounds on $A(\lambda, \alpha)$ and $B(\lambda, \alpha)$. Assuming ϕ and $\widehat{\phi}$ decay sufficiently fast for the Poisson summation arguments to apply, we need to examine the quantity

$$A(\lambda, \alpha, x) = \left| \delta_{0\alpha} - \sum_{\nu \in \mathbb{Z}} e^{2\pi ix\nu} \left[V_\alpha(-iD) \widehat{\psi}_\lambda \right](2\pi\nu) \right|.$$

From (3.8) for $0 \leq \alpha < k$

$$\alpha! A(\lambda, \alpha, x) \leq \left| \delta_{0\alpha} - \widehat{\psi}_\lambda^{(\alpha)}(0) \right| + 2 \sum_{\nu=1}^{\infty} \left| \widehat{\psi}_\lambda^{(\alpha)}(2\pi\nu) \right|$$

$$= 2 \sum_{\nu=1}^{\infty} \left| \sum_{\ell=0}^{\alpha} \binom{\alpha}{\ell} b_\ell \widehat{\phi}_\lambda^{(\alpha-\ell)}(2\pi\nu) \right|$$

$$\leq 2 \sum_{\ell=0}^{\alpha} \binom{\alpha}{\ell} |b_\ell| \sum_{\nu=1}^{\infty} \left| \widehat{\phi}_\lambda^{(\alpha-\ell)}(2\pi\nu) \right|. \qquad (3.9)$$

Example A. (The Gaussian). Take $\phi_\lambda(x) = \dfrac{\lambda}{\sqrt{2\pi}} e^{-\lambda^2 x^2/2}$, $x \in \mathbb{R}$. Then

$$\frac{1}{\widehat{\phi}_\lambda(t)} = e^{t^2/(2\lambda^2)} = 1 + \frac{t^2}{2\lambda^2} + \cdots + \frac{t^{2\beta}}{\beta! 2^\beta \lambda^{2\beta}} + \cdots$$

so that the constants b_α of Lemma 3.4 and (3.9) are given by

$$b_\ell = \begin{cases} \dfrac{\ell!}{(\frac{\ell}{2})!\, 2^{\ell/2} \lambda^\ell}, & \ell \text{ even}, \\[2ex] 0, & \ell \text{ odd}. \end{cases}$$

Assume that λ is such that $e^{-2\pi^2/\lambda^2} \le \min\{1/2, e^{-k+1}\}$. Then using (3.1) gives, for $0 \le \alpha < k$,

$$\alpha! A(\lambda, \alpha, x) \le 2 \sum_{\ell=0}^{\alpha} \binom{\alpha}{\ell} |b_\ell| \sum_{\nu=1}^{\infty} \left| \widehat{\phi}_\lambda^{(\alpha-\ell)}(2\pi\nu) \right|$$

$$\le 2 \sum_{\ell=0}^{\alpha} \binom{\alpha}{\ell} \frac{\ell!}{(\frac{\ell}{2})! 2^{\ell/2}} \lambda^{-\ell} c_{\alpha-\ell} \lambda^{-2(\alpha-\ell)} e^{-2\pi^2/\lambda^2}$$

$$\le C\alpha!\, \lambda^{-2\alpha} e^{-2\pi^2/\lambda^2},$$

where

$$C = (2/\alpha!) \sum_{\ell=0}^{\alpha} \binom{\alpha}{\ell} \frac{\ell!}{(\frac{\ell}{2})! 2^{\ell/2}} c_{\alpha-\ell}.$$

It follows immediately that

$$A(\lambda, \alpha) \le C\lambda^{-2\alpha} e^{-2\pi^2/\lambda^2}, \quad 0 \le \alpha < k. \tag{3.10}$$

Lemma 3.5. *Let $k \in \mathbb{N}$ and define $m = k - 2$ if k is even and $m = k - 1$ otherwise. Let ϕ_λ be defined by $\phi_\lambda(x) = \dfrac{\lambda}{\sqrt{2\pi}} e^{-\lambda^2 x^2/2}$, $x \in \mathbb{R}$, $\lambda > 0$. Let ψ_λ be defined as in (3.7). Then*

$$B(\lambda, k) = \sup_{x \in [0,1]} \sum_{\nu \in \mathbb{Z}} \left| \psi_\lambda(x - \nu) V_k(\nu - x) \right| = \mathcal{O}(\lambda^{-m-k}), \quad \text{as } \lambda \to 0.$$

Proof: Set $\beta = m/2$. Fix $x \in [0,1]$. From (3.7) and the binomial theorem

$$
\sum_{\nu \in \mathbb{Z}} \left| \psi_\lambda(x-\nu)(\nu-x)^k \right| \leq \sum_{j=-\beta}^{\beta} |\mu_j| \sum_{\nu \in \mathbb{Z}} |x-\nu|^k \phi_\lambda(x-\nu-j)
$$

$$
= \sum_{j=-\beta}^{\beta} |\mu_j| \sum_{\nu \in \mathbb{Z}} |x-\nu+j|^k \phi_\lambda(x-\nu)
$$

$$
= \sum_{j=-\beta}^{\beta} |\mu_j| \sum_{\nu \in \mathbb{Z}} \sum_{\ell=0}^{k} \binom{k}{\ell} |j|^{k-\ell} |x-\nu|^\ell \phi_\lambda(x-\nu)
$$

$$
\leq \|\mu\|_\infty \sum_{j=-\beta}^{\beta} \sum_{\ell=0}^{k} \binom{k}{\ell} |j|^{k-\ell} \sum_{\nu \in \mathbb{Z}} |x-\nu|^\ell \phi_\lambda(x-\nu).
$$

Suppose now that $0 \leq \ell \leq k+1$ and ℓ is even. Then by the Poisson summation formula and (3.2),

$$
\sum_{\nu \in \mathbb{Z}} |x-\nu|^\ell \phi_\lambda(x-\nu) = \sum_{\nu \in \mathbb{Z}} (x-\nu)^\ell \phi_\lambda(x-\nu)
$$

$$
= \ell! \sum_{\nu \in \mathbb{Z}} e^{2\pi i x \nu} \left[V_\ell(-iD)\widehat{\phi}_\lambda \right](2\pi\nu)
$$

$$
\leq c_\ell \lambda^{-\ell} \left(\epsilon_\ell + \lambda^{-\ell} e^{-2\pi^2/\lambda^2} \right),
$$

providing λ is sufficiently small. By decreasing λ if necessary, we can assume that for each even ℓ, $0 \leq \ell \leq k+1$, there is a number d_ℓ such that

$$
\sum_{\nu \in \mathbb{Z}} |x-\nu|^\ell \phi_\lambda(x-\nu) \leq d_\ell \lambda^{-\ell}.
$$

Suppose now that ℓ is odd, $0 \leq \ell \leq k$ and $\ell = 2\beta+1$. Once again for all sufficiently small λ, we have, by the Cauchy–Schwarz inequality,

$$
\sum_{\nu \in \mathbb{Z}} |x-\nu|^{2\beta+1} \phi_\lambda(x-\nu)
$$

$$
\leq \left(\sum_{\nu \in \mathbb{Z}} |x-\nu|^{2\beta} \phi_\lambda(x-\nu) \right)^{\frac{1}{2}} \left(\sum_{\nu \in \mathbb{Z}} |x-\nu|^{2\beta+2} \phi_\lambda(x-\nu) \right)^{\frac{1}{2}}
$$

$$
\leq \left(d_{2\beta} \lambda^{-2\beta} \right)^{\frac{1}{2}} \left(d_{2\beta+2} \lambda^{-2\beta-2} \right)^{\frac{1}{2}}
$$

$$
= \left(d_{2\beta} d_{2\beta+2} \right)^{\frac{1}{2}} \lambda^{-\ell}.
$$

It now follows from Lemma 3.4(iv) and our estimate of $\|b\|_\infty$, that

$$
B(\lambda, k) \leq \frac{\|\mu\|_\infty}{k!} \sum_{j=-\beta}^{\beta} \sum_{\ell=0}^{k} \binom{k}{\ell} |j|^{k-\ell} \sup_{x \in [0,1]} \sum_{\nu \in \mathbb{Z}} |x - \nu|^\ell \phi_\lambda(x - \nu)
$$
$$
\leq D\lambda^{-k-m},
$$

where D is a suitable constant, which establishes the required result. ∎

Theorem 3.6. *Let $k \in \mathbb{N}$, and let $\phi_\lambda \in C(\mathbb{R}^n)$ be given by $\phi_\lambda(x) = \left(\lambda/\sqrt{2\pi}\right)^n e^{-\lambda^2 x^2/2}$, $x \in \mathbb{R}^n$, $\lambda > 0$. Let $J = \{z \in \mathbb{Z}^n : \|z\|_\infty \leq [(k-1)/2]\}$ and define $\psi_\lambda \in C(\mathbb{R}^n)$ by*

$$
\psi_\lambda(x) = \sum_{z \in J} \mu_z \phi_\lambda(x - z), \quad x \in \mathbb{R}^n,
$$

where $\mu_z = \mu_z(\lambda) \in \mathbb{R}$. Also define the quasi-interpolation operator L_λ on $W_\infty^k(\mathbb{R}^n)$ by

$$
(L_\lambda f)(x) = \sum_{\nu \in \mathbb{Z}^n} f(\nu)\psi_\lambda(x - \nu), \quad x \in \mathbb{R}^n, \ f \in W_\infty^k(\mathbb{R}^n).
$$

Then there exists a choice of the coefficients $\{\mu_z : z \in J\}$ and constants $\delta > 0$ and $C > 0$ such that if $\lambda = \sqrt{2\pi^2/(k|\ln h|)}$ then with $K_h = S_{1/h} L_\lambda S_h$

$$
\|f - K_h f\|_\infty \leq Ch^k |\ln h|^{\frac{k}{2} + [\frac{k-1}{2}]} \|f\|_{k,\infty}
$$

for all $f \in W_\infty^k(\mathbb{R}^n)$ and $0 < h < \delta$.

Proof: In view of the discussion of the lifting of univariate results via a tensor product construction presented at the end of Section 2, it suffices to consider the univariate case. From Corollary 2.2, for $f \in W_\infty^k(\mathbb{R}) \cap C^k(\mathbb{R})$,

$$
\|f - K_h f\|_\infty \leq \sum_{\alpha=0}^{k-1} A(\lambda, \alpha) h^\alpha \|f^{(\alpha)}\|_\infty + B(\lambda, k) h^k \|f^{(k)}\|_\infty.
$$

Let $m = 2[(k-1)/2]$ and choose ψ_λ in accord with (3.7). Then for sufficiently small $\lambda > 0$, (3.10) and Lemma 3.5 show that

$$
A(\lambda, \alpha) = \mathcal{O}(\lambda^{-2\alpha} e^{-2\pi^2/\lambda^2}), \ 0 \leq \alpha < k, \quad \text{and} \quad B(\lambda, k) = \mathcal{O}(\lambda^{-m-k})
$$

as $\lambda \to 0$. Now choose λ as specified in the statement of the theorem. Then $e^{-2\pi^2/\lambda^2} = h^k$ and $\lambda^{-2} = \mathcal{O}(|\ln h|)$ as $h \to 0$. Hence there exist constants $\delta > 0$ and $C > 0$ such that

$$\|f - S_{1/h}L_\lambda S_h f\|_\infty \leq \sum_{\alpha=0}^{k-1} C|\ln h|^\alpha h^k h^\alpha \|f^{(\alpha)}\|_\infty + C|\ln h|^{\frac{m}{2}+\frac{k}{2}} h^k \|f^{(k)}\|_\infty$$

$$\leq C h^k |\ln h|^{\left[\frac{k-1}{2}\right]+\frac{k}{2}} h^k \sum_{\alpha=0}^{k} \|f^{(\alpha)}\|_\infty, \quad 0 < h < \delta.$$

If f is not in $C^k(\mathbb{R}^n)$, but merely in $W_\infty^k(\mathbb{R}^n)$, then we can obtain the result by smoothing. This procedure is well-documented in a number of places. We refer the reader to Jia and Lei [3] as an example of what can be done. ∎

Before moving on to our final example, we make one short observation about Theorem 3.6. As that result is stated, one has to know the order of smoothness k before one can determine the correct relationship between λ and h. Making instead a choice of λ which is independent of k has the desirable effect that the resulting approximating subspaces are independent of the smoothness, k. One suitable choice is

$$\lambda^2 = \frac{1}{|\ln h|(\ln|\ln h|)}.$$

This choice has the consequence that as $h \to 0$ the constants $A(\lambda, \alpha)$ in Theorem 3.6 become smaller. Also, for any $\epsilon > 0$ we have

$$\ln|\ln h| < |\ln h|^\epsilon \quad \text{as } h \to 0.$$

With the aid of this inequality, the constant $B(\lambda, k)$ may be bounded by $|\ln h|^k$ as $h \to 0$. A look at the above analysis then shows that there is a constant $C > 0$ such that, for all sufficiently small $h > 0$,

$$\|f - K_h f\|_\infty \leq C h^k |\ln h|^k \|f\|_{k,\infty}.$$

Example B. (The Exponential). Take $\phi_\lambda(x) = (\lambda/2)e^{-\lambda|x|}$, $x \in \mathbb{R}$, $\lambda > 0$. Then $\widehat{\phi}_\lambda(t) = \lambda^2/(\lambda^2+t^2)$. The Poisson summation formula arguments of the latter part of Section **2** are again valid, and so the useful formula

$$\sum_{\nu \in \mathbb{Z}} (x-\nu)^\ell \phi_\lambda(x-\nu) = \ell! \sum_{\nu \in \mathbb{Z}} e^{2\pi i x \nu} \left[V_\ell(-iD)\widehat{\phi}_\lambda\right](2\pi\nu)$$

continues to hold for nonnegative $\ell \in \mathbb{Z}$. Noting that

$$\widehat{\phi}'_\lambda(t) = -2t\frac{\lambda^2}{(\lambda^2 + t^2)^2} \quad \text{and} \quad \widehat{\phi}_\lambda^{(2)}(t) = \frac{(6\lambda^2 t^2 - 2\lambda^4)}{(\lambda^2 + t^2)^3},$$

we find, when ϕ_λ is the exponential and $\psi_\lambda = \phi_\lambda$,

$$A(\lambda, \alpha) = \mathcal{O}(\lambda^2) \quad \text{for } \alpha = 0 \text{ or } 1,$$

and

$$B(\lambda, 1) = \mathcal{O}(\lambda^{-1}) \quad \text{and} \quad B(\lambda, 2) = \mathcal{O}(\lambda^{-2}),$$

where all the order relationships hold as $\lambda \to 0$. Proceeding as in the proof of Theorem 3.6 for the Gaussian, we can substitute these bounds into Corollary 2.2 and obtain an error estimate in the univariate case. This univariate estimate can then be lifted by tensor product arguments to a multivariate estimate. This yields:

Theorem 3.7. *Let $\phi_\lambda \in C(\mathbb{R}^n)$ be given by $\phi_\lambda(x) = (\lambda/2)^n e^{-\lambda\|x\|_1}$, $x \in \mathbb{R}^n$, $\lambda > 0$. Define the quasi-interpolation operator L_λ on $W^1_\infty(\mathbb{R}^n)$ by*

$$(L_\lambda f)(x) = \sum_{\nu \in \mathbb{Z}^n} f(\nu)\phi_\lambda(x - \nu), \quad x \in \mathbb{R}^n.$$

Let $K_h = S_{1/h}L_\lambda S_h$. Then there exist constants $\delta > 0$ and $C > 0$ such that if $\lambda = h^{1/3}$ and $f \in W^1_\infty(\mathbb{R}^n)$, then

$$\|f - K_h f\|_\infty \le Ch^{2/3}\|f\|_{1,\infty}, \quad 0 < h < \delta.$$

Furthermore, if $\lambda = h^{1/2}$ and $f \in W^2_\infty(\mathbb{R}^n)$, then

$$\|f - K_h f\|_\infty \le Ch\|f\|_{2,\infty}, \quad 0 < h < \delta.$$

Acknowledgements. R.K. Beatson was partially supported by a grant from the Science and Engineering Research Council of Great Britain. W.A. Light was partially supported by the University of Canterbury, New Zealand.

References

1. de Boor, C. and A. Ron, Fourier Analysis of the approximation power of principal shift-invariant spaces, CMS Technical Report No. 92-01, University of Wisconsin, Madison, 1992.
2. Buhmann, M. D. and N. Dyn, Spectral convergence of multiquadric interpolation, Report NA 10, DAMTP, University of Cambridge, 1991.
3. Jia, R. Q. and J. Lei, Approximation by multiinteger translates of functions having non-compact support, J. Approx. Th., to appear.
4. Powell, M. J. D., The theory of radial basis function approximation in 1990 in *Advances in Numerical Analysis II: Wavelets, Subdivision Algorithms and Radial Functions*, W. Light (ed.), Oxford University Press, Oxford, 1992, 105–210.
5. Stein E. M. and G. Weiss, *Introduction to Fourier Analysis on Euclidean Spaces*, Princeton University Press, Princeton, 1971.

Will Light
Dept. of Mathematics
University of Leicester
Leicester, ENGLAND
pwl@leicester.ac.uk

R. K. Beatson
Dept. of Mathematics
Univ. of Canterbury
Canterbury, New Zealand
rkb@math.canterbury.ac.nz

Numerical Methods of Approximation Theory, Vol. 9
Dietrich Braess and Larry L. Schumaker (eds.), pp. 41–52.
International Series of Numerical Mathematics, Vol. 105

ISBN 3-7643-2746-4.

Estimating the Condition Number for Multivariate Interpolation Problems

P. Binev and K. Jetter

Dedicated to the memory of Lothar Collatz

Abstract. We estimate the condition number in multivariate interpolation problems where the interpolation space stems from shifted versions of a single basis function ϕ. The estimate depends on the properties of the Fourier transform $\phi^\wedge$. In case the interpolation points are a subset of a regular grid, the bound is based on properties of the symbol of ϕ (which is a periodization of $\phi^\wedge$). For scattered data interpolation, we assume that ϕ is a positive definite function where the Fourier transform can be properly bounded. As examples, we consider interpolation by box splines or by certain radial basis functions.

§1. Introduction

We give estimates for the condition numbers of matrices of type

$$A = (\phi(x_k - x_\ell))_{k,\ell=1}^N. \qquad (1.1)$$

Such matrices arise in certain interpolation problems; here $\phi : \mathbb{R}^d \to \mathbb{R}$ is a so-called *basis function*, and $x_1, \ldots, x_N \in \mathbb{R}^d$ are the interpolation (or collocation) points, taken to be distinct. We want to interpolate given data at these points by some function from the interpolation space

$$\mathrm{span}\{\phi(\cdot - x_\ell); \quad \ell = 1, \ldots, N\}. \qquad (1.2)$$

A particularly interesting example along these lines is so-called *radial basis interpolation* where $\phi(x) = g(|x|^2)$ with $|x|$ the Euclidean length of the vector x and $g : [0, \infty) \to \mathbb{R}$ a univariate function [9].

The following hypotheses will be assumed throughout the paper: ϕ is a continuous and even function $(\phi(-x) = \phi(x))$, and it is the inverse Fourier transform of an $L_1(\mathbb{R}^d)$-function $\phi^\wedge$:

$$\phi(x) = \frac{1}{(2\pi)^d} \int_{\mathbb{R}^d} \phi^\wedge(\xi)\, e^{ix\cdot\xi}\, d\xi, \quad x \in \mathbb{R}^d, \tag{1.3}$$

where $x \cdot \xi$ is the usual dot product of the vectors x and ξ. Thus, in terms of the Fourier transform $\phi^\wedge$, our assumptions read (with $\phi^\wedge : \mathbb{R}^d \to \mathbb{R}$)

$$\phi^\wedge \in L_1(\mathbb{R}^d) \quad \text{and} \quad \phi^\wedge(\xi) = \phi^\wedge(-\xi), \quad \xi \in \mathbb{R}^d. \tag{1.4}$$

It is clear that these assumptions imply that the collocation matrix is real-symmetric, hence denoting by $\sigma(A)$ the spectrum of A we have the following condition number for A (in case A is regular):

$$\text{cond}(A) = \lambda_{\max}/\lambda_{\min}, \tag{1.5}$$

where $\lambda_{\max} := \max\{|\lambda|; \lambda \in \sigma(A)\}$ and $\lambda_{\min} := \min\{|\lambda|; \lambda \in \sigma(A)\}$.

In order to estimate this condition number, we shall apply the following expression for the quadratic form represented by A: For arbitrary numbers $c_1, \ldots, c_N \in \mathbb{R}$, we have

$$\sum_{k,\ell=1}^{N} c_k \phi(x_k - x_\ell) c_\ell = \frac{1}{(2\pi)^d} \int_{\mathbb{R}^d} \phi^\wedge(\xi)\, |\sum_{k=1}^{N} c_k e^{ix_k\cdot\xi}|^2\, d\xi. \tag{1.6}$$

This identity follows immediately from (1.3). It shows the following result:

Theorem 1.1. *If in addition to (1.4), $\phi^\wedge$ is non-negative, continuous at $\xi = 0$, and $\phi^\wedge(0) > 0$, then the collocation matrix (1.1) is always positive definite.*

In case the interpolation points are from a regular grid, the assumption $\phi^\wedge \geq 0$ can be weakened to $\phi^\sim \geq 0$, where $\phi^\sim$ is the periodization of $\phi^\wedge$ defined in equation (2.4) below. This case is considered in Section 2, and by using standard methods, we get a quite satisfactory result (Theorem 2.1) which can be applied to these examples for bivariate interpolation: Three- and Four-directional box splines, Rabut's thin plate B-splines, and Hardy multiquadrics.

The situation is more challenging in scattered data interpolation, which we examine in Section 3. Here we use a variation of an estimate of Zygmund [11, Section V.9], but we have to make more careful estimates, since we are working in $\mathbb{R}^d$, $d > 1$. For $d = 2$ we derive a rather explicit result (Theorem 3.4).

Our considerations should be compared with results of Narcowich and Ward [7,8], or Baxter [2], which apply to order-one conditionally negative definite functions. In contrast to these results, our method applies to preconditioned functions of that type. Our approach to the problem seems to be very promising, and we hope to have more complete results (with probably even better bounds) in the future.

§2. Interpolation on Regular Grids

There is a standard and well-known procedure to deal with our problem in case the interpolation points come from a regular grid $x_0 + h\mathbb{Z}^d$ (with $h > 0$ the mesh size). By shifting and scaling, we may assume that

$$x_1, \ldots, x_N \in \mathbb{Z}^d. \tag{2.1}$$

Let us assume that in addition to (1.4),

$$\phi^{\wedge} \in C(\mathbb{R}^d), \tag{2.2a}$$

$$\phi^{\wedge}(\xi) = \mathcal{O}(|\xi|^{-d-\varepsilon}) \quad \text{as} \quad |\xi| \to \infty \tag{2.2b}$$

for some $\varepsilon > 0$, and

$$\phi_| := (\phi(\alpha))_{\alpha \in \mathbb{Z}^d} \in \ell_1(\mathbb{Z}^d). \tag{2.2c}$$

Then putting $C = [-\pi, +\pi]^d$, identity (1.6) reads

$$\sum_{k,\ell=1}^{N} c_k \phi(x_k - x_\ell) c_\ell = \frac{1}{(2\pi)^d} \int_C \phi^{\sim}(\xi) \, | \sum_{k=1}^{N} c_k e^{i x_k \cdot \xi} |^2 \, d\xi, \tag{2.3}$$

with

$$\phi^{\sim}(\xi) := \sum_{\alpha \in \mathbb{Z}^d} \phi^{\wedge}(\xi + 2\pi\alpha) \tag{2.4}$$

the 2π-periodization of $\phi^{\wedge}$. Using Poisson's summation formula we see that

$$\phi^{\sim}(\xi) = \sum_{\alpha \in \mathbb{Z}^d} \phi(\alpha) e^{-i\alpha \cdot \xi}. \tag{2.5}$$

Thus, $\phi^{\sim}$ is the symbol of the Toeplitz-Laurent discrete convolution operator on $\ell_2(\mathbb{Z}^d)$ which is defined by the sequence (2.2.c), and hence the range of the symbol is a closed interval,

$$\text{range } \phi^{\sim} = [\lambda, \Lambda]. \tag{2.6}$$

From this we get

$$\lambda \sum_{k=1}^{N} |c_k|^2 \leq \sum_{k,\ell=1}^{N} c_k \phi(x_k - x_\ell) c_\ell \leq \Lambda \sum_{k=1}^{N} |c_k|^2, \tag{2.7}$$

and hence the following result is obtained:

Theorem 2.1. *If ϕ satisfies the assumptions (1.4) and (2.2), and if the collocation points $x_1, \ldots, x_N$ are taken from the lattice $\mathbb{Z}^d$, then the condition number of the collocation matrix (1.1) is uniformly bounded (independent of the number and the location of the points x_i), under the condition that the symbol $\phi^{\sim}$ does not vanish on $\mathbb{R}^d$.*

Example 1. (Three-directional box splines). Here, for given $k, \ell, m \in \mathbb{N}$, the three-directional box spline $\phi = M_{k,\ell,m} : \mathbb{R}^2 \to \mathbb{R}$ is defined by

$$M_{k,\ell,m}^{\wedge}(\xi_1, \xi_2) := (\text{sinc } \frac{\xi_1}{2})^k (\text{sinc } \frac{\xi_2}{2})^\ell (\text{sinc } \frac{\xi_1 + \xi_2}{2})^m$$

with $\text{sinc } t = \sin t / t$. Hence the assumptions (1.4) and (2.2) are satisfied (note that $M_{k,\ell,m}$ is compactly supported). It was shown in [4] that $\phi^{\sim}(\xi) \neq 0$, $\xi \in \mathbb{R}^d$ (an easier proof than the original one can be found in [3]). Moreover

$$\text{range } M_{k,\ell,m}^{\sim} = [\lambda, 1]$$

with $0 < \lambda = \lambda(k, \ell, m)$, whence A is always positive definite and $\text{cond}(A) \leq 1/\lambda$. It should be noted that λ tends to zero as k, ℓ, m increase.

    The special cases $k = \ell = 2$, $m \in \{1, 2\}$ were dealt with by Arge et al. [1] who have performed numerical tests with scaled three-directional box splines $M_{k,\ell,m}(\cdot/h)$ where interpolation on a rectangular subset of $h\mathbb{Z}^2$ was considered. Let us also note that the tensor product case (where $m = 0$) can be dealt with in an even simpler way.

Example 2. (Four-directional box splines). In this case, for given $k, \ell, m, n \in \mathbb{N}$, we have $\phi = M_{k,\ell,m,n} : \mathbb{R}^2 \to \mathbb{R}$ defined by

$$M_{k,\ell,m,n}^{\wedge}(\xi_1, \xi_2) := M_{k,\ell,m}^{\wedge}(\xi_1, \xi_2) \, (\text{sinc } \frac{\xi_1 - \xi_2}{2})^n.$$

While, in general, $\phi^\sim$ may have sign changes, the results of [6] show that in case $k = \ell$, $m = n$, we have

$$\text{range } M^\sim_{k,k,m,m} = [0,1],$$

and hence identity (2.3) shows that the collocation matrix A is always positive definite. But the condition number tends to infinity as the set of interpolation points approaches the full lattice $\mathbb{Z}^2$. This is again in accordance with the results in [1] for $k = \ell \in \{1,2\}$ and $m = n = 1$. A stable way of computing cardinal interpolants in these cases was proposed in [6].

Example 3. (Rabut's thin plate B-spline). Here, $\phi : \mathbb{R}^2 \to \mathbb{R}$ is defined by

$$\phi^\wedge(\xi_1, \xi_2) = \left(\frac{4 - 2\cos\xi_1 - 2\cos\xi_2}{\xi_1^2 + \xi_2^2} \right)^2,$$

which is a special case of Rabut's polyharmonic B-splines [10]. Obviously, $\phi^\wedge(\xi) > 0$ except for $\xi \in 2\pi\mathbb{Z}^2 \setminus \{0\}$; hence $\phi^\sim(\xi) > 0$ for all $\xi \in \mathbb{R}^2$, and we are in the situation where

$$\text{range } \phi^\sim = [\lambda, \Lambda] \quad \text{with} \quad 0 < \lambda < \Lambda.$$

Hence, A is positive definite with uniformly bounded condition numbers.

Example 4. (preconditioned Hardy multiquadric). With $-\nabla$ the discrete Laplacian, setting $\phi(x) := (-\nabla)^{3/2}\sqrt{|x|^2 + 1}$, we get [5, with $d = 2$, $m = 1$, $\beta = \frac{1}{2}$]

$$\phi^\wedge(\xi_1, \xi_2) = 4\sqrt{\pi} \left(\frac{4 - 2\cos\xi_1 - 2\cos\xi_2}{\xi_1^2 + \xi_2^2} \right)^{3/2} \int_0^\infty e^{-(\xi_1^2 + \xi_2^2)\frac{\tau}{4}} \frac{e^{-1/\tau}}{\tau^{5/2}} d\tau.$$

From this we again conclude (as in Example 3) that A is positive definite with uniformly bounded condition numbers.

The latter example should be compared with the approach taken by Baxter [2]. He deals directly with $\varphi(x) = \sqrt{|x|^2 + 1}$, which is an order-one conditionally negative definite function, and he derives the estimate

$$\left| \sum_{k,\ell=1}^{N} c_k \varphi(x_k - x_\ell) c_\ell \right| \geq \lambda \sum_{k=1}^{N} |c_k|^2$$

subject to the condition that $\sum_{k=1}^{N} c_k = 0$; here $\lambda = |\varphi^\sim(\pi e)|$ for $e = (1,1)^T$, where $\varphi^\sim$ is defined as in (2.4), and $\varphi^\wedge$ is the Fourier transform in the generalized sense.

§3. Scattered Data Interpolation

We now turn to the case where the interpolation points are irregularly distributed. There is some way to apply the results of Section 2, viz. let $h > 0$ be such that

$$\operatorname{dist}(x_k, h\mathbb{Z}^d) = 0, \quad k = 1, \ldots, N, \tag{3.1}$$

where $\operatorname{dist}(x, Y) = \inf\{|x - y|; y \in Y\}$, choose ϕ as in Section 2, and put

$$\phi_h := \phi(\cdot/h). \tag{3.2}$$

It is then obvious that $\operatorname{cond}(A_h)$ can be estimated as before, where

$$A_h = (\phi_h(x_k - x_\ell))_{k,\ell=1}^N. \tag{3.3}$$

We note that in the case of the multiquadric, equation (3.2) means that the constant part in the multiquadric is adjusted to the placement of the knots $x_1, \ldots, x_N$ (a method which is frequently used in order to improve the numerical performance).

On the other hand, although

$$q := \min\{|x_k - x_\ell|_\infty; \quad k, \ell = 1, \ldots, N, \quad k \neq \ell\} \tag{3.4}$$

may be big, it could happen that h in (3.1) will be very small and hence ϕ_h given by (3.2) may be a very local function, a fact which is not at all desirable in applications. For this reason it seems to be more favorable to estimate the condition number of $A = (\phi(x_k - x_\ell))_{k,\ell=1}^N$ in terms of the mesh constant q given in (3.4), i.e., we would like to have estimates of type

$$\lambda(q) \sum_{k=1}^N |c_k|^2 \leq \sum_{k,\ell=1}^N c_k \phi(x_k - x_\ell)c_\ell \leq \Lambda(q) \sum_{k=1}^N |c_k|^2 \tag{3.5}$$

generalizing (2.7). Here, $\lambda(q)$ and $\Lambda(q)$ should not depend on N. (Hence, infinite samples of points would be also admitted. On the other hand, c.f. Example 2, this could lead to $\lambda(q) = 0$ for small q.)

Let us assume that (1.4) holds and $\phi^\wedge(\xi) \geq 0$, $\xi \in \mathbb{R}^d$. The basic idea for proving the lower estimate in (3.5) is to find a function $\psi(x)$ for which the inequality (with ϕ replaced by ψ) holds and, in addition,

$$0 \leq \psi^\wedge(\xi) \leq \phi^\wedge(\xi), \quad \xi \in \Omega,$$

(with Ω some neighborhood of the origin). Then the desired estimate is a consequence of (1.6). Here only the case $d = 2$ is considered, since this is often the case in applications, and will avoid notational problems. Similar results hold for $d > 2$, but with worse constants.

The following tensor product analogue of the test function from [11, p. 223] is used:

$$\psi(y) = \psi(y_1, y_2) = \frac{\cos \frac{3}{2}\pi y_1}{1 - 9y_1^2} \cdot \frac{\cos \frac{3}{2}\pi y_2}{1 - 9y_2^2}.$$

Then

$$\psi^\wedge(\xi) = \begin{cases} \frac{\pi^2}{9} \cos \frac{\xi_1}{3} \cos \frac{\xi_2}{3}, & \text{for } \xi \in [-\frac{3}{2}\pi, \frac{3}{2}\pi]^2 \\ 0 & \text{otherwise} \end{cases} . \tag{3.6}$$

For $y = (y_1, y_2) \in \mathbb{R}^2$ we set $|y|_\infty = \max\{|y_1|, |y_2|\}$, and suppose

$$\rho := 2 - (2 - \frac{\pi}{3\sqrt{3}})^2 > .0528 > \frac{1}{19}. \tag{3.7}$$

Lemma 3.1. *Let $Y \subset \mathbb{R}^2 \setminus (-1, 1)^2$, and assume that $\inf\{|y - z|_\infty; y, z \in Y, y \neq z\} \geq 1$. Then*

$$\sum_{y \in Y} |\psi(y)| \leq (2 - \frac{\pi}{3\sqrt{3}})^2 - 1 = 1 - \rho.$$

Proof: We divide $\mathbb{R}^2 \setminus (-1, 1)^2$ into four sets: $(-1, \infty) \times [1, \infty)$; $(-\infty, -1] \times (-1, \infty)$; $(-\infty, 1) \times (-\infty, -1]$ and $[1, \infty) \times (-\infty, 1)$. Due to the symmetry, it is enough to consider only the first one. Let $Q_j = [j_1, j_1 + 1) \times [j_2, j_2 + 1)$ for $j = (j_1, j_2) \in \mathbb{N}^2$ and $R_p = (-1, 1) \times [p, p + 1)$ for $p \in \mathbb{N}$. Then we have $(-1, \infty) \times [1, \infty) = (\cup_{j \in \mathbb{N}^2} Q_j) \cup (\cup_{p \in \mathbb{N}} R_p)$ and $Q_j \cap Y$ has at most one element, while $R_p \cap Y$ has at most two. Set

$$\gamma(x) := \frac{\cos \frac{3}{2}\pi x}{1 - 9x^2}, \quad x \in \mathbb{R}^2;$$

$$f(0) := 1, \quad f(p) := \frac{1}{9p^2 - 1}, \quad p \in \mathbb{Z} \setminus 0;$$

$$F(j) = F(j_1, j_2) := f(j_1)f(j_2), \quad j \in \mathbb{Z}^2.$$

Then the convexity of $\frac{1}{9x^2 - 1}$ in $[1, \infty)$ gives

$$|\gamma(p + \alpha)| \leq \frac{1}{9(p + \alpha)^2 - 1} \leq (1 - \alpha)f(p) + \alpha f(p) \quad \text{for} \quad p \in \mathbb{N}, \quad \alpha \in [0, 1). \tag{3.8}$$

For $y \in Q_j \cap Y$, we have $y = (j_1 + \alpha_1, j_2 + \alpha_2)$ with $j \in \mathbb{N}^2$, $\alpha \in [0,1)^2$, and therefore

$$
\begin{aligned}
|\psi(y)| = |\gamma(j_1 + \alpha_1)||\gamma(j_2 + \alpha_2)| \\
\leq (1 - \alpha_1)(1 - \alpha_2)F(j_1, j_2) + \alpha_1(1 - \alpha_2)F(j_1, j_2 + 1) \quad (3.9) \\
+ (1 - \alpha_1)\alpha_2 F(j_1 + 1, j_2) + \alpha_1 \alpha_2 F(j_1 + 1, j_2 + 1).
\end{aligned}
$$

Let $y, z \in R_p \cap Y$. Then $y = (\alpha_1, p + \alpha_2)$ and $z = (-\beta_1, p + \beta_2)$, where $\alpha, \beta \in [0,1)^2$ and $\alpha_1 + \beta_1 \geq 1$. A direct computation gives that for any α_1, $\beta_1 \geq 0$ and $\alpha_1 + \beta_1 \geq 1$, we have

$$
|\gamma(\alpha_1)| + |\gamma(-\beta_1)| \leq \frac{9}{8} = f(0) + f(1) \leq f(0) + (\alpha_1 + \beta_1)f(1).
$$

Due to $|\gamma(x)| \leq 1$ we thus may put $\delta := \max\{0, |\gamma(\alpha_1)| - \alpha_1 f(1)\} \in [0,1]$ in order to have

$$
|\gamma(\alpha_1)| \leq \delta f(0) + \alpha_1 f(1); \quad |\gamma(-\beta_1)| \leq (1 - \delta)f(0) + \beta_1 f(1).
$$

Using these inequalities, (3.8) and $f(p) \geq f(p+1)$, we get

$$
\begin{aligned}
|\psi(y)| + |\psi(z)| \leq F(0, p) + \alpha_1(1 - \alpha_2)F(1, p) + \alpha_1 \alpha_2 F(1, p+1) \\
+ \beta_1(1 - \beta_2)F(-1, p) + \beta_1 \beta_2 F(-1, p+1). \quad (3.10)
\end{aligned}
$$

If $R_p \cap Y$ has only one element y, then obviously $|\psi(y)| \leq F(0, p)$.

We apply inequalities of type (3.9) or (3.10) for all points from Y and take their sum. Then on the right-hand side we get a linear combination of $F(j)$, $j \in \mathbb{Z}^2 \setminus \{0\}$. Due to the assumption on the set Y, it is not hard to see that the sum of the coefficients of $F(j)$ is not greater than 1, for each j. Hence,

$$
\sum_{y \in Y} |\psi(y)| \leq \sum_{j \in \mathbb{Z}^2 \setminus 0} F(j) = \sum_{j \in \mathbb{Z}^2} F(j) - 1 = \left\{ \sum_{p \in \mathbb{Z}} f(p) \right\}^2 - 1.
$$

Now

$$
\sum_{p=1}^{\infty} f(p) = \sum_{p=1}^{\infty} \frac{1}{9p^2 - 1} = \frac{1}{2} \sum_{p=1}^{\infty} \left(\frac{1}{3p - 1} - \frac{1}{3p + 1} \right) = \frac{1}{2}\left(1 - \frac{\pi}{3\sqrt{3}}\right)
$$

and therefore

$$\sum_{y \in Y} |\psi(y)| \leq (2 - \frac{\pi}{3\sqrt{3}})^2 - 1. \quad \blacksquare$$

Lemma 3.2. *Let $q > 0$, with q as in (3.4), and assume that for some constant $K = K(q) > 0$,*

$$\phi^\wedge(\xi) \geq K \cos \frac{\xi_1}{3q} \cos \frac{\xi_2}{3q}, \quad \xi \in [-\frac{3}{2}\pi q, \frac{3}{2}\pi q]^2. \tag{3.11}$$

Then

$$\sum_{k,\ell} c_k \phi(x_k - x_\ell) c_\ell \geq \lambda(q) \sum_k |c_k|^2$$

with $\lambda(q) := \rho K \frac{9q^2}{\pi^2}$ and ρ as in (3.7).

Proof: The substitution $\zeta = \frac{\xi}{q}$ shows that we may put $q = 1$. Then $\phi^\wedge(\xi) \geq K \frac{9}{\pi^2} \psi^\wedge(\xi)$, and therefore from (1.6)

$$\frac{\pi^2}{9} \sum_{k,\ell} c_k \phi(x_k - c_\ell) c_\ell \geq \frac{K}{(2\pi)^2} \int_{\mathbb{R}^2} \psi^\wedge(\xi) |\sum_k c_k e^{ix_k \cdot \xi}|^2 d\xi$$

$$= K \sum_{k,\ell} c_k \psi(x_k - x_\ell) c_\ell$$

$$\geq K \left\{ \sum_k |c_k|^2 \psi(0) - \sum_{k,\ell, k \neq \ell} |c_k||c_\ell||\psi(x_k - x_\ell)| \right\}$$

$$\geq K \sum_k |c_k|^2 \left\{ \psi(0) - \sum_{\ell \neq k} |\psi(x_k - x_\ell)| \right\}.$$

Applying Lemma 3.1 to the sum $\sum_{\ell \neq k} |\psi(x_k - x_\ell)|$ we get the result. $\quad \blacksquare$

Lemma 3.3. *Let q be from (3.4), and ρ from (3.7). Then*

$$\int_{[-q\pi, q\pi]^2} |\sum_k c_k e^{ix_k \cdot \xi}|^2 d\xi \leq 144 q^2 (2 - \rho) \sum_k |c_k|^2.$$

Proof: It is enough to prove the lemma for $q = 1$. Since $\psi^\wedge(\xi) \geq \frac{1}{4} \cdot \frac{\pi^2}{9}$ for $\xi \in [-\pi, \pi]^2$, we get

$$\int_{[-\pi,\pi]^2} |\sum_k c_k e^{ix_k \cdot \xi}|^2 d\xi \leq \frac{36}{\pi^2} \int_{\mathbb{R}^2} \psi^\wedge(\xi) |\sum_k c_k e^{ix_k \cdot \xi}|^2 d\xi$$

$$= 144 \sum_{k,\ell} c_k \psi(x_k - x_\ell) c_\ell$$

$$\leq 144 \sum_k |c_k|^2 \{\psi(0) + \sum_{\ell \neq k} |\psi(x_k - x_\ell)|\}.$$

An application of Lemma 3.1 finishes the proof. $\blacksquare$

In order to derive the upper estimate in (3.5), we need a condition of type (2.2b). We suppose that there exist constants M_0 and M_1 such that for some $n > d$

$$|\phi^\wedge(\xi)| \leq M_0 \quad \text{for} \quad |\xi|_\infty \leq q\pi, \tag{3.12a}$$

$$|\phi^\wedge(\xi)| \leq \frac{M_1}{|\xi|_\infty^n} \quad \text{for} \quad |\xi|_\infty \geq q\pi. \tag{3.12b}$$

The following theorem is the main result of this section.

Theorem 3.4. *Let $\phi^\wedge \geq 0$ satisfy (1.4), (3.11) and (3.12), with q as in (3.4). Then (3.5) holds true with $\lambda(q)$ as defined in Lemma 3.2 and*

$$\Lambda(q) = \frac{36q^2}{\pi^2} (2 - \frac{\pi}{3\sqrt{3}})^2 \left\{ M_0 + \frac{M_1}{(q\pi)^n} \sum_{m=1}^\infty \frac{8m}{(2m-1)^n} \right\}, \tag{3.13}$$

and hence

$$\operatorname{cond}(A) \leq \frac{\Lambda(q)}{\lambda(q)} \leq \frac{148}{K} \left\{ M_0 + \frac{8M_1}{(q\pi)^n} (1 + \frac{2n-1}{3^n(n-2)}) \right\}. \tag{3.14}$$

Proof: Let us first derive an estimate for $\Lambda(q)$. We put

$$D_0 = \{\xi; |\xi|_\infty \leq q\pi\} \quad \text{and} \quad D_m = \{\xi; q\pi(2m-1) \leq |\xi|_\infty \leq q\pi(2m+1)\}.$$

Using (3.12b) we get

$$|\phi^\wedge(\xi)| \leq \frac{1}{(2m-1)^n} \frac{M_1}{(q\pi)^n} \quad \text{for} \quad \xi \in D_m. \tag{3.15}$$

Taking into account that D_m contains exactly $8m$ sets of type $j + D_0$, $j \in \mathbb{Z}^2$, and applying (3.12a), (3.15) and Lemma 3.3, we get

$$\sum_{k,\ell} c_k \phi(x_k - x_\ell) c_\ell = \frac{1}{(2\pi)^2} \int_{\mathbb{R}^2} \phi^\wedge(\xi) |\sum_k c_k e^{ix_k \cdot \xi}|^2 d\xi$$

$$= \frac{1}{(2\pi)^2} \sum_{j \in \mathbb{Z}^2} \int_{j+D_0} \phi^\wedge(\xi) |\sum_k (c_k e^{ix_k \cdot j}) e^{ix_k \cdot (\xi - j)}|^2 d\xi,$$

$$\leq \Lambda(q) \sum_k |c_k|^2$$

with $\Lambda(q)$ as defined in (3.13). Combining this with Lemma 3.2, we get (3.5), and using

$$\sum_{m=1}^\infty \frac{8m}{(2m-1)^n} = 4 \sum_{m=1}^\infty \left(\frac{1}{(2m-1)^{n-1}} + \frac{1}{(2m-1)^n} \right)$$

$$\leq 4\left(2 + \frac{4}{3^n} + \frac{2}{(n-2)3^{n-1}}\right) = 8\left(1 + \frac{2n-1}{3^n(n-2)}\right),$$

we find from (1.6) that

$$\Lambda(q) \leq 4(2-\rho)\left\{M_0 + \frac{8M_1}{(q\pi)^n}\left(1 + \frac{2n-1}{3^n(n-2)}\right)\right\}\frac{9q^2}{\pi^2}.$$

Finally, (3.14) follows from $\frac{4(2-\rho)}{\rho} < 148$ by using (3.7). ∎

Example 3 Continued. (Rabut's thin plate B-spline). As an application of Theorem 3.4, we consider the splines from Example 3. Choose $q = 1$. Then $K = \frac{1}{6}$, $M_0 = 1$, $n = 4$ and $M_1 = 16$. Thus

$$\mathrm{cond}(A) \leq 900\left\{1 + \frac{128}{\pi^4}\left(1 + \frac{7}{162}\right)\right\} < 2134.$$

References

1. Arge, E., M. Daehlen, and A. Tveito, Box spline interpolation: a computational study, preprint, 1991.
2. Baxter, B., Norm estimates for inverses of Toeplitz distance matrices, Report # NA 16, DAMTP, University of Cambridge, September, 1991.

3. Binev, P.G. and K. Jetter, Euler splines from 3-directional box splines, in *Constructive Theory of Functions, Varna 1991*, K. Ivanov et al. (eds.), to appear.

4. de Boor, C., K. Höllig, and S.D. Riemenschneider, Bivariate cardinal interpolation by splines on a three-direction mesh, Ill. J. Math. **29** (1985), 533–566.

5. Chui, C.K., K. Jetter, and J.D. Ward, Cardinal interpolation with differences of tempered functions, Computers Math. Applic., to appear.

6. Jetter, K. and J. Stöckler, Algorithms for cardinal interpolation using box splines and radial basis functions, Numer. Math. **60** (1991), 97–114.

7. Narcowich, F. J. and J. D. Ward, Norms of inverses and condition numbers for matrices associated with scattered data, J. Approx. Theory **64** (1991), 69–94.

8. Narcowich, F. J. and J. D. Ward, Norms of inverses for matrices associated with scattered data, in *Curves and Surfaces*, P.-J. Laurent, A. Le Méhauté and L. L. Schumaker (eds.), Academic Press, Boston, 1991, 341–348.

9. Powell, M.J.D., The theory of radial basis function approximation in 1990, in *Advances in Numerical Analysis II: Wavelets, Subdivision Algorithms and Radial Functions*, W. Light (ed.), Oxford University Press, Oxford, 1992, 105–210.

10. Rabut, Ch., B-splines polyharmoniques cardinales: interpolation, quasi-interpolation, filtrage, Thèse, Université Paul Sabatier, Toulouse, 1990.

11. Zygmund, A., *Trigonometric Series*, Vol. I, 2nd ed., Cambridge University Press, Cambridge, 1959.

Peter Binev
Dept. of Mathematics
University of Sofia
Sofia 1156, Bulgaria

Kurt Jetter
Dept. of Mathematics
University of Duisburg
4100 Duisburg, Germany
hn277je@unidui.uni-duisburg.de

Research supported by DAAD grant # 314/102/002/1 and NATO grant # CRG 900 158.

Numerical Methods of Approximation Theory, Vol. 9
Dietrich Braess and Larry L. Schumaker (eds.), pp. 53–75.
International Series of Numerical Mathematics, Vol. 105

Wavelets on a Bounded Interval

Charles K. Chui and Ewald Quak

Dedicated to the memory of Lothar Collatz

Abstract. The aim of this paper is to present two different approaches to the study of multiresolution analysis and wavelets on a bounded interval. Recently, Meyer obtained orthonormal wavelets on a bounded interval by restricting Daubechies' scaling functions and wavelets to $[0, 1]$ and applying the Gram-Schmidt procedure to orthonormalize the restrictions. Our own approach – presented in the second part of the paper – is based on the semi-orthogonal Chui-Wang spline-wavelets. In this case we no longer have orthogonality in one scale, but there are explicit formulae for these wavelets.

§1. Introduction

Let us first recall the notions of scaling functions and multiresolution analysis as introduced by Meyer [11] and Mallat [10]. For a function $\phi \in L^2(\mathbb{R})$, let a reference subspace V_0 be generated as the L^2-closure of the linear span of the integer translates of ϕ, namely:

$$V_0 := \mathrm{clos}_{L^2}\langle\phi(\cdot - k)\colon\ k \in \mathbb{Z}\rangle,$$

and consider the other subspaces $V_j := \mathrm{clos}_{L^2}\langle\phi_{j,k}\colon\ k \in \mathbb{Z}\rangle$, $j \in \mathbb{Z}$, where

$$\phi_{j,k}(x) := 2^{j/2}\phi(2^j x - k), \qquad j, k \in \mathbb{Z}.$$

Definition 1.1. *A function $\phi \in L^2(\mathbb{R})$ is said to generate a multiresolution analysis (MRA) if it generates a nested sequence of closed subspaces V_j that satisfy*

(i) $\qquad \cdots V_{-1} \subset V_0 \subset V_1 \subset \cdots$;

(ii) $\qquad \mathrm{clos}_{L^2}\left(\bigcup_{j \in \mathbb{Z}} V_j \right) = L^2(\mathbb{R})$;

(iii) $\qquad \bigcap_{j \in \mathbb{Z}} V_j = \{0\}$;

(iv) $\qquad f(x) \in V_j \Leftrightarrow f(x + 2^{-j}) \in V_j \Leftrightarrow f(2x) \in V_{j+1}, \quad j \in \mathbb{Z}; \quad$ and

(v) $\qquad \{\phi(\cdot - k)\}_{k \in \mathbb{Z}}$ *forms a Riesz basis for V_0, i.e.,*

there are constants A and B, with $0 < A \leq B < \infty$, such that

$$A\|\{c_k\}\|_{\ell^2}^2 \leq \left\| \sum_{k=-\infty}^{\infty} c_k \phi(\cdot - k) \right\|_2^2 \leq B\|\{c_k\}\|_{\ell^2}^2,$$

for all sequences $\{c_k\}$ for which $\|\{c_k\}\|_{\ell^2}^2 = \sum_{k=-\infty}^{\infty} |c_k|^2 < \infty$. If ϕ generates an MRA, then ϕ is called a scaling function. In case different integer translates of ϕ are orthogonal ($\phi(\cdot - k) \perp \phi(\cdot - \tilde{k})$ for $k \neq \tilde{k}$), the scaling function is called an orthogonal scaling function.

Since the subspaces V_j are nested, there exist complementary orthogonal subspaces W_j, such that

$$V_{j+1} = V_j \oplus W_j, \qquad j \in \mathbb{Z}.$$

This gives rise to an orthogonal decomposition of $L^2(\mathbb{R})$, namely

$$L^2(\mathbb{R}) = \bigoplus_{j \in \mathbb{Z}} W_j;$$

(here and in the following $\oplus$ denotes orthogonal sums).

Definition 1.2. *A function $\psi \in L^2(\mathbb{R})$ is called a wavelet if it generates the complementary orthogonal subspaces W_j of an MRA; that is, $W_j = \mathrm{clos}_{L^2}\langle \psi_{j,k} \colon k \in \mathbb{Z}\rangle$, $j \in \mathbb{Z}$, where $\psi_{j,k}(x) := 2^{j/2}\psi(2^j x - k)$, $j, k \in \mathbb{Z}$. Obviously, $\psi_{j,k} \perp \psi_{\tilde{j},\tilde{k}}$ for $j \neq \tilde{j}$. If $\langle \psi_{j,k}, \psi_{\tilde{j},\tilde{k}}\rangle = \delta_{j,\tilde{j}}\delta_{k,\tilde{k}}$, where $\langle f, g\rangle =$*

$\int_{-\infty}^{\infty} f(x)\overline{g(x)}\,dx$ is the standard inner product, then ψ is called an *orthonormal wavelet*.

In the following, two different sets of scaling functions (and wavelets) are presented which will be adapted to the case of a bounded interval; *i.e.*, $L^2[0,1]$ instead of $L^2(\mathbb{R})$.

Example 1.3. Since $\phi \in V_0 \subset V_1$, there is a bi-infinite sequence $\{p_k\}$ such that

$$\phi(x) = \sum_{k=-\infty}^{\infty} p_k \phi(2x - k).$$

Daubechies [8] was the first to give finite sequences $\{p_k\}$ for the construction of compactly supported orthogonal scaling functions ϕ (and wavelets ψ). More precisely, for every $N \in \mathbb{N}$, there is an orthogonal scaling function ϕ_N^D (henceforth called Daubechies' scaling function), having support $[0, 2N - 1]$. *E.g.*, for $N = 2$

$$p_0^{(2)} = \frac{1 + \sqrt{3}}{4}, p_1^{(2)} = \frac{3 + \sqrt{3}}{4}, p_2^{(2)} = \frac{3 - \sqrt{3}}{4}, p_3^{(2)} = \frac{1 - \sqrt{3}}{4}$$

and $p_k^{(2)} = 0$ otherwise. Furthermore, as $\psi \in W_0 \subset V_1$, we have some sequence $\{q_k\}$, such that

$$\psi(x) = \sum_{k=-\infty}^{\infty} q_k \phi(2x - k).$$

For o.n. scaling functions, we may choose $q_k = (-1)^k \bar{p}_{-k+1}$, $k \in \mathbb{Z}$, so that the corresponding o.n. Daubechies wavelet ψ_N^D is also supported on an interval of length $2N - 1$. ∎

Example 1.4. Cardinal B-splines N_m, for $m \in \mathbb{N}$, are also suitable scaling functions which are recursively defined by integral convolution, namely:

$$N_m(x) := \int_0^1 N_{m-1}(x - t)\,dt,$$

where $N_1(x) := \chi_{[0,1)}(x)$ is the characteristic function of the interval $[0,1)$.

For $\phi = N_m$, the subspace V_0 consists of all polynomial splines of order m that have continuous derivatives up to order $m - 2$ and are square-integrable on $\mathbb{R}$:

$$V_0 := \{f \in C^{m-2} \cap L^2(\mathbb{R}): f|_{[k,k+1)} \in \Pi_{m-1}, k \in \mathbb{Z}\}$$

and analogously,

$$V_j := \{f \in C^{m-2} \cap L^2(\mathbb{R}) :\ f|_{[\frac{k}{2^j}, \frac{k+1}{2^j})} \in \Pi_{m-1}, k \in \mathbb{Z}\},$$

where Π_{m-1} denotes the space of all polynomials of degree at most $m - 1$, and C^{m-2} the space of all $m - 2$ times continuously-differentiable functions on $\mathbb{R}$. Clearly, for $m \geq 2$, these scaling functions are not orthogonal. Here, two possible choices for wavelets were found by Chui and Wang [5,6].

Firstly, the so-called interpolatory wavelets are defined by setting

$$\psi_m^I(x) := L_{2m}^{(m)}(2x - 1),$$

where L_{2m} is the $(2m)^{\text{th}}$ order fundamental cardinal spline, which is uniquely determined by the infinitely many conditions

$$L_{2m}(k) = \delta_{k,0}, \qquad k \in \mathbb{Z}.$$

The wavelets ψ_m^I are supported on all of $\mathbb{R}$, but are exponentially decaying for $|x| \to \infty$.

On the other hand, there are compactly supported wavelets with minimal support $[0, 2m - 1]$, the so-called B-wavelets ψ_m, given by

$$\psi_m(x) := \sum_n q_n N_m(2x - n),$$

where

$$q_n := \frac{(-1)^n}{2^{m-1}} \sum_{\ell=0}^{m} \binom{m}{\ell} N_{2m}(n + 1 - \ell), \quad n = 0, \ldots, 3m - 2,$$

and $q_n = 0$ otherwise. ∎

Restriction to a bounded interval, say $[0, 1]$, entails some changes in the concepts of a multiresolution analysis. There can no longer be a bi-infinite sequence of nested subspaces. Instead, we have to start from an initial subspace $V_0^{[0,1]}$ and investigate spaces $V_j^{[0,1]}$, only for $j \geq 0$:

$$V_0^{[0,1]} \subset V_1^{[0,1]} \subset \cdots$$

such that

$$\text{clos}_{L^2[0,1]}\left(\bigcup_{j\geq 0} V_j\right) = L^2[0,1].$$

In this case, we deal with the complementary orthogonal subspaces $W_j^{[0,1]}$ satisfying

$$V_{j+1}^{[0,1]} = V_j^{[0,1]} \oplus W_j^{[0,1]}, \qquad j \in \mathbb{Z}_+,$$

and the orthogonal decomposition of $L^2[0,1]$, namely

$$L^2[0,1] = V_0^{[0,1]} \oplus W_0^{[0,1]} \oplus W_1^{[0,1]} \oplus \cdots$$
$$= V_0^{[0,1]} \bigoplus_{j \in \mathbb{Z}_+} W_j^{[0,1]}.$$

The behaviour at the endpoints of the interval $[0,1]$ must now be taken into account and special boundary functions (scaling functions and wavelets) have to be introduced, as the relevant families of functions in $V_j^{[0,1]}$ and $W_j^{[0,1]}$ can no longer be controlled just by dilation and translation alone. The aim is to produce Riesz bases for the spaces $V_j^{[0,1]}$ consisting of a finite set of suitable translates $\phi_{j,k}$ of the original scaling function and a finite set of specially constructed *boundary scaling functions* $\phi_{j,k}^b$ as well as bases of the complementary subspaces $W_j^{[0,1]}$ consisting of a finite set of translates of the wavelet function $\psi_{j,k}$ and a finite set of special *boundary wavelets* $\psi_{j,k}^b$.

One approach is to use Daubechies' scaling functions and wavelets of Example 1.3, taking care that the newly contructed bases are again orthonormal. This was first carried out by Meyer in [12], whose approach (without proofs) is described in Section 2. Our approach, as laid out in Section 3, is based on the non-orthogonal spline wavelets of Example 1.4 and makes use of the concept of multiple knots to construct boundary scaling functions and wavelets.

§2. Meyer's Approach

Meyer has derived an orthonormal basis for $L^2[0,1]$ using Daubechies' scaling functions and wavelets on $L^2(\mathbb{R})$. Our presentation follows closely the approach in [12], omitting the detailed proofs for which the reader is referred to Meyer's paper. Let $N \in \mathbb{N}$ be fixed for the rest of this section so that we can omit the index N for convenience when investigating the scaling function ϕ_N^D with support $[0, 2N - 1]$.

The subspaces $V_j^{[0,1]}$ are given by the restrictions of functions in V_j to the interval $[0,1]$:

$$V_j^{[0,1]} := \{g \in L^2[0,1]: \ g = f|_{[0,1]}, f \in V_j\}. \tag{2.1}$$

Due to the compact support of ϕ^D, $V_j^{[0,1]}$ is a finite-dimensional linear space. More precisely,

$$\operatorname{supp} \phi_{j,k}^D = \left[\frac{k}{2^j}, \frac{k+2N-1}{2^j}\right] \tag{2.2}$$

which implies

$$\operatorname{supp} \phi_{j,k}^D \cap (0,1) \neq \emptyset \quad \text{iff} \quad k \in I(j), \tag{2.3}$$

where the index set $I(j)$ is defined as

$$I(j) := \{k \in \mathbb{Z}: \ -2N+2 \leq k \leq 2^j - 1\}, \quad j \in \mathbb{Z}_+, \tag{2.4}$$

enabling us to write

$$V_j^{[0,1]} = \langle \phi_{j,k}^D: \ k \in I(j) \rangle, \quad j \in \mathbb{Z}_+. \tag{2.5}$$

This spanning set for $V_j^{[0,1]}$ is actually linearly independent, *i.e.*, it forms a basis of the space $V_j^{[0,1]}$. We have

Theorem 2.1. *For* $j \in \mathbb{Z}_+$, *the functions* $\phi_{j,k}^D$, $k \in I(j)$, *form a basis of* $V_j^{[0,1]}$, *so that*

$$\dim V_j^{[0,1]} = 2^j + 2N - 2.$$

It should be noted that depending on j, two cases have to be distinguished. Let $j_0 := \min\{j \in \mathbb{Z}: \ 2^j \geq 4N - 4\}$. For $j \geq j_0$, we see that the length of $\operatorname{supp} \phi_{j,k}^D$ is less than $1/2$ and therefore the index set $I(j)$ can be split into three disjoint sets, depending on whether $\phi_{j,k}^D(0) \neq 0$, $\phi_{j,k}^D(1) \neq 0$ or $\operatorname{supp} \phi_{j,k}^D \subseteq [0,1]$, as follows:

$$I(j) = I_1(j) \ \dot\cup \ I_2(j) \ \dot\cup \ I_3(j), \ \text{where}$$

$$I_1(j) = \{k \in I(j): \ \phi_{j,k}^D(0) \neq 0\} = \{k \in \mathbb{Z}: \ -2N+2 \leq k \leq -1\}; \tag{2.6}$$

$$I_3(j) = \{k \in I(j): \ \phi_{j,k}^D(1) \neq 0\} = \{k \in \mathbb{Z}: \ 2^j - 2N + 2 \leq k \leq 2^j - 1\} \ \text{and}$$

$$I_2(j) = \{k \in I(j): \ \operatorname{supp} \phi_{j,k}^D \subseteq [0,1]\} = \{k \in \mathbb{Z}: \ 0 \leq k \leq 2^j - 2N + 1\}.$$

This means that we can distinguish those $\phi^D_{j,k}$ ($k \in I_1(j)$) that have been cut only at 0, or only at 1 ($k \in I_3(j)$), and those that have not been altered at all by the restriction to the interval [0,1] ($k \in I_2(j)$).

In the case $j < j_0$, the influence of the two endpoints 0 and 1 can no longer be separated as before and the treatment has to be different.

Theorem 2.1 gives us a basis of the space $V_j^{[0,1]}$, but the restriction of the functions in V_j to [0,1] destroys at least some of the orthogonality relations. Now it has to be investigated how an o.n. basis of $V_j^{[0,1]}$ can be obtained.

For $0 \leq j < j_0$, all functions $\phi^D_{j,k}$, $k \in I(j)$, have to be orthonormalized following the classical Gram-Schmidt procedure to obtain an o.n. basis $\phi^b_{j,k}$ of $V_j^{[0,1]}$.

For $j \geq j_0$, the decomposition (2.6) can be used again. We obtain that $\phi^D_{j,k} \perp \phi^D_{j,\tilde{k}}$ for $k \in I_1(j)$ and $\tilde{k} \in I_3(j)$ as their supports are disjoint. Furthermore, it is also true that $\phi^D_{j,k} \perp \phi^D_{j,\tilde{k}}$ for $k \in I(j)$ and $\tilde{k} \in I_2(j)$. So, to produce an o.n. basis of $V_j^{[0,1]}$, one must only orthonormalize the functions $\phi^D_{j,k}$, $k \in I_1(j)$, among themselves using Gram-Schmidt and also the functions $\phi^D_{j,k}$, $k \in I_3(j)$, among themselves. These operations are dilation invariant, and produce new *boundary functions at 0 and 1* to complete an o.n. basis of $V_j^{[0,1]}$, namely:

$$2^{j/2}\phi^{b_0}_1(2^j x), 2^{j/2}\phi^{b_0}_2(2^j x), \ldots, 2^{j/2}\phi^{b_0}_{2N-2}(2^j x) \quad \text{for} \quad 0$$

and

$$2^{j/2}\phi^{b_1}_1(2^j(1-x)), 2^{j/2}\phi^{b_1}_2(2^j(1-x)), \ldots, 2^{j/2}\phi^{b_1}_{2N-2}(2^j(1-x)) \quad \text{for} \quad 1.$$

Summarizing the above results, we have

Proposition 2.2. *For $j \geq j_0$, the set of functions*

$$\begin{cases} 2^{j/2}\phi^{b_0}_i(2^j x), & i = 1, \ldots, 2N-2; \\ 2^{j/2}\phi^D(2^j x - k), & k = 0, \ldots, 2^j - 2N + 1; \\ 2^{j/2}\phi^{b_1}_i(2^j(1-x)), & i = 1, \ldots, 2N-2, \end{cases}$$

forms an o.n. basis of $V_j^{[0,1]}$.

An interesting additional result is

Lemma 2.3. *The linear space $V_0^{[0,1]}$ of dimension $2N - 1$ contains the space Π_{N-1} of polynomials of degree at most $N - 1$.*

The wavelets to be constructed in the following are orthogonal to $V_0^{[0,1]}$ and, as a consequence of Lemma 2.4, have vanishing moments up to degree $N - 1$.

For the construction of the corresponding wavelets, it is necessary to find an o.n. basis of the orthogonal complement $W_j^{[0,1]}$ of $V_j^{[0,1]}$ relative to $V_{j+1}^{[0,1]}$. As we are now working with finite dimensional subspaces, we can deduce that $W_j^{[0,1]}$ has dimension 2^j. The support of ψ^D has length $2N - 1$; so the space of all restrictions of functions in W_j to the interval $[0,1]$ is spanned by $2^j + 2N - 2$ functions $\psi_{j,k}^D$, not all of which can be in $W_j^{[0,1]}$. The following lemma basically establishes that those $\psi_{j,k}^D$ with more than half their support lying outside the interval $[0,1]$, have to be discarded.

Lemma 2.4. *The functions $\psi^D(2^j x - k)$, for $-2N + 2 \leq k \leq -N$ and for $2^j - N + 1 \leq k \leq 2^j - 1$, restricted to $[0,1]$, belong to $V_j^{[0,1]}$.*

Lemma 2.4 leaves us with $\psi_{j,k}^D$, $-N + 1 \leq k \leq 2^j - N$, to consider. In fact, we have the following

Theorem 2.5. *For all $j \in \mathbb{Z}_+$, a basis of $V_{j+1}^{[0,1]}$ consists of the basis $\phi_{j,k}^D$, $k \in I(j)$, of $V_j^{[0,1]}$ and the functions $\psi_{j,k}^D$, $-N + 1 \leq k \leq 2^j - N$.*

Still, the decomposition found in Theorem 2.5 is not an orthonormal one, the loss of orthogonality being caused by the restriction to $[0, 1]$. An o.n. basis of $W_j^{[0,1]}$ can be found in the following way, again distinguishing between two cases.

For $0 \leq j < j_0$, set $h_{j,k} := \psi_{j,k}^D - \sum_{\tilde{k} \in I(j)} c_{j,\tilde{k}}^{(k)} \phi_{j,\tilde{k}}^b$, $-N + 1 \leq k \leq 2^j - N$,

and determine the coefficients $c_{j,\tilde{k}}^{(k)}$ so that $h_{j,k} \in W_j^{[0,1]}$; i.e., $h_{j,k} \perp \phi_{j,\ell}^b$. In the second stage, the $h_{j,k}$ have to be orthonormalized to give an o.n. basis $\psi_{j,k}^b$, $-N + 1 \leq k \leq 2^j - N$, of $W_j^{[0,1]}$.

Again, for $j \geq j_0$, the situation can be described more clearly. For $0 \leq k \leq 2^j - 2N + 1$, supp $\psi_{j,k}^D \subseteq [0,1]$ and $\psi_{j,k}^D \perp V_j^{[0,1]}$; i.e. $\psi_{j,k} \in W_j^{[0,1]}$. This takes care of $2^j - 2N + 2$ basis functions. To specify the missing $2N - 2$ wavelets, $N - 1$ boundary wavelets corresponding to 0 and $N - 1$ boundary wavelets corresponding to 1 have to be constructed. We start with the wavelets $\psi_{j,k}^D$, $-N + 1 \leq k \leq -1$, in W_j. A close inspection of the respective supports shows that $\psi_{j,k}^D \perp \phi_{\tilde{k}}^{b_1}$ for $\tilde{k} = 1, \ldots, 2N - 2$, and $\psi_{j,k}^D \perp \phi_{j,\tilde{k}}^D$, $\psi_{j,k}^D \perp \psi_{j,\tilde{k}}^D$ for $\tilde{k} \geq 0$.

Set $2^{j/2}h_k(2^j x) := \psi_{j,k}^D(x) - \sum_{\tilde{k}=1}^{2N-2} c_{j,\tilde{k}}^{(k)} \phi_{\tilde{k}}^{b_0}(2^j x)$ and, as above, determine the coefficients $c_{j,\tilde{k}}^{(k)}$ by imposing the orthogonality conditions $2^{j/2}h_k(2^j \cdot) \perp \phi_i^{b_0}(2^j \cdot)$. Finally, these h_k are orthonormalized to construct $N-1$ *0-boundary wavelets* $2^{j/2}\psi_1^{b_0}(2^j x), \ldots, 2^{j/2}\psi_{N-1}^{b_0}(2^j x)$. As for the boundary scaling functions, these operations are dilation-invariant. The construction for the *1-boundary wavelets* $2^{j/2}\psi_1^{b_1}(2^j(1-x)), \ldots, 2^{j/2}\psi_{N-1}^{b_1}(2^j(1-x))$ is done similarly. This finally results in

Proposition 2.6. *For $j \geq j_0$, the set of functions*

$$\begin{cases} 2^{j/2}\psi_i^{b_0}(2^j x), & i = 1, \ldots, N-1; \\ 2^{j/2}\psi^D(2^j x - k), & k = 0, \ldots, 2^j - 2N + 1; \\ 2^{j/2}\psi_i^{b_1}(2^j(1-x)), & i = 1, \ldots, N-1, \end{cases}$$

forms an o.n. basis of $W_j^{[0,1]}$.

Remark. It has recently been brought to our attention that Cohen and Daubechies [7], based on the scaling function ϕ^D and the wavelet ψ^D, pursue a different approach to a multiresolution analysis of $L^2[0,1]$. They construct spaces $\widetilde{V}_j^{[0,1]}$ with dimension 2^j, $j \geq 0$. To this end, they only include those functions $\phi_{j,k}^D$ whose supports are completely in the interval $[0,1]$; *i.e.*, for which the restriction to $[0,1]$ makes no difference. As there are exactly $2^j - 2N + 2$ such functions, another $2N - 2$ functions have to be added to achieve dimension 2^j. This is done in such a way as to ensure that all of the polynomials of degree $N - 1$ are included in the spaces $\widetilde{V}_j^{[0,1]}$.

<h2 align="center">§3. Spline-wavelets for $L^2[0,1]$</h2>

The theory of spline-wavelets for $L^2(\mathbb{R})$ was developed by Chui and Wang in a series of papers [4-6]. These wavelets are not orthogonal, just semi-orthogonal, but this allows for symmetry and antisymmetry and for deriving explicit formulae for the spline-wavelets. For a detailed exposition see the monograph [2]. The spline approach also allows us to consider vastly irregular knot sequences, not just $2^j\mathbb{Z}$. First results on existence of wavelets for irregular simple knot sequences were established by Buhmann and Micchelli [1], and explicit formulae using determinants by Lyche and Mørken [9].

A big advantage of the spline approach is that it readily adapts to the case of the bounded interval $[0,1]$ by introducing multiple knots at the endpoints.

So no cut-off due to restriction of functions on $\mathbb{R}$ to $[0, 1]$ is necessary and after suitable adaptation at the endpoints, most of the concepts of the *B*-spline MRA of $L^2(\mathbb{R})$ can be carried over to $[0, 1]$. The relevant proofs mostly use well-known ingredients of the classical spline theory, some of which are nevertheless repeated here for the reader's convenience.

Let $m \in \mathbb{N}$ be fixed throughout this section. Usually the index m will be suppressed for notational convenience.

Definition 3.1. *For $j \in \mathbb{Z}_+$, let a knot sequence on $[0,1]$ be given by* $\mathbf{t}^{(j)} :=$ $\mathbf{t}_m^{(j)} := \{t_k^{(j)}\}_{k=-m+1}^{2^j+m-1}$, *with*

$$t_{-m+1}^{(j)} = t_{-m+2}^{(j)} = \cdots = t_0^{(j)} = 0 \quad \text{(knot of multiplicity } m\text{)};$$

$$t_k^{(j)} = k2^{-j}, \quad k = 1, \ldots, 2^j - 1;$$

$$t_{2^j}^{(j)} = t_{2^j+1}^{(j)} = \cdots = t_{2^j+m-1}^{(j)} = 1 \quad \text{(knot of multiplicity } m\text{)}.$$

The (polynomial) spline space of order m (degree $m - 1$) for the knot sequence $\mathbf{t}^{(j)}$ is defined as

$$S_{m,\mathbf{t}^{(j)}} := S_{m,j} := S_{m,j}^{m-2}$$
$$:= \{s \in C^{m-2}[0, 1]: \ s_{|(t_k^{(j)}, t_{k+1}^{(j)})} \in \Pi_{m-1}, \quad k = 0, \ldots, 2^j - 1\}.$$

Note that the multiple knots at 0 and 1, being endpoints, do not diminish the overall order of smoothness of the elements of $S_{m,j}$. It is clear that a nested sequence of subspaces $V_j^{[0,1]}$ is given by

$$V_j^{[0,1]} := S_{m,j} \tag{3.1}$$

and $V_0^{[0,1]} = \Pi_{m-1}$. By standard spline theory one can establish the following (see for example Schumaker [13]):

Theorem 3.2. *A basis for $V_j^{[0,1]}$ is given by the B-splines $B_{i,m,j}$, $i = -m + 1, \ldots, 2^j - 1$ and thus*

$$\dim V_j^{[0,1]} = 2^j + m - 1. \quad\blacksquare$$

Here,

$$B_{i,m,j}(x) := (t_{i+m}^{(j)} - t_i^{(j)})[t_i^{(j)}, t_{i+1}^{(j)}, \ldots, t_{i+m}^{(j)}]_t(t - x)_+^{m-1}$$

and $[\cdot, \ldots, \cdot]_t$ is the m-th divided difference of $(t - x)_+^{m-1}$ with respect to the variable t. The support of $B_{i,m,j}$ is $[t_i^{(j)}, t_{i+m}^{(j)}]$. For $i = -m + 1, \ldots, -1$, the knot sequence defining $B_{i,m,j}$ contains a multiple knot at 0, and for $i = 2^j - m + 1, \ldots, 2^j - 1$, a multiple knot at 1. So these are the *boundary scaling functions* (for $2^j \geq m - 1$ there are $m - 1$ for 0 and $m - 1$ different ones for 1). The inner ones ($i = 0, \ldots, 2^j - m$ for $2^j \geq m$) are just dilations and translations of the cardinal B-spline $N_m(x) = m[0, 1, \ldots, m]_t(t - x)_+^{m-1}$ used as the scaling function for $L^2(\mathbb{R})$, namely:

$$B_{i,m,j}(x) = N_m(2^j x - i), \qquad i = 0, \ldots, 2^j - m.$$

Note again that these scaling functions are not orthogonal.

To find suitable wavelet functions spanning $W_j^{[0,1]}$ in the orthogonal decompositions $V_{j+1}^{[0,1]} = V_j^{[0,1]} \oplus W_j^{[0,1]}$, we use an argument that identifies the wavelet space $W_j^{[0,1]}$ with a subspace of a spline space of order $2m$, an argument which originates from the properties of the Chui-Wang interpolatory wavelets [5], and also forms the basis for the reasoning in Chui and Wang [4], Buhmann and Micchelli [1], etc.

For each $m \in \mathbb{N}$, define the spline space

$$\widetilde{S}_{2m, \mathbf{t}_m^{(j+1)}} := \langle B_{i, 2m, \mathbf{t}_m^{(j+1)}} : i = -m + 1, \ldots, 2^{j+1} - m - 1 \rangle \qquad (3.2)$$

(i.e., the span of all B-splines of order $2m$ which are defined on the refined knot sequence $\mathbf{t}_m^{(j+1)}$ of order m), and its subspace

$$\widetilde{S}_{2m, \mathbf{t}_m^{(j+1)}}^0 := \{ s \in \widetilde{S}_{2m, \mathbf{t}_m^{(j+1)}} : s(t_k^{(j)}) = 0, k = 0, \ldots, 2^j \} \qquad (3.3)$$

of all splines in $\widetilde{S}_{2m, \mathbf{t}_m^{(j+1)}}$ that vanish on the coarse knot sequence $\mathbf{t}_m^{(j)}$.

Now $\widetilde{S}_{2m, \mathbf{t}_m^{(j+1)}}^0$ can be identified with $W_j^{[0,1]}$ by taking m derivatives.

Theorem 3.3. *For each $m \in \mathbb{N}$, the m-th order differential operator D^m maps the space $\widetilde{S}_{2m, \mathbf{t}_m^{(j+1)}}^0$ one-to-one onto the wavelet space $W_j^{[0,1]}$.*

Proof: Firstly, we have to show that for $s \in \widetilde{S}_{2m, \mathbf{t}_m^{(j+1)}}^0$

$$D^m s = s^{(m)} \in W_j^{[0,1]}, \;\; i.e., \;\; (i) \; s^{(m)} \in V_{j+1}^{[0,1]} \;\; \text{and} \;\; (ii) \; s^{(m)} \perp V_j^{[0,1]}.$$

As $s \in C^{2m-2}[0,1]$ is a spline with knot sequence $\mathbf{t}_m^{(j+1)}$, (i) is clear. To show (ii), we regard, for $i = -m+1, \ldots, 2^j - 1$,

$$\int_0^1 s^{(m)}(x) B_{i,m,j}(x) dx. \tag{3.4}$$

The multiplicity of 0 and 1 in $\mathbf{t}_m^{(j+1)}$ is m; therefore $s \in C^{m-1}(\mathbb{R})$, which implies $s^{(\ell)}(0) = s^{(\ell)}(1) = 0$ for $\ell = 0, 1, \ldots, m-1$. Thus, integrating by parts $m-1$ times yields

$$
\begin{aligned}
\int_0^1 s^{(m)}(x) B_{i,m,j}(x) dx &= \int_0^1 (-1)^{m-1} s'(x) B_{i,m,j}^{(m-1)}(x) dx \\
&= (-1)^{m-1} \sum_{k=0}^{2^j-1} \int_{t_k^{(j)}}^{t_{k+1}^{(j)}} s'(x) c_k dx \\
&= (-1)^{m-1} \sum_{k=0}^{2^j-1} c_k [s(t_{k+1}^{(j)}) - s(t_k^{(j)})] = 0,
\end{aligned}
\tag{3.5}
$$

as $B_{i,m,j}^{(m-1)}$ is piecewise constant ($B_{i,m,j}^{(m-1)} = c_k$ on $(t_k^{(j)}, t_{k+1}^{(j)})$) and s vanishes at the original knots $t_k^{(j)}$.

Next, $s^{(m)} \equiv 0$ on $[0,1]$ implies $s \in \Pi_{m-1}$, but $s^{(\ell)}(0) = 0$, for $\ell = 0, 1, \ldots, m-1$, gives $s \equiv 0$, and so D^m is a one-to-one mapping. Finally, $\dim \widetilde{S}_{2m,\mathbf{t}_m^{(j+1)}} = 2^{j+1} - 1$. Every $s \in \widetilde{S}_{2m,\mathbf{t}_m^{(j+1)}}$ vanishes at 0 and 1, but the other conditions $s(t_k^{(j)})$, $k = 1, \ldots, 2^j - 1$, are independent, so that $\dim \widetilde{S}^0_{2m,\mathbf{t}_m^{(j+1)}} = 2^j$. On the other hand, as $\dim V_j^{[0,1]} = 2^j + m - 1$, $\dim W_j^{[0,1]}$ must be 2^j, and the mapping is onto. ∎

In Chui-Wang [5], an interpolatory wavelet on $\mathbb{R}$ is defined as the m-th derivative of the fundamental cardinal spline L_{2m} for which $L_{2m}(k) = \delta_{0k}$, $k \in \mathbb{Z}$. Analogously, $W_j^{[0,1]}$ is spanned by the m-th derivatives of the fundamental splines $L_{i,2m}$ in $\widetilde{S}^0_{2m,\mathbf{t}_m^{(j+1)}}$, for which $L_{i,2m}((2k-1)2^{-j-1}) = \delta_{ik}$, $i, k = 1, \ldots, 2^j$.

Theorem 3.4. *The wavelet space $W_j^{[0,1]}$ is spanned by the functions $\psi_{i,m}^I = L_{i,2m}^{(m)}, i = 1, \ldots, 2^j$, where for $k = 1, \ldots, 2^{j+1} - 1$, $L_{i,2m} \in \widetilde{S}^0_{2m,\mathbf{t}_m^{(j+1)}}$ satisfy*

$$L_{i,2m}(t_k^{(j+1)}) = \begin{cases} 1 & \text{for } k = 2i - 1 \\ 0 & \text{otherwise.} \end{cases}$$

Let $\mathcal{B}$ be the $(2^{j+1} - 1) \times (2^{j+1} - 1)$ *so-called collocation matrix with entries* $b_{k,i} = B_{i-m,2m,\mathbf{t}_m^{(j+1)}}(t_k^{(j+1)})$, *for* $i, k = 1, \ldots, 2^{j+1} - 1$, *and let* $B_{k,\ell}$ *be its minors that are obtained by deleting the k-th row and the ℓ-th column. Then, the $L_{i,2m}$'s can be computed as*

$$L_{i,2m}(x) = \sum_{\ell=1}^{2^{j+1}-1} B_{2i-1,\ell} B_{\ell-m,2m,\mathbf{t}_m^{(j+1)}}(x) / \det \mathcal{B}.$$

Proof: As the splines $B_{i-m,2m,\mathbf{t}_m^{(j+1)}}$ and the knot sequence $\mathbf{t}_m^{(j+1)}$ satisfy the Schoenberg-Whitney condition for spline interpolation [13], $\det \mathcal{B}$ is nonzero. As a function of x, suppressing the second and third indices of the B-splines, we see that the expression

$$\det \begin{pmatrix} B_{-m+1}(t_1^{(j+1)}) & \cdots & B_{2^{j+1}-1-m}(t_1^{(j+1)}) \\ \vdots & \vdots & \vdots \\ B_{-m+1}(x) & \cdots & B_{2^{j+1}-1-m}(x) \\ \vdots & \vdots & \vdots \\ B_{-m+1}(t_{2^{j+1}-1}^{(j+1)}) & \cdots & B_{2^{j+1}-1-m}(t_{2^{j+1}-1}^{(j+1)}) \end{pmatrix} \quad i\text{-th row}$$

vanishes at all knots but $t_i^{(j+1)}$. Expanding the above determinant with respect to the i-th row and normalizing appropriately gives the representation of Theorem 3.4. Clearly, the $L_{i,2m}$'s are linearly independent and are elements of $\widetilde{S}^0_{2m,\mathbf{t}_m^{(j+1)}}$ and consequently form a basis of this space. By Theorem 3.3, the m-th derivatives form a basis of $W_j^{[0,1]}$. $\blacksquare$

Certainly, the basis functions for $W_j^{[0,1]}$ in Theorem 3.4 have support on all of the interval $[0,1]$ and depend on all the functions $B_{i,2m,\mathbf{t}_m^{(j+1)}}$.

In order to construct a basis of $W_j^{[0,1]}$ with localized supports and a clear distinction between *boundary* and *inner wavelets*, the Chui-Wang approach for compactly supported wavelets in $L^2(\mathbb{R})$ must be suitably adapted. For $2^j \geq 2m-1$, there are $2^j - 2m + 2$ CW wavelets whose supports are completely in $[0,1]$. As in Meyer's approach, additional $m - 1$ *boundary wavelets at 0* and $m - 1$ *boundary wavelets at 1* have to be constructed to obtain a basis of $W_j^{[0,1]}$.

More precisely, a closer inspection of the basis of $\widetilde{S}_{2m,\mathbf{t}_m^{(j+1)}}$ reveals that for $i = -m+1, \ldots, -1$, the splines $B_{i,2m,\mathbf{t}_m^{(j+1)}}$ have a multiple knot at 0, and for

$i = 2^{j+1} - 2m + 1, \ldots, 2^{j+1} - m - 1$, a multiple knot at 1. The rest are translates of the $2m$-th order cardinal B-spline $B_{i,2m,\mathbf{t}_m^{(j+1)}}(x) = N_{2m}(2^{j+1}x - i)$, $i = 0, \ldots, 2^{j+1} - 2m$.

If $2^j \geq 2m - 1$, there exists at least one *inner wavelet* whose support lies completely in $[0, 1]$ and which can be constructed as in Chui-Wang [6].

Lemma 3.5. *For all $j \in \mathbb{N}$ such that $2^j \geq 2m - 1$, there exist $2^j - 2m + 2$ linearly independent inner wavelets $\psi_{j,i}$, $i = 0, \ldots, 2^j - 2m + 1$, in $W_j^{[0,1]}$ which are given by*

$$\psi_{j,i}(x) = \frac{1}{2^{2m-1}} \sum_{k=0}^{2m-2} (-1)^k N_{2m}(k + 1) B^{(m)}_{2i+k,2m,\mathbf{t}_m^{(j+1)}}(x).$$

In fact,

$$\psi_{j,i}(x) = \psi_m(2^j x - i),$$

where ψ_m is the m-th order Chui-Wang B-wavelet.

Proof: Again, using Theorem 3.3, we just have to show that

$$\Psi_{j,i}(x) = \sum_{k=0}^{2m-2} (-1)^k N_{2m}(k + 1) B_{2i+k,2m,\mathbf{t}_m^{(j+1)}}(x), i = 0, \ldots, 2^j - 2m + 1,$$

$$(3.6)$$

are linearly independent elements of $\widetilde{S}^0_{2m,\mathbf{t}_m^{(j+1)}}$. We rewrite $\Psi_{j,i}$ as

$$\Psi_{j,i}(x) = \sum_{k=0}^{2m-2} (-1)^k N_{2m}(k + 1) N_{2m}(2^{j+1}x - 2i - k) \qquad (3.7)$$

to see that supp $\Psi_{j,i} = \left[\frac{2i}{2^{j+1}}, \frac{2i+4m-2}{2^{j+1}} \right]$. From the definition, it is clear that $\Psi_{j,i} \in \widetilde{S}_{2m,\mathbf{t}_m^{(j+1)}}$. To show that $\Psi_{j,i} \in \widetilde{S}^0_{2m,\mathbf{t}_m^{(j+1)}}$, it suffices to show, due to the support of $\Psi_{i,j}$, that

$$\Psi_{j,i}(2(i + \ell) \cdot 2^{-j-1}) = 0 \quad \text{for} \quad \ell = 1, \ldots, 2m - 2. \qquad (3.8)$$

We obtain

$$\sum_{k=0}^{2m-2} (-1)^k N_{2m}(k + 1) N_{2m}(2(i + \ell) - 2i - k)$$

$$= \sum_{k=0}^{2m-2} (-1)^k N_{2m}(k + 1) N_{2m}(2\ell - k). \qquad (3.9)$$

For $\ell = 1, \ldots, m - 1$, this is equal to

$$\sum_{k=0}^{2\ell-1} (-1)^k N_{2m}(k + 1) N_{2m}(2\ell - k) \tag{3.10}$$

$$= \sum_{k=0}^{\ell-1} (-1)^k N_{2m}(k + 1) N_{2m}(2\ell - k) + \sum_{k=\ell}^{2\ell-1} (-1)^k N_{2m}(k + 1) N_{2m}(2\ell - k)$$

$$= \sum_{k=0}^{\ell-1} (-1)^k N_{2m}(k + 1) N_{2m}(2\ell - k) + \sum_{\tilde{k}=0}^{\ell-1} (-1)^{\tilde{k}+1} N_{2m}(2\ell - \tilde{k}) N_{2m}(\tilde{k} + 1)$$

$$= \quad 0$$

after an index transformation $\tilde{k} \leftrightarrow 2\ell - k - 1$ in the second term. For $\ell = m, \ldots, 2m - 2$, we obtain

$$\sum_{2(\ell-m)+1}^{2m-2} (-1)^k N_{2m}(k + 1) N_{2m}(2\ell - k)$$

$$= \sum_{k=2\ell-2m+1}^{\ell-1} (-1)^k N_{2m}(k + 1) N_{2m}(2\ell - k) \tag{3.11}$$

$$+ \sum_{k=\ell}^{2m-2} (-1)^k N_{2m}(k + 1) N_{2m}(2\ell - k) = 0$$

as above, by an index transformation $\tilde{k} \leftrightarrow 2\ell - k - 1$ in the second term. The linear independence follows directly from the overlapping supports of $\Psi_{j,i}$. $\blacksquare$

Having found $2^j - 2m + 2$ inner wavelets, we have to find $m - 1$ additional *boundary wavelets* at 0 and at 1, respectively. We will show how to construct those for 0, and the others are obtained by an index transformation $i \leftrightarrow 2^j - 2m + 1 - i$, using the same approach as for 0.

For $i = -m + 1, \ldots, -1$, set

$$\Psi_{j,i}(x) := \sum_{k=-m+1}^{-1} \alpha_{i,k} B_{k,2m,\mathbf{t}_m^{(j+1)}}(x)$$

$$+ \sum_{k=0}^{2m-2+2i} (-1)^k N_{2m}(k + 1 - 2i) B_{k,2m,\mathbf{t}_m^{(j+1)}}(x); \tag{3.12}$$

i.e., $\Psi_{j,i}$ is a linear combination of the $m-1$ *boundary splines in* $\widetilde{S}_{2m,\mathbf{t}_m^{(j+1)}}$ and $2m+2i-1$ *inner B-splines* $B_{k,2m,\mathbf{t}_m^{(j+1)}}(x) = N_{2m}(2^{j+1}x - k)$ with simple knots. The coefficients of the inner splines are just the ones used for the inner wavelets, while the coefficients $\alpha_{i,k}$ for the boundary splines have to be chosen to achieve $\Psi_{j,i} \in \widetilde{S}_{2m,\mathbf{t}_m^{(j+1)}}^0$. Note that supp $\Psi_{j,i} = [0,(4m-2+2i)2^{-j-1}]$. For $2\ell \cdot 2^{-j-1}$, $\ell = m, m+1, \ldots, 2m-2+i$, it holds for the boundary B-splines, $k = -m+1, \ldots, -1$, that $B_{k,2m,\mathbf{t}_m^{(j+1)}}(2\ell \cdot 2^{-j-1}) = 0$, so that

$$\begin{aligned}
\Psi_{j,i}(2\ell \cdot 2^{-j-1}) &= \sum_{k=0}^{2m-2+2i} (-1)^k N_{2m}(k+1-2i)N_{2m}(2\ell - k) \\
&= \sum_{k=2\ell-2m+1}^{2m-2+2i} (-1)^k N_{2m}(k+1-2i)N_{2m}(2\ell - k) \\
&= \sum_{k=2\ell-2m+1}^{\ell+i} (-1)^k N_{2m}(k+1-2i)N_{2m}(2\ell - k) \\
&\quad + \sum_{k=\ell+i+1}^{2m-2+2i} (-1)^k N_{2m}(k+1-2i)N_{2m}(2\ell - k) = 0
\end{aligned} \qquad (3.13)$$

by an index transformation as before, this time $\tilde{k} \leftrightarrow 2\ell - k + 2i - 1$. For the rest of the knots, $2\ell \cdot 2^{-j-1}$, $\ell = 1, \ldots, m-1$, the conditions $\Psi_{j,i}(2\ell \cdot 2^{-j-1}) = 0$ lead to the system of equations

$$\begin{aligned}
\sum_{k=-m+1}^{-1} &\alpha_{i,k} B_{k,2m,\mathbf{t}_m^{(j+1)}}(2\ell \cdot 2^{-j-1}) = \\
&- \sum_{k=0}^{2m-2+2i} (-1)^k N_{2m}(k+1-2i)N_{2m}(2\ell - k).
\end{aligned} \qquad (3.14)$$

As supp $B_{-\ell,2m,\mathbf{t}_m^{(j+1)}} = [0,(2m-\ell)2^{-j-1}]$ and $2\ell \cdot 2^{-j-1} \in (0,(2m-(m-\ell))2^{-j-1})$, i.e., $2\ell \cdot 2^{-j-1}$ is in the interior of the support of $B_{-m+\ell}$, the conditions of the Schoenberg-Whitney interpolation theorem [13] are satisfied, the $(m-1) \times (m-1)$ matrix

$$\widetilde{B} = (\tilde{b}_{\ell,k})_{1\le \ell,k \le m-1} = (B_{-k,2m,\mathbf{t}_m^{(j+1)}}(t_\ell^{(j)}))_{1\le \ell,k \le m-1} \qquad (3.15)$$

is non-singular, and the coefficient vectors $\alpha_i = (\alpha_{i,-1}, \ldots, \alpha_{i,-m+1})^{\mathsf{T}}$ can be computed uniquely as

$$\alpha_i = \widetilde{B}^{-1} r_i, \qquad (3.16)$$

where $r_i = \left(- \sum\limits_{k=0}^{2m-2+2i} (-1)^k N_{2m}(k+1-2i) N_{2m}(2\ell - k) \right)_{\ell=1,\ldots,m-1}^{\top}$, $i = -m+1,\ldots,-1$. Note that only the right-hand sides differ for the various $\Psi_{j,i}$, while the coefficient matrix is always the same. Actually, as $B_{-k,2m,\mathbf{t}_m^{(j+1)}}(x) = B_{-k,2m,\mathbf{t}_m^{(j)}}(2x)$, the matrix has to be computed only once, namely for the smallest j such that $2^j \geq 2m - 1$.

A more careful study of the supports of the splines $B_{-k,2m,\mathbf{t}_m^{(j+1)}}$ for $k = 1,\ldots,m-1$ reveals the structure of $\widetilde{B}$ (here we only demonstrate it for even m; the one for odd m is similar), namely:

$$
\begin{pmatrix}
B_{-1}(2 \cdot 2^{-j-1}) & & \cdots & & B_{-m+1}(2 \cdot 2^{-j-1}) \\
B_{-1}(4 \cdot 2^{-j-1}) & & \cdots & & B_{-m+1}(4 \cdot 2^{-j-1}) \\
\vdots & & \vdots & & \vdots \\
B_{-1}\left(\frac{m}{2} 2^{-j-1}\right) & & \cdots & & B_{-m+1}\left(\frac{m}{2} 2^{-j-1}\right) \\
B_{-1}\left(\left(\frac{m}{2}+1\right) 2^{-j-1}\right) & B_{-m+3}\left(\left(\frac{m}{2}+1\right) 2^{-j-1}\right) & 0 & & 0 \\
B_{-1}\left(\left(\frac{m}{2}+2\right) 2^{-j-1}\right) & B_{-m+5}\left(\left(\frac{m}{2}+2\right) 2^{-j-1}\right) & 0 & \cdots & 0 \\
\vdots & & \vdots & & \vdots \\
B_{-1}((2m-4)2^{-j-1}) & B_{-3}((2m-4)2^{-j-1}) & 0 & \cdots & 0 \\
B_{-1}((2m-2)2^{-j-1}) & 0 & & 0 &
\end{pmatrix}
$$

Summarizing the above results, we obtain

Lemma 3.6. *For $j \in \mathbb{N}$, such that $2^j \geq 2m-1$, there exist $m-1$ boundary wavelets for the endpoint 0 which can be constructed as*

$$
\psi_{j,i}(x) = \frac{1}{2^{2m-1}} \sum_{k=-m+1}^{-1} (\widetilde{B}^{-1} r_i)_k B^{(m)}_{k,2m,\mathbf{t}_m^{(j+1)}}(x)
$$

$$
+ \frac{1}{2^{2m-1}} \sum_{k=0}^{2m-2+2i} (-1)^k N_{2m}(k+1-2i) B^{(m)}_{k,2m,\mathbf{t}_m^{(j+1)}}(x)
$$

with supports $[0, (2m-1+i)2^{-j}]$ for $i = -m+1,\ldots,-1$. ∎

Due to the inherent symmetry in the spline constructions for the knot sequences $\mathbf{t}_m^{(j)}$, we get the *boundary wavelets for 1* as $\psi_{j,2^j-2m+1-i}(x) := \psi_{j,i}(1-x)$. The final result is

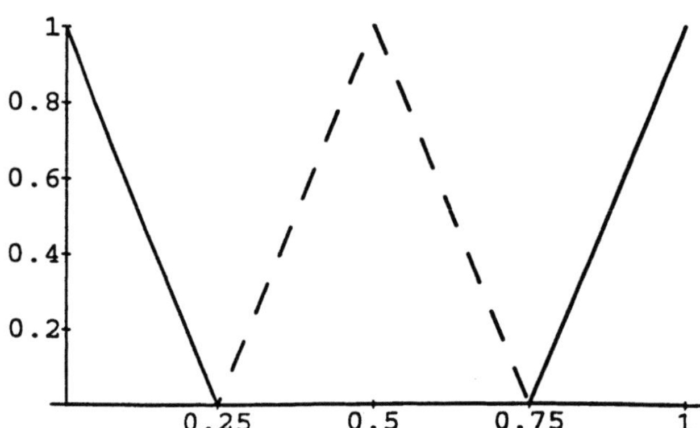

Figure 1. $B_{-1,2,\mathbf{t}^{(2)}}, B_{1,2,\mathbf{t}^{(2)}}, B_{3,2,\mathbf{t}^{(2)}}$.

Theorem 3.7. *If* $2^j \geq 2m - 1$, *the* 2^j *dimensional space* $W_j^{[0,1]}$ *is spanned by the inner wavelets* $\psi_{j,i}$, $i = 0, \ldots, 2^j - 2m + 1$, *of Lemma 4.5, the boundary wavelets* $\psi_{j,i}$ *for* $0, i = -m + 1, \ldots, -1$, *of Lemma 4.6, and the boundary wavelets* $\psi_{j,i}$ *for* $1, i = 2^j - 2m + 2, \ldots, 2^j - m$. ∎

Remark. For the finitely many j, such that $2^j < 2m - 1$, there are no inner wavelets inherited from the MRA of $L^2(\mathbb{R})$. The effects of the endpoints 0 and 1 cannot be separated. As 2^j is small in this case, the interpolatory wavelets in Theorem 3.4 can be used without too much computational effort.

Example 3.8. The figures display the results of the procedures for constructing linear spline-wavelets ($m = 2$) for the knot sequence $\mathbf{t}^{(2)} = \{0, 0, 1/4, 1/2, 3/4, 1, 1\}$ and the refined knot sequence $\mathbf{t}^{(3)} = \{0, 0, 1/8, 1/4, 3/8, 1/2, 5/8, 3/4, 7/8, 1, 1\}$. Figure 1 shows the *cardinal B-spline* $B_{1,2,\mathbf{t}^{(2)}}$ and the *boundary splines* $B_{-1,2,\mathbf{t}^{(2)}}$ and $B_{3,2,\mathbf{t}^{(2)}}$, while Figure 2 shows the basis elements of the space $\widetilde{S}_{4,\mathbf{t}^{(3)}}$. The four interpolating elements of $\widetilde{S}^0_{4,\mathbf{t}^{(3)}}$ as described in Theorem 3.4 are given in Figure 3 and the corresponding interpolatory wavelets $\psi^I_{2,i}$, $i = 1, 2, 3, 4$, in Figure 4. Finally, Figure 5 shows the elements $\Psi_{2,i}$ of $\widetilde{S}^0_{4,\mathbf{t}^{(3)}}$ and in Figure 6 the wavelets $\psi_{2,0}$ and $\psi_{2,1}$ are *inner Chui-Wang* wavelets in the sense of Lemma 3.5, while $\psi_{2,-1}$ and $\psi_{2,2}$ are *boundary wavelets* constructed by following Lemma 3.6.

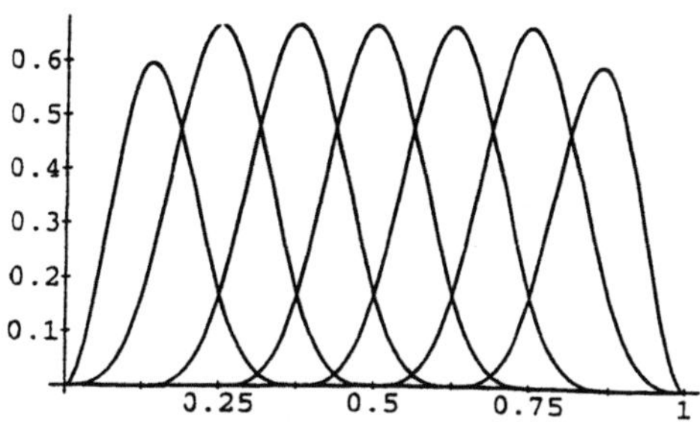

Figure 2. Basis splines of $\widetilde{S}_{4,\mathbf{t}^{(3)}}$.

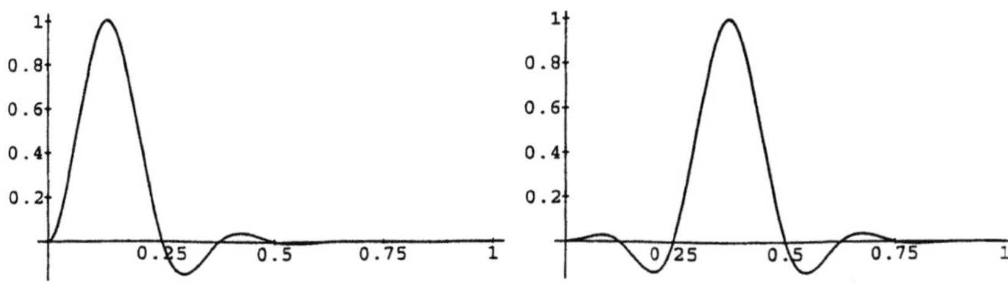

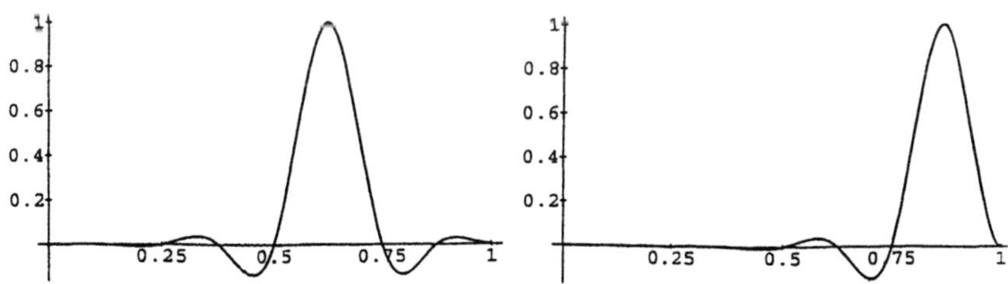

Figure 3. Interpolating elements of $\widetilde{S}^0_{4,\mathbf{t}^{(3)}}$.

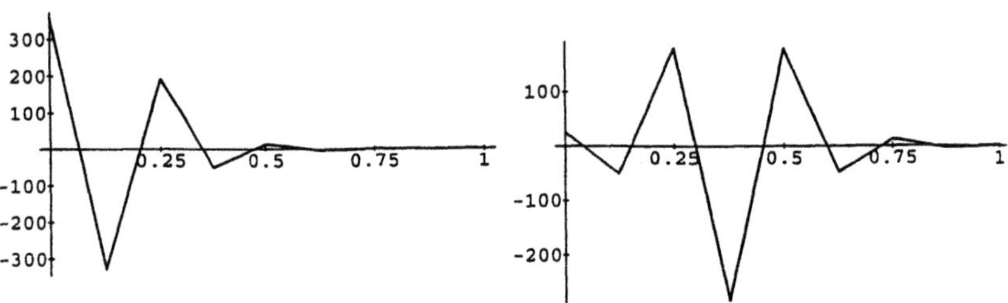

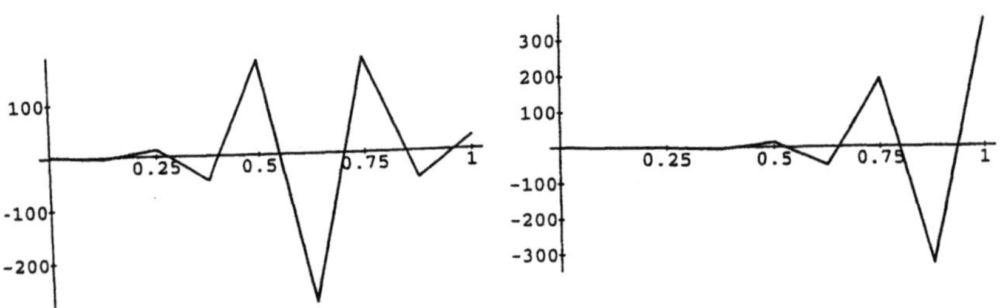

Figure 4. $\psi_{2,i}^{I}, i = 1, 2, 3, 4.$

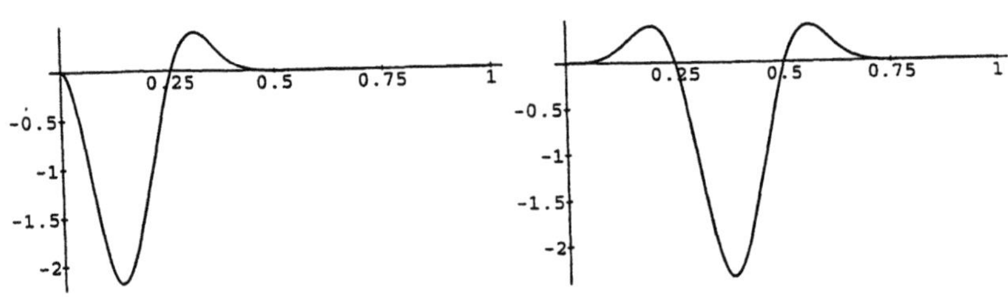

Figure 5a. The elements $\Psi_{2,-1}, \Psi_{2,0}$ of $\widetilde{S}_{4,\mathbf{t}^{(3)}}^{0}.$

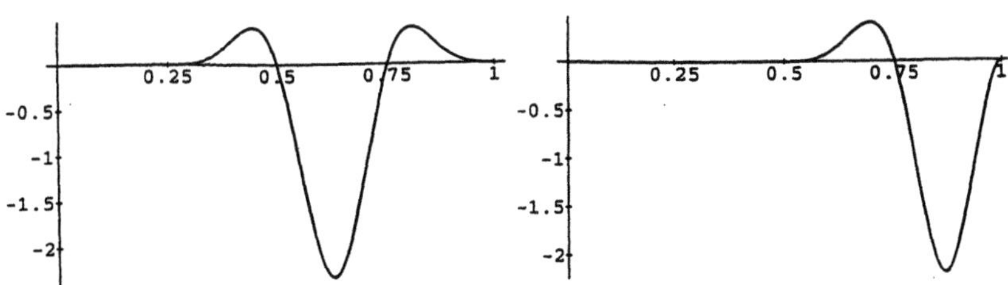

Figure 5b. The elements $\Psi_{2,1}, \Psi_{2,2}$ of $\widetilde{S}^0_{4,\mathbf{t}(3)}$.

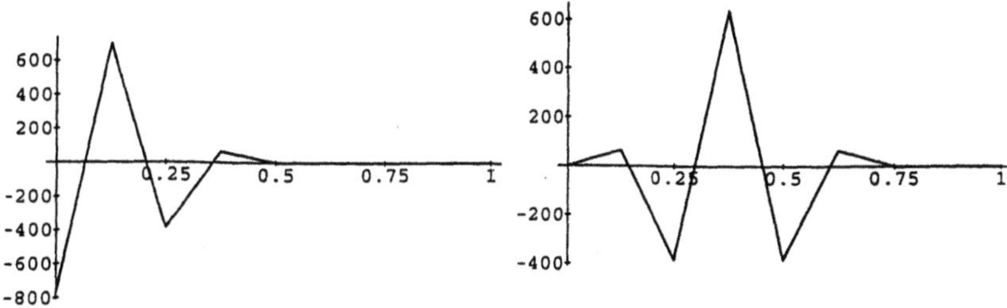

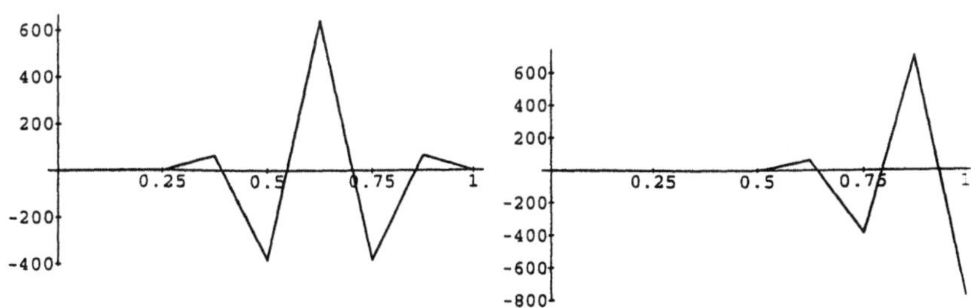

Figure 6. The wavelets $\psi_{2,i}, i = -1, 0, 1, 2.$

§4. Final Remarks

As the splines and spline-wavelets are only semi-orthogonal, we need their
so-called dual splines and wavelets, as developed by Chui-Wang [6] for $L^2(\mathbb{R})$,
to derive the representations of $f \in L^2[0,1]$ as

$$
f(x) = \sum_{\substack{j \in \mathbb{N} \\ k=0,\ldots,2^j-1}} \langle f, \widetilde{\psi}_{j,k} \rangle \psi_{j,k}(x) + \sum_{k=0}^{m-1} \langle f, \widetilde{B}_{k,m,\mathbf{t}_m^{(0)}} \rangle B_{k,m,\mathbf{t}_m^{(0)}}(x)
$$

or

$$
f(x) = \sum_{\substack{j \in \mathbb{N} \\ k=0,\ldots,2^j-1}} \langle f, \psi_{j,k} \rangle \widetilde{\psi}_{j,k}(x) + \sum_{k=0}^{m-1} \langle f, B_{k,m,\mathbf{t}_m^{(0)}} \rangle \widetilde{B}_{k,m,\mathbf{t}_m^{(0)}}(x).
$$

The explicit formulae for the dual B-splines and wavelets as well as the decomposition and reconstruction sequences for a finite interval are the subject
of an upcoming paper [3]. From the duality conditions $\langle \psi_{j,k}, \widetilde{\psi}_{\tilde{j},\tilde{k}} \rangle = \delta_{j,\tilde{j}}\delta_{k,\tilde{k}}$
for the relevant indices (j,k) and $(\tilde{j},\tilde{k})$, it is clear, however, that these dual
functions must be supported on the entire interval $[0,1]$ and must be different
for different indices. Consequently, the corresponding decomposition relations
are described by full (finite) matrices, while the matrices for the reconstruction
are banded.

A further question that has to be settled is the numerical stability of the
bases. The Meyer bases of Section 2 have been identified as numerically unstable [7]. For the spline functions of Section 3, numerical stability/instability
has yet to be investigated.

Acknowledgements. The research of C. K. Chui was supported by NSF
Grant DMS-89-01345, ARO Grant DAAL 03-90-G-0091, and Texas Higher
Education Grants TATP 32134 and TARP 32135.

References

1. Buhmann, M. D. and C. A. Micchelli, Spline prewavelets for non-uniform
   knots, Numer. Math. , to appear.
2. Chui, C. K., *An Introduction to Wavelets*, Academic Press, Boston, 1992.
3. Chui, C. K. and E. Quak, Dual spline-wavelets on a bounded interval, in
   preparation.

4. Chui, C. K. and J. Z. Wang, An analysis of cardinal spline-wavelets, J. Approx. Th. , to appear.

5. Chui, C. K. and J. Z. Wang, A cardinal spline approach to wavelets, Proc. Amer. Math. Soc. **113** (1991), 785–793.

6. Chui. C. K. and J. Z. Wang, On compactly supported wavelets and a duality principle, Trans. Amer. Math. Soc. , to appear.

7. Cohen, A., Private communication.

8. Daubechies, I., Orthonormal bases of compactly supported wavelets, Comm. Pure and Appl. Math. **41** (1988), 909–996.

9. Lyche, T. and K. Mørken, Spline-wavelets of minimal support, in *Numerical Methods of Approximation Theory, Vol. 9*, Dietrich Braess and Larry L. Schumaker (eds.), Birkhäuser Verlag, Basel, 1992, 177–194.

10. Mallat, S. G., Multiresolution approximations and and wavelet orthonormal bases of $L^2(\mathbb{R})$, Trans. Amer. Math. Soc. **315** (1989), 69–87.

11. Meyer, Y., *Ondelettes et Opérateurs*, Hermann, Paris, 1990.

12. Meyer, Y., Ondelettes sur l'intervalle, Rev. Mat. Iberoamericana, **7** (1991), 115–143.

13. Schumaker, L. L., *Spline Functions: Basic Theory*, Wiley-Interscience, New York, 1981.

Charles K. Chui
Center for Approximation Theory
Texas A&M University
College Station, Texas 77843-3368
USA
cchui@tamu.edu

Ewald Quak
Center for Approximation Theory
Texas A&M University
College Station, Texas 77843-3368
USA
egq4460@tamvenus.bitnet

Numerical Methods of Approximation Theory, Vol. 9
Dietrich Braess and Larry L. Schumaker (eds.), pp. 77–95.
International Series of Numerical Mathematics, Vol. 105

77

Quasi-Kernel Polynomials and Convergence Results for Quasi-Minimal Residual Iterations

Roland W. Freund

Dedicated to the memory of Lothar Collatz

Abstract. Recently, Freund and Nachtigal [9] have proposed a novel polynomial-based iteration, the quasi-minimal residual algorithm (QMR), for solving general nonsingular non-Hermitian linear systems. Motivated by the QMR method, in [6] we have introduced the general concept of quasi-kernel polynomials, and we have shown that the QMR algorithm is based on a particular instance of quasi-kernel polynomials. In this paper, we continue our study of quasi-kernel polynomials. In particular, we derive bounds for the norms of quasi-kernel polynomials. These results are then applied to obtain convergence theorems both for the QMR method, and for a transpose-free variant of QMR, the TFQMR algorithm.

§1. Introduction

Many iterative algorithms for solving large nonsingular systems of linear equations

$$Ax = b \tag{1}$$

are based on polynomials. For example, the classical conjugate gradient algorithm [13] for Hermitian positive definite matrices A and its various extensions

to non-Hermitian linear systems all fall into the category of polynomial-based
methods. For a survey of iterative methods, we refer the reader to [7] and the
references given there.

Typically, polynomial-based algorithms with optimal convergence properties correspond to polynomials that are characterized as solutions of constrained least squares approximation problems of the form

$$\min_{\psi \in \mathcal{P}_n : \, \psi(0)=1} \langle \psi, \psi \rangle. \tag{2}$$

Here and in the sequel, we use the notation

$$\mathcal{P}_n := \{\psi(\lambda) \equiv \gamma_0 + \gamma_1 \lambda + \cdots + \gamma_n \lambda^n \, : \, \gamma_0, \gamma_1, \ldots, \gamma_n \in \mathbb{C}\} \tag{3}$$

for the set of all complex polynomials of degree at most n. Moreover, $\langle \cdot, \cdot \rangle$ in
(2) is an inner product that is positive definite on $\mathcal{P}_n$. The exact solutions of
problems of the form (2) are normalized kernel polynomials, and they can be
expressed in terms of the orthogonal polynomials associated with the inner
product $\langle \cdot, \cdot \rangle$.

For the special case of Hermitian matrices A, only inner products on
the real line arise, and then the corresponding kernel and orthogonal polynomials satisfy three-term recurrences. As a result, optimal polynomial-based
algorithms for Hermitian linear systems can be implemented using only three-
term vector recursions. The situation is less satisfactory for the case of general
non-Hermitian matrices A. Here, inner products in the complex plane arise,
and the corresponding complex kernel and orthogonal polynomials no longer
satisfy short recurrences. Consequently, optimal polynomial-based methods
for non-Hermitian linear systems in general involve long vector recurrences.
Usually, it is too expensive to run the full versions of these optimal algorithms.
Instead, restarted or truncated variants are used, which often leads to rather
slow convergence.

Recently, Freund and Nachtigal [9] proposed the quasi-minimal residual
algorithm (QMR) for solving general nonsingular non-Hermitian linear systems (1). The QMR method produces approximate solutions of (1) that are
characterized by a quasi-optimality condition, namely a quasi-minimization
of the residual norm. The point is that the QMR algorithm—unlike optimal
polynomial-based iteration schemes—can be implemented using only short
vector recurrences.

Motivated by the QMR method, in [6] we have introduced the general
concept of quasi-kernel polynomials, and we have shown that the QMR algorithm is based on a particular instance of quasi-kernel polynomials. Moreover,

in [6] some general theory for quasi-kernel polynomials was given, such as recurrence relations and a characterization of roots of quasi-kernel polynomials as generalized eigenvalues.

In this paper, we continue our study of quasi-kernel polynomials. In particular, we derive bounds for the norms of quasi-kernel polynomials. These results are then applied to obtain convergence theorems both for the original QMR method, and for a transpose-free variant of QMR, the TFQMR algorithm [5].

The remainder of the paper is organized as follows. In §2 we briefly recall the definition of quasi-kernel polynomials. In §3 we present bounds for the norms of quasi-kernel polynomials. We then turn to non-Hermitian linear systems (1), and in §4 we describe the general setting of polynomial-based methods for the iterative solution of such systems. Finally, in §5 and §6 we consider the QMR and TFQMR algorithms, respectively, and we give convergence results for both methods.

Throughout the paper, all vectors and matrices are allowed to have real or complex entries. As usual, $M^T = [\,m_{kj}\,]$ and $M^H = [\,\overline{m_{kj}}\,]$ denote the transpose and the conjugate transpose, respectively, of the matrix $M = [\,m_{jk}\,]$. We use the notation $\sigma_{\max}(M)$ and $\sigma_{\min}(M)$ for the largest and smallest singular value, respectively, of the matrix M. The vector norm $\|x\|_2 = \sqrt{x^H x}$ is always the Euclidean norm, $\|M\|_2 = \sigma_{\max}(M)$ is the corresponding matrix norm, and for nonsingular square matrices M, $\kappa_2(M) = \sigma_{\max}(M)/\sigma_{\min}(M)$ is the Euclidean condition number. Moreover, for square matrices M, we denote by $\lambda(M)$ the set of all eigenvalues of M. In addition to (3), we use the notation $\mathcal{P}_\infty$ for the set of all complex polynomials. The symbol 0 will be used for the number zero, the zero matrix, and the polynomial $\psi(\lambda) \equiv 0$; its actual meaning and, in the case of the zero matrix, its dimension will be apparent from the context. Finally, I_n denotes the $n \times n$ identity matrix.

§2. Quasi-Kernel Polynomials

We consider constrained approximation problems of the form (2), where $\langle \cdot, \cdot \rangle$ is assumed to be a given positive semidefinite inner product on $\mathcal{P}_\infty$, i.e., for all $\varphi_1,\ \varphi_2,\ \varphi,\ \psi \in \mathcal{P}_\infty$ and $\gamma_1,\ \gamma_2 \in \mathbb{C}$:

$$\langle \gamma_1 \varphi_1 + \gamma_2 \varphi_2, \psi \rangle = \gamma_1 \langle \varphi_1, \psi \rangle + \gamma_2 \langle \varphi_2, \psi \rangle,$$

$$\overline{\langle \varphi, \psi \rangle} = \langle \psi, \varphi \rangle,$$

$$\langle \varphi, \varphi \rangle \geq 0.$$

Notice that we do not require $\langle \varphi, \varphi \rangle > 0$ for $\varphi \neq 0$, i.e., $\langle \cdot, \cdot \rangle$ is not assumed to be positive definite. It will be convenient to rewrite (2) in the form

$$\min_{\psi \in \mathcal{P}_n : \, \psi(0)=1} \|\psi\|. \tag{4}$$

Here and in the sequel,

$$\|\varphi\| := \langle \varphi, \varphi \rangle^{1/2}, \quad \varphi \in \mathcal{P}_\infty,$$

always denotes the seminorm induced by $\langle \cdot, \cdot \rangle$.

If $\langle \cdot, \cdot \rangle$ is positive definite on $\mathcal{P}_n$, then there always exists a unique optimal solution ψ_n^{K} of (4). It is well known (see, e.g., [21, Chapter XVI], [1, Chapter I], [20], or [6, Corollary 3.3]) that ψ_n^{K} is just the suitably normalized nth *kernel polynomial* associated with $\langle \cdot, \cdot \rangle$. Furthermore, ψ_n^{K} is given by

$$\psi_n^{K}(\lambda) \equiv \frac{\sum_{j=0}^{n} \overline{\varphi_j(0)} \varphi_j(\lambda)}{\sum_{j=0}^{n} |\varphi_j(0)|^2}, \tag{5}$$

where $\varphi_j \in \mathcal{P}_j$, $j = 0, 1, \ldots, n$, is any set of *orthonormal polynomials* with respect to $\langle \cdot, \cdot \rangle$, i.e.,

$$\langle \varphi_j, \varphi_k \rangle = \begin{cases} 0, & \text{if } j \neq k, \\ 1, & \text{if } j = k. \end{cases}$$

In the following, we will refer to ψ_n^{K} as a *true nth kernel polynomial*.

Roughly speaking, quasi-kernel polynomials are approximations to true kernel polynomials, which are obtained from a set of arbitrary basis polynomials, rather than a set of orthonormal polynomials. More precisely, the setting is as follows. Let $\Pi_J := \{\varphi_j\}_{j=0}^{J}$, where J can either be a finite integer or equal to ∞, be a given set of polynomials

$$\varphi_j \in \mathcal{P}_j \quad \text{with} \quad \deg \varphi_j = j, \tag{6}$$

which span $\mathcal{P}_J$. We note that, by (6), each polynomial $\lambda \varphi_{j-1}$, $1 \leq j \leq J$, has a unique representation of the form

$$\lambda \varphi_{j-1}(\lambda) \equiv \sum_{k=0}^{j} h_{k+1,j} \varphi_k(\lambda), \quad \text{where} \quad h_{j+1,j} \neq 0. \tag{7}$$

In the sequel, we always assume that $n \in \{1, 2, \ldots, J\}$ if J is finite, and $n \in \{1, 2, \ldots\}$ if $J = \infty$. Moreover, we denote by

$$H_n = \begin{bmatrix} h_{11} & h_{12} & \cdots & & h_{1n} \\ h_{21} & \ddots & & & \vdots \\ 0 & \ddots & & \ddots & \vdots \\ \vdots & \ddots & h_{n,n-1} & & h_{nn} \\ 0 & \cdots & & 0 & h_{n+1,n} \end{bmatrix} \tag{8}$$

the $(n+1) \times n$ upper *Hessenberg* matrix, which contains the recurrence coefficients $h_{k+1,j}$ from (7) as entries. We remark that, in view of (7), all subdiagonal elements $h_{j+1,j}$ in (8) are different from zero, i.e., H_n is an *unreduced* upper Hessenberg matrix.

It follows from (6) that any polynomial $\psi_n \in \mathcal{P}_n$ with $\psi_n(0) = 1$ can be parametrized in the form

$$\psi_n(\lambda) \equiv 1 - \lambda \sum_{j=0}^{n-1} \zeta_{j+1} \varphi_j(\lambda), \quad \text{where} \quad \zeta_1, \zeta_2, \ldots, \zeta_n \in \mathbb{C}. \tag{9}$$

In [6] we have shown that each polynomial (9) satisfies

$$\begin{aligned}
\|\psi_n\| &= \left((d_n - H_n z)^H G_n (d_n - H_n z) \right)^{1/2} \\
&\leq \left(\|G_n\|_2 \right)^{1/2} \|d_n - H_n z\|_2.
\end{aligned} \tag{10}$$

Here

$$z := \begin{bmatrix} \zeta_1 \\ \vdots \\ \zeta_n \end{bmatrix} \in \mathbb{C}^n \tag{11}$$

is the vector containing the free parameters in (9) as coefficients,

$$G_n := \left[\langle \varphi_j, \varphi_k \rangle \right]_{j,k=0,1,\ldots,n} \tag{12}$$

is the Gram matrix of $\varphi_0, \varphi_1, \ldots, \varphi_n$, and

$$d_n := [1/\varphi_0 \quad 0 \quad \cdots \quad 0]^T \in \mathbb{C}^{n+1}. \tag{13}$$

Note that, by (6), φ_0 is a nonzero constant.

The estimate (10) is the basis for the definition of quasi-kernel polynomials. We choose the parameters $\zeta_1, \zeta_2, \ldots, \zeta_n$ in (9) such that, instead of $\|\psi_n\|$, only the second factor in the upper bound in (10) is minimal. This leads to the following definition.

Definition 1. [6, Definition 3.1]. *Let* $z_n \in \mathbb{C}^n$ *be the solution of the least squares problem*

$$\|d_n - H_n z_n\|_2 = \min_{z \in \mathbb{C}^n} \|d_n - H_n z\|_2. \tag{14}$$

The polynomial ψ_n given by (9) and (11) (with $z := z_n$) is called the nth quasi-kernel polynomial (corresponding to the inner product $\langle \cdot, \cdot \rangle$ and derived from Π_J).

Recall that H_n is an unreduced $(n+1) \times n$ Hessenberg matrix, and thus it has full rank n. This guarantees that (14) always has a unique solution z_n.

In [6] we have established some general theory for quasi-kernel polynomials, such as recurrence relations and a characterization of roots of quasi-kernel polynomials as generalized eigenvalues. For example, in [6, Theorem 3.2], it is shown that ψ_n satisfies the same relation (5) as true kernel polynomials, where the φ_j's are now the given basis polynomials (6). In particular, if the polynomials in Π_J are orthonormal, then the quasi-kernel polynomial ψ_n coincides with the true kernel polynomial.

§3. Bounds for the Norms of Quasi-Kernel Polynomials

In this section, we present some bounds for the norm $\|\psi_n\|$ of the nth quasi-kernel polynomial ψ_n.

First, note that, from (10) and (14), one obtains the estimates

$$
\min_{\psi \in \mathcal{P}_n \colon \psi(0)=1} \|\psi\| \leq \|\psi_n\|
$$
$$
\leq (\|G_n\|_2)^{1/2} \min_{z \in \mathbb{C}^n} \|d_n - H_n z\|_2. \tag{15}
$$

For later use, we remark that the first factor of the upper bound in (15) satisfies the inequality

$$
(\|G_n\|_2)^{1/2} \leq \sqrt{n+1} \max_{j=0,\dots,n} \|\varphi_j\|, \tag{16}
$$

which follows from

$$
\|G_n\|_2 = \max_{y \in \mathbb{C}^{n+1} \colon \|y\|_2=1} y^H G_n y
$$

and the Cauchy-Schwarz inequality.

Clearly, the Gram matrix G_n defined in (12) is positive definite if the inner product $\langle \cdot, \cdot \rangle$ is positive definite on $\mathcal{P}_n$. In this case, the condition number $\kappa_2(G_n)$ of G_n is a measure for how far $\|\psi_n\|$ can be from the minimal value $\|\psi_n^K\|$ of (4).

Theorem 2. *Assume that the inner product $\langle \cdot, \cdot \rangle$ is positive definite on $\mathcal{P}_n$, and let ψ_n^{K} be the true nth kernel polynomial. Then:*

$$\|\psi_n^{\mathrm{K}}\| \leq \|\psi_n\| \leq (\kappa_2(G_n))^{1/2}\, \|\psi_n^{\mathrm{K}}\|. \tag{17}$$

In particular, $\psi_n = \psi_n^{\mathrm{K}}$ if $G_n = I_n$.

Proof: The polynomial ψ_n^{K} is the optimal solution of the approximation problem on the left-hand side of (15). This implies the first inequality in (17).

Let ψ_n^{K} be represented in the form (9), and let $z_n^{\mathrm{K}} \in \mathbb{C}^n$ be the corresponding parameter vector (11). Using the first relation in (10), it follows that

$$\begin{aligned}
\|\psi_n^{\mathrm{K}}\| &= \left((d_n - H_n z_n^{\mathrm{K}})^H G_n (d_n - H_n z_n^{\mathrm{K}})\right)^{1/2} \\
&\geq (\sigma_{\min}(G_n))^{1/2}\, \|d_n - H_n z_n^{\mathrm{K}}\|_2.
\end{aligned} \tag{18}$$

On the other hand, from (15), we have

$$\begin{aligned}
\|\psi_n\| &\leq (\sigma_{\max}(G_n))^{1/2} \min_{z \in \mathbb{C}^n} \|d_n - H_n z\|_2 \\
&\leq (\sigma_{\max}(G_n))^{1/2}\, \|d_n - H_n z_n^{\mathrm{K}}\|_2.
\end{aligned} \tag{19}$$

Finally, by combining the estimates (18) and (19), we obtain the second inequality in (17). ∎

Next, we show that the upper bound in (15) can be related to certain constrained approximation problems. This result will be used in §5 and §6 to derive convergence theorems for QMR and TFQMR, respectively.

We use the following setting. Let $m \leq J$ be an arbitrary, but fixed integer. We denote by $\lambda_1, \lambda_2, \ldots, \lambda_l \in \mathbb{C}$ the distinct zeros of the mth basis polynomial φ_m, and μ_j is the multiplicity of λ_j. Note that

$$\varphi_m(\lambda) \equiv \gamma \prod_{j=1}^{l} (\lambda - \lambda_j)^{\mu_j} \quad \text{for some} \quad \gamma \in \mathbb{C},\ \gamma \neq 0.$$

Furthermore, we denote by

$$H := [\, I_m \quad 0\,]\, H_m \tag{20}$$

the $m \times m$ matrix obtained by deleting the last row of the mth Hessenberg matrix H_m in (8). Using (7) and (20), one easily verifies the identity

$$\lambda \Phi(\lambda) \equiv \Phi(\lambda) H + [\, 0 \quad \cdots \quad 0 \quad h_{m+1,m} \varphi_m(\lambda)\,], \tag{21}$$

where $\Phi(\lambda)$ is a row vector defined by

$$\Phi(\lambda) := [\,\varphi_0(\lambda) \quad \cdots \quad \varphi_{m-1}(\lambda)\,].$$

By setting $\lambda = \lambda_j$ in (21), we deduce that the zeros of φ_m are just the eigenvalues of H, i.e.,

$$\{\lambda_1, \lambda_2, \ldots, \lambda_l\} = \lambda(H), \tag{22}$$

and each λ_j is an eigenvalue of algebraic multiplicity μ_j. Next, we recall that H_m (and thus H) is an unreduced upper Hessenberg matrix. This implies that, for each λ_j, the matrix $H - \lambda_j I_m$ has rank $m - 1$, and hence the eigenspace corresponding to λ_j is of dimension 1. As a result, there is exactly one Jordan block for each eigenvalue of H. Therefore, H has the Jordan canonical form

$$J = SHS^{-1}, \tag{23}$$

where S is a nonsingular $m \times m$ matrix and

$$J = \mathrm{diag}(J(\lambda_1), J(\lambda_2), \ldots, J(\lambda_l)) \tag{24}$$

is a block diagonal matrix with Jordan blocks

$$J(\lambda_j) = \lambda_j I_{\mu_j} + N_j, \quad N_j := \begin{bmatrix} 0 & 1 & 0 & \cdots & 0 \\ \vdots & 0 & 1 & \ddots & \vdots \\ \vdots & \ddots & \ddots & \ddots & 0 \\ \vdots & \ddots & \ddots & 0 & 1 \\ 0 & \cdots & \cdots & \cdots & 0 \end{bmatrix} \in \mathbb{R}^{\mu_j \times \mu_j}. \tag{25}$$

After these preliminaries, our result can now be stated as follows.

Theorem 3. *Let $m \leq J$ be a fixed integer, and let H be the $m \times m$ matrix defined by (20) with spectrum (22). Let $J = SHS^{-1}$ be the Jordan normal form (23), (24), with Jordan blocks (25). Then, for $n = 1, 2, \ldots, m - 1$, the nth quasi-kernel polynomial ψ_n satisfies*

$$\|\psi_n\| \leq \frac{\kappa_2(S)}{|\varphi_0|} \left(\|G_n\|_2\right)^{1/2} \varepsilon_n, \tag{26}$$

where

$$\varepsilon_n := \min_{\psi \in \mathcal{P}_n:\, \psi(0)=1} \ \max_{j=1,\ldots,l} \|\psi(J(\lambda_j))\|_2. \tag{27}$$

Moreover, *if H is diagonalizable, then:*

$$\varepsilon_n = \min_{\psi \in \mathcal{P}_n:\, \psi(0)=1} \ \max_{\lambda \in \lambda(H)} |\psi(\lambda)|. \tag{28}$$

Proof: Let $n \in \{1, 2, \ldots, m-1\}$. By using (13) and by setting $u := \varphi_0 z$, we obtain from (15) the estimate

$$\|\psi_n\| \leq \frac{1}{|\varphi_0|} \left(\|G_n\|_2\right)^{1/2} \vartheta_n, \tag{29}$$

where

$$\vartheta_n := \min_{u \in \mathbb{C}^n} \|e_1^{(n+1)} - H_n u\|_2, \quad e_1^{(n+1)} := [1 \quad 0 \quad \cdots \quad 0]^T \in \mathbb{R}^{n+1}. \tag{30}$$

In view of (20), H_n is the $(n+1) \times n$ leading submatrix of H, i.e.,

$$H = \begin{bmatrix} H_n & * \\ 0 & * \end{bmatrix},$$

and we have the relation

$$H \begin{bmatrix} u \\ 0 \end{bmatrix} = \begin{bmatrix} H_n u \\ 0 \end{bmatrix} \quad \text{for all} \quad u \in \mathbb{C}^n. \tag{31}$$

Next, using the fact that H is an unreduced upper Hessenberg matrix, we deduce that

$$\left\{ \begin{bmatrix} u \\ 0 \end{bmatrix} : u \in \mathbb{C}^n \right\} = \{\phi(H)e_1^{(m)} \mid \phi \in \mathcal{P}_{n-1}\}. \tag{32}$$

By means of (31) and (32), the quantity ϑ_n in (30) can be rewritten in the form

$$\vartheta_n = \min_{\phi \in \mathcal{P}_{n-1}} \|e_1^{(m)} - \phi(H)e_1^{(m)}\|_2 = \min_{\psi \in \mathcal{P}_n:\, \psi(0)=1} \|\psi(H)e_1^{(m)}\|_2. \tag{33}$$

With (23)–(25) and since $\|e_1^{(m)}\|_2 = 1$, it follows that, for all $\psi \in \mathcal{P}_\infty$,

$$\|\psi(H)e_1^{(m)}\|_2 \leq \kappa_2(S) \|\psi(J)\|_2 = \kappa_2(S) \max_{j=1,\ldots,l} \|\psi(J(\lambda_j))\|_2. \tag{34}$$

By combining (29), (33), and (34), we obtain the desired estimate (26), (27).

Finally, if H is diagonalizable, then all Jordan blocks $J(\lambda_j) = [\,\lambda_j\,]$ are of size 1, and (27) reduces to (28). ∎

We remark that the norms in (27) can be estimated by means of the inequality

$$\|\psi(J(\lambda_j))\|_2 \le \sum_{i=0}^{\mu_j-1} \frac{1}{i!} \left| \psi^{(i)}(\lambda_j) \right|, \quad \psi \in \mathcal{P}_\infty, \tag{35}$$

which is an immediate consequence of the standard relation

$$\psi(J(\lambda_j)) = \sum_{i=0}^{\mu_j-1} \frac{1}{i!} \psi^{(i)}(\lambda_j) N_j^i.$$

Here N_j is the nilpotent matrix defined in (25), and $\psi^{(i)}$ denotes the ith derivative of ψ.

§4. Polynomial-Based Matrix Iterations

We now return to systems of linear equations (1). From now on, it is always assumed that A in (1) is a nonsingular non-Hermitian $N \times N$ matrix.

Many iterative algorithms for solving linear systems (1) are *polynomial-based methods*: they produce approximations x_n to $A^{-1}b$ of the form

$$x_n = x_0 + \phi_n(A)r_0 \quad \text{with} \quad \phi_n \in \mathcal{P}_{n-1}, \quad n = 1, 2, \ldots . \tag{36}$$

Here $x_0 \in \mathbb{C}^N$ is any initial guess for the solution of (1), and $r_0 := b - Ax_0$. We remark that the residual vector $r_n := b - Ax_n$ corresponding to the nth iterate (36) of any polynomial-based algorithm is given by

$$r_n = \psi_n(A)r_0, \quad \text{where} \quad \psi_n(\lambda) \equiv 1 - \lambda\phi_n(\lambda). \tag{37}$$

Note that

$$\psi_n \in \mathcal{P}_n \quad \text{and} \quad \psi_n(0) = 1. \tag{38}$$

Generally, any ψ_n that satisfies (38) is called an nth *residual polynomial*.

The goal in designing a polynomial-based method is to choose at each step the residual polynomial ψ_n in (37) such that some norm of r_n is as small as possible. A standard approach [18] is to require that the Euclidean norm of r_n is minimal, i.e.,

$$\|r_n\|_2 = \min_{\psi \in \mathcal{P}_n:\, \psi(0)=1} \|\psi(A)r_0\|_2. \tag{39}$$

By setting

$$\langle \varphi, \psi \rangle := r_0^H (\psi(A))^H \varphi(A) r_0 \quad \text{and} \quad \|\psi\| := \langle \psi, \psi \rangle^{1/2}, \tag{40}$$

we see that (39) is an instance of a constrained polynomial approximation problem of the form (4). We remark that other strategies (see [17,19]) for choosing the residual polynomial ψ_n in (37) also lead to problems of the form (4), e.g., with inner products given by

$$\langle \varphi, \psi \rangle := \iint_G \varphi(\lambda) \overline{\psi(\lambda)} \omega(\lambda) \, d\sigma \quad \text{or} \quad \langle \varphi, \psi \rangle := \int_C \varphi(\lambda) \overline{\psi(\lambda)} \omega(\lambda) \, |d\lambda| \, .$$

Typically, $G \subset \mathbb{C}$ is a compact set containing the spectrum $\lambda(A)$ of A or some approximation to $\lambda(A)$, and C is a curve bounding such a set G.

In the remainder of this paper, we always assume that $\langle \cdot, \cdot \rangle$ is the inner product defined by (40). Note that the corresponding polynomial norm $\|\cdot\|$ is equivalent to the Euclidean vector norm in $\mathbb{C}^N$, in the sense that

$$\|\psi\| = \|v\|_2 \quad \text{for all} \quad v = \psi(A) r_0 \quad \text{with} \quad \psi \in \mathcal{P}_\infty. \tag{41}$$

Finally, we call a polynomial-based method *optimal* if its residual vectors satisfy (39). We recall (see, e.g., [6]) that an optimal iteration scheme terminates after a finite number $J_{\mathrm{opt}}(\leq N)$ of steps with the exact solution $x_{J_{\mathrm{opt}}} = A^{-1}b$ of (1). Furthermore, the residual vectors r_n^{opt} of an optimal polynomial-based method are unique, and they are given by

$$r_n^{\mathrm{opt}} = \psi_n^K(A) r_0, \quad n = 1, 2, \ldots, J_{\mathrm{opt}}, \tag{42}$$

where ψ_n^K is the nth true kernel polynomial associated with $\langle \cdot, \cdot \rangle$.

§5. Convergence Results for the QMR Algorithm

Faber and Manteuffel [2] have shown that—except for a very special class of non-Hermitian matrices—optimal polynomial-based methods for general non-Hermitian linear systems can be implemented only with long recurrences, which involve vectors from all previous iterations. Their result motivated the development of a quasi-optimal polynomial-based iteration scheme, the QMR algorithm [9], which can be implemented using only short vector recurrences. In this section, we apply the results from §3 to obtain two convergence theorems for the QMR algorithm.

The QMR iteration uses the Lanczos process [14] with look-ahead [16,8] to generate basis vectors of the form

$$v_j = \phi_j(A)r_0, \quad j = 0, 1, \ldots, J_{\mathrm{L}}. \tag{43}$$

Here

$$\phi_j \in \mathcal{P}_j \quad \text{with} \quad \deg \phi_j = j, \quad j = 0, 1, \ldots, J_{\mathrm{L}}, \tag{44}$$

are the so-called Lanczos polynomials, and J_{L} is the termination index of the look-ahead Lanczos algorithm. We note that $J_{\mathrm{L}} \leq J_{\mathrm{opt}}$. For a detailed description of the Lanczos process, we refer the reader to [8] and the references given there.

In [6] we have shown that the QMR algorithm is based on quasi-kernel polynomials. More precisely, the QMR residual vectors are given by

$$r_n^{\mathrm{QMR}} = \psi_n(A)r_0, \quad n = 1, 2, \ldots, J, \tag{45}$$

where $J := J_{\mathrm{L}}$, and ψ_n is the nth quasi-kernel polynomial corresponding to the inner product (40) and derived from the normalized Lanczos polynomials

$$\varphi_j = \begin{cases} \dfrac{\phi_j}{\|\phi_j\|}, & j = 0, 1, \ldots, J_{\mathrm{L}} - 1, \\[2mm] \phi_{J_{\mathrm{L}}}, & j = J_{\mathrm{L}}. \end{cases} \tag{46}$$

For later use, we remark that the condition number of the Gram matrix G_n associated with the basis polynomials (46) can be expressed in terms of the extreme singular values of the matrix of normalized Lanczos vectors

$$V_n := \left[\frac{v_0}{\|v_0\|_2} \quad \frac{v_1}{\|v_1\|_2} \quad \cdots \quad \frac{v_n}{\|v_n\|_2} \right].$$

More precisely, for $n = 0, 1, \ldots, J_{\mathrm{L}} - 1$, we have

$$(\kappa_2(G_n))^{1/2} = \frac{\sigma_{\max}(V_n)}{\sigma_{\min}(V_n)}. \tag{47}$$

The relation (47) is an immediate consequence of the identity

$$y^H G_n y = \|V_n y\|^2, \quad y \in \mathbb{C}^{n+1},$$

which follows from (12), (40), (43), (46), and (41).

Next, we apply Theorem 2, and we rewrite, by means of (41) and (47), the bounds in (17) in terms of the residual vectors (42) and (45). This gives the following result.

Theorem 4. *For $n = 1, 2, \ldots, J_{\mathrm{L}} - 1$, the* QMR *residual vectors satisfy*

$$\|r_n^{\mathrm{opt}}\|_2 \leq \|r_n^{\mathrm{QMR}}\|_2 \leq \frac{\sigma_{\max}(V_n)}{\sigma_{\min}(V_n)} \|r_n^{\mathrm{opt}}\|_2. \tag{48}$$

In particular, $r_n^{\mathrm{QMR}} = r_n^{\mathrm{opt}}$ if the columns of V_n are orthonormal.

We remark that a different proof for (48), without using quasi-kernel polynomials, was given by Nachtigal [15, Section 4.3].

Next, we derive upper bounds for $\|r_n^{\mathrm{QMR}}\|_2$ by applying Theorem 3 (with $m = J_{\mathrm{L}}$) for the special case of the QMR quasi-kernel polynomials.

In the sequel, the same notation as in §3 is used. We set $m := J_{\mathrm{L}}$, and let

$$H := [\, I_{J_{\mathrm{L}}} \quad 0 \,] \, H_{J_{\mathrm{L}}} \tag{49}$$

be the Hessenberg matrix of recurrence coefficients for the basis polynomials (46). As before, we denote by $\lambda_1, \lambda_2, \ldots, \lambda_l$ the distinct zeros of φ_m. We remark that, by (46), the polynomial $\varphi_m = \phi_{J_{\mathrm{L}}}$ is just the last Lanczos polynomial. It is well known (see [16,12]) that each zero of $\phi_{J_{\mathrm{L}}}$ is also an eigenvalue of A. Therefore, together with (22), we have

$$\{\lambda_1, \lambda_2, \ldots, \lambda_l\} = \lambda(H) \subseteq \lambda(A). \tag{50}$$

Furthermore, we will need the relation

$$|\varphi_0| = 1/\|r_0\|_2, \tag{51}$$

which follows from (46) (with j=0) and (41), and the inequality

$$(\|G_n\|_2)^{1/2} \leq \sqrt{n+1}, \quad n = 1, 2, \ldots, m - 1, \tag{52}$$

which follows from (16) and (46).

Finally, by applying Theorem 3, we obtain the following convergence result for the QMR algorithm.

Theorem 5. *Let H be the $J_{\mathrm{L}} \times J_{\mathrm{L}}$ matrix (49) corresponding to the normalized Lanczos polynomials (46), and let (50) be the spectrum of H. Let $J = SHS^{-1}$ be the Jordan normal form (23), (24) of H, with Jordan blocks (25). Then, for $n = 1, 2, \ldots, J_{\mathrm{L}} - 1$, the* QMR *residual vectors satisfy*

$$\frac{\|r_n^{\mathrm{QMR}}\|_2}{\|r_0\|_2} \leq \kappa_2(S) \sqrt{n+1}\, \varepsilon_n, \tag{53}$$

where

$$\varepsilon_n := \min_{\psi \in \mathcal{P}_n: \psi(0)=1} \ \max_{j=1,\ldots,l} \|\psi(J(\lambda_j))\|_2.$$

Moreover, if H is diagonalizable, then:

$$\varepsilon_n \leq \min_{\psi \in \mathcal{P}_n: \psi(0)=1} \ \max_{\lambda \in \lambda(A)} |\psi(\lambda)|. \tag{54}$$

Proof: The inequality (53) is obtained by rewriting (26) in terms of the QMR residuals (45) instead of ψ_n, and by using the relations (51) and (52). The estimate (54) follows from (28) and (50). $\blacksquare$

We remark that, for the special case of diagonalizable H, the bound (53), (54) was first given by Freund and Nachtigal [9, Theorem 6.1].

Of course, the bound in (54) is not very practical since it would require the knowledge of all eigenvalues of A. However, often one knows some compact set $G \subset \mathbb{C}$ such that

$$\lambda(A) \subset G \quad \text{and} \quad 0 \notin G. \tag{55}$$

In this case, we can replace (54) by the estimate

$$\varepsilon_n \leq \min_{\psi \in \mathcal{P}_n: \psi(0)=1} \ \max_{\lambda \in G} |\psi(\lambda)|. \tag{56}$$

The solutions of constrained polynomial approximation problems of the form (56) are known explicitly for some "simple" sets, such as disks [22], certain ellipses [3,4], and line segments [11].

Finally, we remark that the right-hand side in (53) is less than 1, provided that n is large enough. Here, this is shown only for the case that all eigenvalues of A are contained in the right half of the complex plane, i.e.,

$$\lambda(A) \subset \{\lambda \in \mathbb{C} : \ \mathrm{Re}\,\lambda > 0\}. \tag{57}$$

Moreover, for simplicity, we assume that the Lanczos matrix H in (49) is diagonalizable. In view of (57), there exists an ellipse in the right half plane that contains the spectrum of A, i.e., the condition (55) is satisfied for some ellipse G. Let $f_1 \neq f_2$ be the two foci of G. Then G can be represented in the form

$$G = \left\{ \lambda \in \mathbb{C} : \ |\lambda - f_1| + |\lambda - f_2| \leq \frac{|f_1 - f_2|}{2} \left(r + \frac{1}{r} \right) \right\}, \quad \text{with} \quad r \geq 1.$$

In [3, Theorem 2], it was shown that, for such ellipses G, the right-hand side in (56) can be bounded as follows:

$$\min_{\psi \in \mathcal{P}_n: \psi(0)=1} \ \max_{\lambda \in G} |\psi(\lambda)| \leq \frac{r^n + 1/r^n}{R^n + 1/R^n}, \quad n = 1, 2, \ldots . \tag{58}$$

Here, R is the unique solution of

$$\frac{1}{2}\left(R + \frac{1}{R}\right) = \frac{|f_1| + |f_2|}{|f_1 - f_2|}, \quad R > 1.$$

We note that the condition $0 \notin G$ implies that $R > r$. By combining (56) and (58), we obtain

$$\varepsilon_n \leq \frac{r^n + 1/r^n}{R^n + 1/R^n}. \tag{59}$$

Since $1 \leq r < R$, the estimate (59) guarantees that the right-hand side in (53) tends to zero as n increases.

§6. A Convergence Result for the TFQMR Algorithm

The QMR algorithm involves matrix-vector multiplications with the coefficient matrix A, as well as its transpose A^T. This is a disadvantage for certain applications, where matrix-vector products $A \cdot v$, but not $A^T \cdot w$, can be computed efficiently without ever explicitly generating the matrix A. This was the motivation for developing a transpose-free variant of the standard QMR method, the TFQMR algorithm [5,10], which involves matrix-vector products with A only. We remark that the QMR and TFQMR methods are not equivalent, and the two algorithms produce different sets of iterates.

In this section, we present a convergence result for the TFQMR algorithm. It appears that this is the first convergence theorem for a transpose-free Lanczos-based iterative method.

In [6] we have shown that—like the QMR algorithm—the TFQMR method is also based on quasi-kernel polynomials. Its residual vectors are given by

$$r_n^{\text{TFQMR}} = \psi_n(A)r_0, \quad n = 1, 2, \ldots, J,$$

where $J := 2J_{\text{L}}$ and ψ_n is the nth quasi-kernel polynomial corresponding to the inner product (40) and derived from basis polynomials $\varphi_0, \varphi_1, \ldots, \varphi_J$ that are now defined as squares or products of Lanczos polynomials (44). More precisely, the basis polynomials are defined as follows:

$$\varphi_j = \begin{cases} \dfrac{\phi_k^2}{\|\phi_k^2\|}, & \text{if } j = 2k,\ k = 0, 1, \ldots, J_{\mathrm{L}} - 1, \\[2ex] \dfrac{\phi_{k-1}\phi_k}{\|\phi_{k-1}\phi_k\|}, & \text{if } j = 2k - 1,\ k = 1, 2, \ldots, J_{\mathrm{L}} - 1, \\[2ex] \dfrac{\phi_{J_{\mathrm{L}}-1}\phi_{J_{\mathrm{L}}}}{\|\phi_{J_{\mathrm{L}}-1}\phi_{J_{\mathrm{L}}}\|}, & \text{if } j = 2J_{\mathrm{L}} - 1 \text{ and } \|\phi_{J_{\mathrm{L}}-1}\phi_{J_{\mathrm{L}}}\| \neq 0, \\[2ex] \phi_{J_{\mathrm{L}}-1}\phi_{J_{\mathrm{L}}}, & \text{if } j = 2J_{\mathrm{L}} - 1 \text{ and } \|\phi_{J_{\mathrm{L}}-1}\phi_{J_{\mathrm{L}}}\| = 0, \\[2ex] \phi_{J_{\mathrm{L}}}^2, & \text{if } j = 2J_{\mathrm{L}}. \end{cases} \tag{60}$$

We remark that, as in the previous section, J_{L} denotes the termination index of the look-ahead Lanczos algorithm.

Again, we apply Theorem 3, but this time with $m := 2J_{\mathrm{L}}$. Let

$$H' := [\, I_{J_{2\mathrm{L}}} \quad 0 \,] H_{J_{2\mathrm{L}}}, \tag{61}$$

where $H_{J_{2\mathrm{L}}}$ now denotes the Hessenberg matrix (8) containing the recurrence coefficients of the basis polynomials (60). Here, the notation H' is used to distinguish the matrix (61) from the Hessenberg matrix (49) associated with the QMR algorithm.

Recall from §3 that the eigenvalues of H' are just the distinct zeros of the polynomial φ_m. Now, from (60) (with $j = 2J_{\mathrm{L}}$), we have that

$$\varphi_m = \phi_{J_{\mathrm{L}}}^2 \tag{62}$$

is the square of the last Lanczos polynomial. It follows that

$$\lambda(H') = \{\lambda_1, \lambda_2, \ldots, \lambda_l\},$$

where, as in §5, the λ_j's denote the distinct zeros of $\phi_{J_{\mathrm{L}}}$. Hence, in view of (50), the eigenvalues of the TFQMR and QMR Hessenberg matrices H' and H coincide, and we have

$$\lambda(H') = \lambda(H) \subseteq \lambda(A). \tag{63}$$

However, their algebraic multiplicities are different. Indeed, by (62), each eigenvalue λ_j of H' has even algebraic multiplicity $2\mu_j$, while λ_j is an eigenvalue of H with algebraic multiplicity $\mu_j (\geq 1)$. In particular, the blocks in the Jordan normal form (23), (24) of H' are now of the form

$$J(\lambda_j) = \lambda_j I_{2\mu_j} + N_j, \quad N_j := \begin{bmatrix} 0 & 1 & 0 & \cdots & 0 \\ \vdots & 0 & 1 & \ddots & \vdots \\ \vdots & \ddots & \ddots & \ddots & 0 \\ \vdots & & \ddots & \ddots & 0 & 1 \\ 0 & \cdots & \cdots & \cdots & 0 \end{bmatrix} \in \mathbb{R}^{2\mu_j \times 2\mu_j}. \quad (64)$$

Finally, we note that, for the basis polynomials (60), the relations (51) and (52) hold too.

Therefore, we can proceed as in the previous section, and by applying Theorem 3, we obtain the following convergence result for the TFQMR algorithm.

Theorem 6. *Let H' be the $2J_{\mathrm{L}} \times 2J_{\mathrm{L}}$ matrix (61) corresponding to the basis polynomials (60). Then, the eigenvalues $\lambda_1, \lambda_2, \ldots, \lambda_l$ of H' are the zeros of the last Lanczos polynomial $\phi_{J_{\mathrm{L}}}$, and each λ_j is an eigenvalue of even algebraic multiplicity $2\mu_j$. Let $J = SH'S^{-1}$ be the Jordan normal form (23), (24), with Jordan blocks (64). Then, for $n = 1, 2, \ldots, 2J_{\mathrm{L}} - 1$, the TFQMR residual vectors satisfy*

$$\frac{\|r_n^{\mathrm{TFQMR}}\|_2}{\|r_0\|_2} \leq \kappa_2(S)\sqrt{n+1}\,\varepsilon_n,$$

where

$$\varepsilon_n := \min_{\psi \in \mathcal{P}_n: \psi(0)=1} \max_{j=1,\ldots,l} \|\psi(J(\lambda_j))\|_2. \quad (65)$$

Moreover, if all zeros of the last Lanczos polynomial $\phi_{J_{\mathrm{L}}}$ are simple, then:

$$\varepsilon_n \leq \min_{\psi \in \mathcal{P}_n: \psi(0)=1} \max_{\lambda \in \lambda(A)} \left(|\psi(\lambda)| + |\psi'(\lambda)|\right). \quad (66)$$

We remark that the estimate (66) follows from (65) by using (63) and inequality (35) (with μ_j replaced by 2).

References

1. Chihara, T. S., *An Introduction to Orthogonal Polynomials*, Gordon and Breach, New York, 1978.
2. Faber, V. and T. Manteuffel, Necessary and sufficient conditions for the existence of a conjugate gradient method, SIAM J. Numer. Anal. **21** (1984), 352–362.

3. Fischer, B. and R. W. Freund, On the constrained Chebyshev approximation problem on ellipses, J. Approx. Th. **62** (1990), 297–315.

4. Fischer, B. and R. W. Freund, Chebyshev polynomials are not always optimal, J. Approx. Th. **65** (1991), 261–272.

5. Freund, R. W., A transpose-free quasi-minimal residual algorithm for non-Hermitian linear systems, SIAM J. Sci. Stat. Comput., to appear.

6. Freund, R. W., Quasi-kernel polynomials and their use in non-Hermitian matrix iterations, J. Comput. Appl. Math., to appear.

7. Freund, R. W., G. H. Golub, and N. M. Nachtigal, Iterative solution of linear systems, Acta Numerica **1** (1992), 57–100.

8. Freund, R. W., M. H. Gutknecht, and N. M. Nachtigal, An implementation of the look-ahead Lanczos algorithm for non-Hermitian matrices, SIAM J. Sci. Stat. Comput., to appear.

9. Freund R. W. and N. M. Nachtigal, QMR: a quasi-minimal residual method for non-Hermitian linear systems, Numer. Math. **60** (1991), 315–339.

10. Freund R. W. and N. M. Nachtigal, TFQMR: a maximal subspace look-ahead QMR algorithm for non-Hermitian linear systems, in preparation.

11. Freund R. W. and St. Ruscheweyh, On a class of Chebyshev approximation problems which arise in connection with a conjugate gradient type method, Numer. Math. **48** (1986), 525–542.

12. Gutknecht, M. H., A completed theory of the unsymmetric Lanczos process and related algorithms, Part I, SIAM J. Matrix Anal. Appl. **13** (1992), 594–639.

13. Hestenes, M. R. and E. L. Stiefel, Methods of conjugate gradients for solving linear systems, J. Res. Nat. Bur. Standards **49** (1952), 409–436.

14. Lanczos, C., An iteration method for the solution of the eigenvalue problem of linear differential and integral operators, J. Res. Nat. Bur. Standards **45** (1950), 255–282.

15. Nachtigal, N. M., A look-ahead variant of the Lanczos algorithm and its application to the quasi-minimal residual method for non-Hermitian linear systems, Ph.D. Dissertation, Massachusetts Institute of Technology, Cambridge, 1991.

16. Parlett, B. N., D. R. Taylor, and Z. A. Liu, A look-ahead Lanczos algorithm for unsymmetric matrices, Math. Comp. **44** (1985), 105–124.

17. Saad, Y., Least squares polynomials in the complex plane and their use for solving nonsymmetric linear systems, SIAM J. Numer. Anal. **24** (1987), 155–169

18. Saad, Y. and M. H. Schultz, GMRES: a generalized minimal residual algorithm for solving nonsymmetric linear systems, SIAM J. Sci. Stat. Comput. **7** (1986), 856–869.

19. Smolarski, D. C. and P. E. Saylor, An optimum iterative method for solving any linear system with a square matrix, BIT **28** (1988), 163–178.
20. Stiefel, E. L., Kernel polynomials in linear algebra and their numerical applications, U.S. National Bureau of Standards, Applied Mathematics Series **49** (1958), 1–22.
21. Szegö, G., *Orthogonal Polynomials*, Fourth Edition, Amer. Math. Soc., Providence, 1975.
22. Varga, R. S., A comparison of the successive overrelaxation method and semi-iterative methods using Chebyshev polynomials, J. Soc. Indust. Appl. Math. **5** (1957), 39–46.

This work was supported by Cooperative Agreement NCC 2-387 between the National Aeronautics and Space Administration and the Universities Space Research Association while the author was in residence at the Research Institute for Advanced Computer Science (RIACS), NASA Ames Research Center, Moffett Field, CA 94035.

Roland W. Freund
AT&T Bell Laboratories
600 Mountain Avenue, Room 2C-420
P.O. Box 636
Murray Hill, NJ 07974-0636
freund@research.att.com

Numerical Methods of Approximation Theory, Vol. 9
Dietrich Braess and Larry L. Schumaker (eds.), pp. 97–115.
International Series of Numerical Mathematics, Vol. 105
Copyright © 1992 by Birkhäuser Verlag, Basel
ISBN 3-7643-2746-4.

Rate of Approximation of Weighted Derivatives by Linear Combinations of SMD Operators

Margareta Heilmann

Dedicated to the memory of Lothar Collatz

Abstract. We consider the Durrmeyer modification of the Szász-Mirakjan operators. The rate of simultaneous approximation by linear combinations is related to the Ditzian-Totik modulus of smoothness. We prove a direct and a corresponding inverse theorem and, partly as a corollary, an equivalence result.

§1. Definition of the Operators and Linear Combinations

In this paper we consider simultaneous approximation by linear combinations of Szász-Mirakjan-Durrmeyer operators in weighted L_p-norms. In the following, these operators will be called SMD operators, and are defined by

$$(S_n f)(x) = \sum_{k=0}^{\infty} p_{nk}(x) n \int_I p_{nk}(t) f(t) dt, \quad n \in \mathbb{N}, \quad x \in I, \qquad (1)$$

where

$$p_{nk}(x) = \frac{(nx)^k}{k!} e^{-nx},$$

for every function f for which the right-hand side of (1) makes sense. Here and in the following, I denotes the interval $[0, \infty)$.

This variant of the Szász-Mirakjan operators (cf. [19]) was first considered in a paper of Mazhar and Totik [17]. This kind of modification of classical operators is due to Durrmeyer [9], who gave the appropriate definition for the Bernstein operators. In [7] Ditzian and Ivanov stated the commutativity of the Bernstein-Durrmeyer operators with the aid of a representation in terms of their eigenfunctions, which was proved by Derriennic [6]. In [12] this result was generalized also to the SMD operators. As already observed in [7,11,13,15], the commutativity will play an important role in this paper.

We consider the following linear combinations S_n^r of the operators S_n. For $r \in \mathbb{N}$, let

$$S_n^r f = \sum_{i=0}^{r-1} \alpha_i(n) S_{n_i} f \tag{2}$$

$$\text{with} \quad n = n_0 < \cdots < n_{r-1} \le An, \qquad \sum_{i=0}^{r-1} |\alpha_i(n)| \le B, \tag{3}$$

where A and B denote positive constants independent of n. We define special linear combinations S_n^{r*} with coefficients

$$\alpha_i^*(n) = n_i^{r-1} \prod_{\substack{j=0 \\ j \ne i}}^{r-1} (n_i - n_j)^{-1}, \quad i = 0, \ldots, r-1, \tag{4}$$

which are independent of n for special choices of n_i. These special linear combinations S_n^{r*} preserve all polynomials of degree at most $r - 1$ (cf. [11, Lemma 5.6]).

§2. Moduli of Smoothness and Appropriate K-functionals

For characterizing the relation between the smoothness of a function $f \in L_p(I)$ and $\varphi^s f \in L_p(I)$, $1 \le p \le \infty$, and the rate of approximation of the derivatives of f in weighted L_p-norms, respectively, we use weighted and nonweighted Ditzian-Totik moduli of smoothness (cf. [8]). In our case, where we choose the step-weight $\varphi(x) = \sqrt{x}$ and assume $t > 0$ to be sufficiently small, they are defined as follows:

$$\omega_\varphi^r(f, t)_p = \sup_{0 < h \le t} \|\Delta_{h\varphi}^r f\|_p,$$

$$\omega_\varphi^r(f, t)_{\varphi^s, p} = \sup_{0 < h \le t} \|\varphi^s \Delta_{h\varphi}^r f\|_p^{[t^*, \infty)} + \sup_{0 < h \le t^*} \|\varphi^s \vec{\Delta}_h^r f\|_p^{[0, 12t^*]}$$

where $t^* = r^2 t^2$. The symmetric and forward differences are given by

$$\Delta^r_{h\varphi(x)} f(x) = \sum_{k=0}^{r} (-1)^k \binom{r}{k} f(x + (\tfrac{r}{2} - k)h\varphi(x)),$$

$$\vec{\Delta}^r_h f(x) = \sum_{k=0}^{r} (-1)^k \binom{r}{k} f(x + (r - k)h)$$

whenever the needed arguments of the function f are contained in the corresponding interval. Otherwise, they are defined to be zero. Ditzian and Totik proved, cf. [8, Chapters 2, 3, 6.1], that these moduli of smoothness are equivalent to appropriate K-functionals given by

$$K^r_\varphi(f, t^r)_p = \inf \left\{ \|f - g\|_p + t^r \|\varphi^r g^{(r)}\|_p ; g, \varphi^r g^{(r)} \in L_p(I) \right\},$$

$$K^r_\varphi(f, t^r)_{\varphi^s, p} = \inf \left\{ \|\varphi^s(f - g)\|_p + t^r \|\varphi^{s+r} g^{(r)}\|_p ; \varphi^s g, \varphi^{s+r} g^{(r)} \in L_p(I) \right\}$$

for the nonweighted and the weighted case respectively. We need the equivalence of the weighted modulus to the modified weighted K-functional (cf. [16])

$$\bar{K}^r_\varphi(f, t^r)_{\varphi^s, p} = \inf \left\{ \|\varphi^s(f - g)\|_p + t^r \|\varphi^{s+r} g^{(r)}\|_p + t^{2r} \|\varphi^s g^{(r)}\|_p ; \right.$$

$$\left. \varphi^s g, \varphi^s g^{(r)}, \varphi^{s+r} g^{(r)} \in L_p(I) \right\}.$$

§3. The Weight-functions and Representations for the Derivatives

In this section we present a collection of useful properties of the weight functions p_{nk} and several representations for the derivatives of $S_n f$. We have

$$\sum_{k=0}^{\infty} p_{nk}(x) = 1, \tag{5}$$

$$\int_I p_{nk}(t)\, dt = \frac{1}{n}, \tag{6}$$

$$\frac{k}{n} p_{nk}(x) = x p_{n,k-1}(x), \tag{7}$$

$$\frac{d}{dx} p_{nk}(x) = n[p_{n,k-1}(x) - p_{nk}(x)], \tag{8}$$

$$\varphi(x)^{2s} p_{nk}(x) p_{n,k+2s}(t) = \beta(k, s) p_{n,k+s}(x) \varphi(t)^{2s} p_{n,k+s}(t),$$

$$\text{where} \quad \frac{s!}{(2s)!} \le \beta(k, s) = \frac{[(k+s)!]^2}{k!(k+2s)!} \le 1. \tag{9}$$

For simplicity, we always define $p_{nk} = 0$ if $k < 0$.

For the derivatives (cf.[13, (3.11)]) we will use the following representations. Let $s \in \mathbb{N}_0$, $x \in I$, $f^{(s)} \in L_p(I)$. Then

$$(S_n f)^{(s)}(x) = n \sum_{k=0}^{\infty} p_{nk}(x) \int_I p_{n,k+s}(t) f^{(s)}(t) \, dt. \tag{10}$$

For $\varphi^{2s} f^{(2s)} \in L_p(I)$, $s \in \mathbb{N}_0$, we get by (9)

$$\varphi(x)^{2s}(S_n f)^{(2s)}(x)$$
$$= n \sum_{k=0}^{\infty} \beta(k,s) p_{n,k+s}(x) \int_I p_{n,k+s}(t) \varphi(t)^{2s} f^{(2s)}(t) dt. \tag{11}$$

Thus, one can change between the representations (10) and (11) as needed.

Sometimes it is convenient to work with positive operators $S_{n,s}$ defined by

$$(S_{n,s} f)(x) = n \sum_{k=0}^{\infty} p_{nk}(x) \int_I p_{n,k+s}(t) f(t) dt, \quad f \in L_p(I). \tag{12}$$

Together with the representation (10) for the derivatives, (12) leads to

$$(S_n f)^{(s)} = S_{n,s} f^{(s)} \tag{13}$$

for $f^{(s)} \in L_p(I)$, and the above mentioned formula (11) holds analogously,

$$\varphi(x)^{2s}(S_{n,2s} f)(x) = n \sum_{k=0}^{\infty} \beta(k,s) p_{n,k+s}(x) \int_I p_{n,k+s}(t) \varphi(t)^{2s} f(t) dt. \tag{14}$$

We define corresponding linear combinations by

$$S_{n,s}^r f = \sum_{i=0}^{r-1} \alpha_i(n) S_{n_i,s} f, \tag{15}$$

with coefficients $\alpha_i(n)$ fulfilling (3), and for which we have as an analogue of (13),

$$(S_n^r f)^{(s)} = S_{n,s}^r f^{(s)}. \tag{16}$$

In particular, we consider linear combinations $S_{n,s}^{r*}$, which we derive from (15) with the special choice of $\alpha_i^*(n)$ defined in (4). It is possible to show that for every $s \in \mathbb{N}_0$, $S_{n,s}^{r*}$ preserves all polynomials of degree at most $r - 1$. For details we refer to [14, Lemma 4.7, Bemerkung 4.3].

§4. Auxiliary Lemmas

For the proofs of the main results, we need some more or less technical lemmas which will be proved in this section.

Definition 1. *Let $s \in \mathbb{N}_0$, $m \in \mathbb{N}$ and $\varphi(t) = \sqrt{t}$. For $u \in I$, we define*

$$H_{n,m,2s}(u) = mn \left\{ \int_I \int_0^u - \int_0^u \int_I \right\} (u-t)^{m-1} \varphi(t)^{2s} \sum_{k=0}^{\infty} p_{nk}(x) p_{nk}(t) \, dt dx,$$
$$(17)$$

$$\widetilde{H}_{n,m,2s}(u) = mn \left\{ \int_I \int_0^u - \int_0^u \int_I \right\} (u-t)^{m-1} \sum_{k=0}^{\infty} p_{nk}(x) p_{n,k+2s}(t) \, dt dx.$$
$$(18)$$

First we establish a representation of $H_{n,m,2s}(u)$ in terms of $H_{n,m+\nu,0}(u)$, $0 \le \nu \le s$, from which we can derive a needed estimate by using results for $H_{n,m+\nu,0}(u)$ which we proved in [13, Lemma 4.10, Korollar 4.11].

Lemma 2. *Let $s \in \mathbb{N}_0$, $m \in \mathbb{N}$. Then $H_{n,m,2s}(u)$ can be written as*

$$H_{n,m,2s}(u) = \sum_{\nu=0}^{s} \binom{s}{\nu} (-1)^{\nu} \varphi(u)^{2(s-\nu)} \frac{m}{m+\nu} H_{n,m+\nu,0}(u).$$

Proof: We consider

$$\begin{aligned}
H_{n,m,2s}(u) =\; & mn \int_I \int_0^u (u-t)^{m-1} \varphi(t)^{2s} \sum_{k=0}^{\infty} p_{nk}(x) p_{nk}(t) \, dt dx \\
& - mn \int_0^u \int_I (u-t)^{m-1} \varphi(t)^{2s} \sum_{k=0}^{\infty} p_{nk}(x) p_{nk}(t) \, dt dx \\
=:\; & I_1 - I_2.
\end{aligned}$$
$$(19)$$

By interchanging the order of integration and using (6) and (5), we get

$$\begin{aligned}
I_1 &= m \int_0^u (u-t)^{m-1} \varphi(t)^{2s} \, dt \\
&= m u^{m+s} \sum_{i=0}^{m-1} \binom{m-1}{i} (-1)^i \frac{1}{s+i+1} = u^{m+s} \frac{1}{\binom{m+s}{s}}.
\end{aligned}$$
$$(20)$$

We used [10, (1.41)] for the last equation. We use the representation $\varphi(t)^{2s} = \sum_{\nu=0}^{s} \binom{s}{\nu}(-1)^\nu u^{s-\nu}(u-t)^\nu$, and consider the second term on the right-hand side of (19)

$$I_2 = \sum_{\nu=0}^{s} \binom{s}{\nu}(-1)^\nu u^{s-\nu} mn \sum_{k=0}^{\infty} \int_0^u p_{nk}(x) \left\{ \int_I (u-t)^{m+\nu-1} p_{nk}(t)\, dt \right\} dx.$$

(21)

Integration by parts leads to

$$m \int_I (u-t)^{m+\nu-1} p_{nk}(t)\, dt$$
$$= \frac{m}{m+\nu} \int_I (u-t)^{m+\nu} p'_{nk}(t)\, dt + \begin{cases} \frac{m}{m+\nu} u^{m+\nu} & \text{if } k = 0 \\ 0 & \text{if } k \neq 0. \end{cases}$$

(22)

Since $n \int_0^u p_{n0}(x)\, dx = \int_0^u p'_{n0}(x)\, dx = 1 - p_{n0}(u)$, by putting (22) in (21), and using again the identity [10, (1.41)], we get

$$I_2 = [1 - p_{n0}(u)] u^{m+s} \sum_{\nu=0}^{s} \binom{s}{\nu}(-1)^\nu \frac{m}{m+\nu}$$
$$+ \sum_{\nu=0}^{s} \binom{s}{\nu}(-1)^\nu u^{s-\nu} n \sum_{k=0}^{\infty} \int_0^u p_{nk}(x)\,dx \frac{m}{m+\nu} \int_I (u-t)^{m+\nu} p'_{nk}(t)\,dt$$
$$= u^{m+s} \frac{1}{\binom{m+s}{s}} - p_{n0}(u) u^{m+s} \sum_{\nu=0}^{s} \binom{s}{\nu}(-1)^\nu \frac{m}{m+\nu}$$
$$+ \sum_{\nu=0}^{s} \binom{s}{\nu}(-1)^\nu u^{s-\nu} n \sum_{k=0}^{\infty} \int_0^u p_{nk}(x)\,dx \frac{m}{m+\nu} \int_I (u-t)^{m+\nu} p'_{nk}(t)\,dt.$$

(23)

With (8) we derive

$$\sum_{k=0}^{\infty} \int_0^u p_{nk}(x)\, dx \int_I (u-t)^{m+\nu} p'_{nk}(t)\, dt$$
$$= \sum_{k=0}^{\infty} \int_0^u n[p_{n,k+1}(x) - p_{nk}(x)]\, dx \int_I (u-t)^{m+\nu} p_{nk}(t)\, dt$$
$$= -\sum_{k=0}^{\infty} \int_0^u p'_{n,k+1}(x)\, dx \int_I (u-t)^{m+\nu} p_{nk}(t)\, dt$$
$$= -\sum_{k=0}^{\infty} p_{n,k+1}(u) \int_I (u-t)^{m+\nu} p_{nk}(t)\, dt$$

as $p_{n,k+1}(0) = 0$. Together with (23) we get

$$I_2 = u^{m+s}\frac{1}{\binom{m+s}{s}} - p_{n,0}(u)u^{m+s}\sum_{\nu=0}^{s}\binom{s}{\nu}(-1)^{\nu}\frac{m}{m+\nu}$$

$$- \sum_{\nu=0}^{s}\binom{s}{\nu}(-1)^{\nu}u^{s-\nu}n\frac{m}{m+\nu}\sum_{k=0}^{\infty}p_{n,k+1}(u)\int_I (u-t)^{m+\nu}p_{nk}(t)\,dt.$$

$$\tag{24}$$

Putting (20) and (24) in (19), the proposed formula follows immediately with the representation (cf.[13, Lemma 4.10])

$$H_{n,m+\nu,0}(u) = n\sum_{k=0}^{\infty}p_{n,k+1}(u)\int_I (u-t)^{m+\nu}p_{nk}(t)\,dt + u^{m+\nu}p_{n0}(u). \qquad\blacksquare$$

From the above mentioned lemma we derive the following corollary.

Corollary 3. *Let* $u \in [1/n,\infty)$. *Then* $H_{n,2m,2s}(u) \le Cn^{-m}\varphi(u)^{2(s+m)}$, *where* C *is a constant independent of* n.

Proof: Using the relations (cf. [13, Korollar 4.11])

$$H_{n,2m,0}(u) = \sum_{i=0}^{m-1} Q_{i,2m}\varphi(u)^{2(m-i)}n^{-m-i}$$

$$H_{n,2m-1,0}(u) = \sum_{i=0}^{m-2} Q_{i,2m-1}\varphi(u)^{2(m-i-1)}n^{-m-i},$$

the representation for $H_{n,2m,2s}(u)$ in Lemma 2, and the fact that $u \in [1/n,\infty)$, we get

$$H_{n,2m,2s}(u) < Cn^{-m}u^{m+s}\left\{\sum_{\substack{\nu=0\\ \nu\text{ even}}}^{s} u^{-\nu/2}n^{-\nu/2} + \sum_{\substack{\nu=1\\ \nu\text{ odd}}}^{s} u^{-(\nu+1)/2}n^{-(\nu+1)/2}\right\}$$

$$\le Cn^{-m}u^{m+s}. \qquad\blacksquare$$

Now we look at $\widetilde{H}_{n,m,2s}(u)$.

Lemma 4. *Let* $s \in \mathbb{N}_0$, $m \in \mathbb{N}$. *Then* $\widetilde{H}_{n,m,2s}(u)$ *can be written as*

$$\widetilde{H}_{n,m,2s}(u) = -m\sum_{k=0}^{2s-1}\int_0^u (u-t)^{m-1}p_{nk}(t)\,dt + m\sum_{\nu=1}^{m-1}u^{m-\nu}n^{-\nu}$$

$$\times \sum_{j=\nu}^{m-1}\binom{m-1}{j}\binom{j}{\nu}\frac{(j+2s)!}{(j-\nu+2s)!}\frac{1}{j-\nu+1}(-1)^{j+1}.$$

Proof: We look at

$$\widetilde{H}_{n,m,2s}(u) = mn \int_I \int_0^u (u-t)^{m-1} \sum_{k=0}^{\infty} p_{nk}(x) p_{n,k+2s}(t)\, dt\, dx$$

$$- mn \int_0^u \int_I (u-t)^{m-1} \sum_{k=0}^{\infty} p_{nk}(x) p_{n,k+2s}(t)\, dt\, dx \qquad (25)$$

$$=: I_1 - I_2.$$

From (6) and (5) we get

$$I_1 = m \int_0^u (u-t)^{m-1} \left\{ \sum_{k=0}^{\infty} p_{nk}(t) - \sum_{k=0}^{2s-1} p_{nk}(t) \right\} dt$$

$$= u^m - m \sum_{k=0}^{2s-1} \int_0^u (u-t)^{m-1} p_{nk}(t)\, dt. \qquad (26)$$

We examine the second term of the right-hand side of (25). Since

$$\int_I (u-t)^{m-1} p_{n,k+2s}(t)\, dt = \sum_{j=0}^{m-1} \binom{m-1}{j} (-1)^j u^{m-1-j} \int_I t^j p_{n,k+2s}(t)\, dt,$$

using (7) to get $t^j p_{n,k+2s}(t) = n^{-j} \left\{ \prod_{l=1}^{j} (k+2s+l) \right\} p_{n,k+2s+j}(t)$, and applying (6) and interchanging the order of summation, we get

$$I_2 = m \sum_{j=0}^{m-1} \binom{m-1}{j} (-1)^j u^{m-1-j} n^{-j} \sum_{k=0}^{\infty} \left\{ \prod_{l=1}^{j} (k+2s+l) \right\} \int_0^u p_{nk}(x)\, dx.$$

Using the Newton form of the interpolation polynomial of $\prod_{l=1}^{j}(x+2s+l)$ at the knots $x_i = i$, $i = 0(1)j$, and calculating the result for $x = k$, we get the identity

$$\prod_{l=1}^{j}(k+2s+l) = \sum_{\nu=0}^{j} \binom{j}{\nu} \frac{(2s+j)!}{(2s+\nu)!} \prod_{l=0}^{\nu-1}(k-l).$$

This together with (7) and (5) leads to

$$I_2 =$$

$$= m \sum_{j=0}^{m-1} \binom{m-1}{j} (-1)^j u^{m-1-j} n^{-j} \sum_{\nu=0}^{j} \binom{j}{\nu} \frac{(2s+j)!}{(2s+\nu)!} \int_0^u (nx)^{\nu} \sum_{k=\nu}^{\infty} p_{n,k-\nu}(x)\, dx$$

$$= m \sum_{j=0}^{m-1} \binom{m-1}{j} (-1)^j u^{m-1-j} n^{-j} \sum_{\nu=0}^{j} \binom{j}{\nu} \frac{(2s+j)!}{(2s+\nu)!} n^{\nu} \frac{1}{\nu+1} u^{\nu+1}.$$

Shifting the indices $(\nu \to j - \nu)$ and interchanging the order of summation gives

$$I_2 = m \sum_{\nu=0}^{m-1} u^{m-\nu} n^{-\nu} \sum_{j=\nu}^{m-1} \binom{m-1}{j} \binom{j}{\nu} \frac{(2s+j)!}{(2s+j-\nu)!} \frac{1}{j-\nu+1} (-1)^j. \quad (27)$$

Calculating the term $\nu = 0$ and using $[10, (1.41)]$ gives

$$mu^m \sum_{j=0}^{m-1} \binom{m-1}{j} \frac{1}{j+1} (-1)^j = u^m.$$

Thus, the proposed formula follows from (26) and (27) together with (25). ∎

In the next corollary, we derive an estimate for $\widetilde{H}_{n,2m,2s}(u)$ needed below.

Corollary 5. *Let* $u \in [0, 1/n]$. *Then* $\widetilde{H}_{n,2m,2s}(u) \le Cn^{-2m}$, *where* C *is a constant independent of* n.

§5. The Direct Result

For the linear combinations S_n^{r*}, we can prove the following direct result.

Theorem 6. *Let* $\varphi(x) = \sqrt{x}$, $s \in \mathbb{N}_0$, *and suppose* f, $\varphi^{2s} f^{(2s)} \in L_p(I)$, $1 \le p < \infty$. *Then*

$$\|\varphi^{2s}(S_n^{r*} f - f)^{(2s)}\|_p \le C\omega_\varphi^{2r}(f^{(2s)}, n^{-1/2})_{\varphi^{2s},p} \quad (28)$$

where C *is a constant independent of* n.

Proof: We use the equivalence of $\omega_\varphi^{2r}(f^{(2s)}, n^{-1/2})_{\varphi^{2s},p}$ to the modified weighted K–functional $\bar{K}_\varphi^{2r}(f^{(2s)}, n^{-r})_{\varphi^{2s},p}$, and prove instead of (28)

$$\|\varphi^{2s}(S_n^{r*} f - f)^{(2s)}\|_p \le C\bar{K}_\varphi^{2r}(f^{(2s)}, n^{-r})_{\varphi^{2s},p} . \quad (29)$$

Using the property (16), we get by applying $[15,\text{ Lemma 3.1}]$ for every g with $\varphi^{2s}g$, $\varphi^{2s}g^{(2r)}$, $\varphi^{2s+2r}g^{(2r)} \in L_p(I)$

$$\|\varphi^{2s}(S_n^{r*} f - f)^{(2s)}\|_p \le C\|\varphi^{2s}(f^{(2s)} - g)\|_p + \|\varphi^{2s}(S_{n,2s}^{r*}g - g)\|_p. \quad (30)$$

We first look at the second term on the right-hand side and prove the estimate

$$\|\varphi^{2s}(S_{n,2s}^{r*}g - g)\|_p \le Cn^{-r}\|\varphi^{2s}(\varphi^2 + n^{-1})^r g^{(2r)}\|_p. \quad (31)$$

Consider the Taylor expansion of the function g, and define

$$g(t) = \sum_{\mu=0}^{r-1} \frac{(t-x)^\mu}{\mu!} g^{(\mu)}(x) + \sum_{\mu=r}^{2r-1} \frac{(t-x)^\mu}{\mu!} g^{(\mu)}(x) + R_{2r}(g,t,x)$$

$$=: g_1(t) + g_2(t) + R_{2r}(g,t,x),$$

where $R_{2r}(g,t,x) = \frac{1}{(2r-1)!} \int_x^t (t-u)^{2r-1} g^{(2r)}(u)\,du$ denotes the integral remainder. Since all polynomials of degree at most $r-1$ are preserved by $S_{n,2s}^{r*}$ for every $s \in \mathbb{N}_0$ (cf.[14, Bemerkung 4.8]), (31) is reduced to

$$\|\varphi^{2s}(S_{n,2s}^{r*} g_2)\|_p \leq C\left\{ n^{-r}\|\varphi^{2s}\varphi^{2r} g^{(2r)}\|_p + n^{-2r}\|\varphi^{2s} g^{(2r)}\|_p \right\} \tag{32}$$

$$\text{and} \quad \|\varphi^{2s}\left[S_{n,2s}^{r*} R_{2r}(g,t,\cdot)\right]\|_p \leq C n^{-r}\|\varphi^{2s}(\varphi^2 + n^{-1})^r g^{(2r)}\|_p. \tag{33}$$

We first prove (32) and split the interval $[0,\infty)$ into $[0,1/n]$ and $[1/n,\infty)$.

We use the following Hardy inequality (cf.[18, Chapter V, Lemma 3.14, (ii)])

$$\left\{ \int_I \left(\int_x^\infty h(y)\,dy \right)^p x^{m-1}\,dx \right\}^{1/p} \leq \frac{p}{m}\left\{ \int_I [yh(y)]^p\, y^{m-1}\,dy \right\}^{1/p} \tag{34}$$

where $h \geq 0$, $p \geq 1$ and $m > 0$.

For $x \in [0,\frac{1}{n}]$ we have proved in [14, Korollar 4.9] that $|S_{n,s}^{r*} f_\mu(x)| \leq C n^{-\mu}$ from which we derive

$$\|\varphi^{2s}(S_{n,2s}^{r*} g_2)\|_p^{[0,1/n]} \leq C \sum_{\mu=r}^{2r-1} n^{-\mu}\|\varphi^{2s} g^{(\mu)}(x)\|_p. \tag{35}$$

Now we apply Hardy's inequality $(2r-\mu)$ times, where the m in (34) has the value $(s+l-1)p+1$ in the l-th step, $l = 1,\dots,2r-\mu$ (i.e., $m > 0$). For h we take $|g^{(\mu+l)}|$. Hence

$$\|\varphi^{2s} g^{(\mu)}\|_p \leq \left\{ \int_I \left(\int_x^\infty |g^{(\mu+1)}(y)|\,dy \right)^p x^{sp}\,dx \right\}^{1/p}$$

$$\leq C\left\{ \int_I |g^{(\mu+1)}(y)|^p y^{(s+1)p}\,dy \right\}^{1/p} \leq \cdots \leq$$

$$\leq C\left\{ \int_I \left(\int_x^\infty |g^{(2r)}(y)|\,dy \right)^p x^{(s+2r-\mu-1)p}\,dx \right\}^{1/p}$$

$$\leq C\left\{ \int_I |g^{(2r)}(y)|^p y^{(s+2r-\mu)p}\,dy \right\}^{1/p}$$

$$= C\|\varphi^{2(s+2r-\mu)} g^{(2r)}\|_p.$$

Together with (35), this implies

$$\|\varphi^{2s}(S^{r*}_{n,2s}g_2)\|_p^{[0,1/n]}$$

$$\leq C \sum_{\mu=r}^{2r-1} n^{-\mu} \left\{ \|\varphi^{2(s+2r-\mu)}g^{(2r)}\|_p^{[0,1/n]} + \|\varphi^{2(s+2r-\mu)}g^{(2r)}\|_p^{[1/n,\infty)} \right\} \quad (36)$$

$$\leq C \left\{ n^{-2r}\|\varphi^{2s}g^{(2r)}\|_p + n^{-r}\|\varphi^{2s}\varphi^{2r}g^{(2r)}\|_p \right\},$$

since $x^{2r-\mu} \leq n^{-2r+\mu}$ for $x \in [0,1/n]$ and $x^{r-\mu} \leq n^{-r+\mu}$ for $x \in [1/n,\infty)$.

For $x \in [\frac{1}{n},\infty)$, we have $|S^{r*}_{n,s}f_\mu(x)| \leq Cn^{-r}x^{\mu-r}$ (cf. [14, Korollar 4.9]). Hence

$$\|\varphi^{2s}(S^{r*}_{n,2s}g_2)\|_p^{[1/n,\infty)} \leq Cn^{-r} \sum_{\mu=r}^{2r-1} \|\varphi^{2(s+\mu-r)}g^{(\mu)}\|_p. \quad (37)$$

Again we use Hardy's inequality (34) $(2r - \mu)$ times, now setting $m = (s + \mu - r - 1 + l)p + 1$ in the l-th step, $l = 1, \ldots, 2r - \mu$ (i.e., $m > 0$ since $\mu \geq r$). This leads to

$$\|\varphi^{2(s+\mu-r)}g^{(\mu)}\|_p \leq \left\{ \int_I |g^{(\mu+1)}(x)|^p x^{(s+\mu-r+1)p}\,dx \right\}^{1/p} \leq \cdots \leq$$

$$\leq C \left\{ \int_I |g^{(2r)}(x)|^p x^{(s+r)p}\,dx \right\}^{1/p} = C\|\varphi^{2s}\varphi^{2r}g^{(2r)}\|_p.$$

Now (37) implies

$$\|\varphi^{2s}(S^{r*}_{n,2s}g_2)\|_p^{[1/n,\infty)} \leq Cn^{-r}\|\varphi^{2s}\varphi^{2r}g^{(2r)}\|_p. \quad (38)$$

Combining (36) and (38) gives (32).

Now we prove the estimate (33) explicitly for $p = 1$ and $p = \infty$. The cases of the other p-norms, $1 < p < \infty$, then follow by the interpolation theorem of Riesz-Thorin [3, Theorem 1.1.1]. By the property (3) in the definition of the linear combinations, it is enough to prove the estimate

$$\|\varphi^{2s}[S_{n,2s}R_{2r}(g,t,\cdot)]\|_p \leq Cn^{-r}\|\varphi^{2s}(\varphi^2 + n^{-1})^r g^{(2r)}\|_p. \quad (39)$$

Using definition (12) and splitting the integral into integrals from 0 to x and from x to ∞, we get

$$|\varphi(x)^{2s}[S_{n,2s}R_{2r}(g,t,x)](x)|$$

$$\leq \frac{n}{(2r-1)!}\varphi(x)^{2s} \sum_{k=0}^{\infty} p_{nk}(x) \int_0^x p_{n,k+2s}(t) \int_t^x (u-t)^{2r-1}|g^{(2r)}(u)|\,du\,dt$$

$$+ \frac{n}{(2r-1)!}\varphi(x)^{2s} \sum_{k=0}^{\infty} p_{nk}(x) \int_x^\infty p_{n,k+2s}(t) \int_x^t (t-u)^{2r-1}|g^{(2r)}(u)|\,du\,dt.$$

$$(40)$$

We consider two cases.

Case 1: $p = \infty$. Using the identity (9) in the first term of (40) and taking into account the boundedness of the $\beta(k, s)$, which essentially means changing from the representation (10) to (11), leads to

$$|\varphi(x)^{2s}[S_{n,2s}R_{2r}(g, t, x)](x)|$$

$$\leq \frac{n}{(2r-1)!} \sum_{k=0}^{\infty} p_{n,k+s}(x) \int_0^x p_{n,k+s}(t)\varphi(t)^{2s} \int_t^x (u-t)^{2r-1}|g^{(2r)}(u)|\,du\,dt$$

$$+ \frac{n}{(2r-1)!}\varphi(x)^{2s} \sum_{k=0}^{\infty} p_{nk}(x) \int_x^{\infty} p_{n,k+2s}(t) \int_x^t (t-u)^{2r-1}|g^{(2r)}(u)|\,du\,dt$$

$$=: \ T_1 + T_2.$$

Taking into consideration that $0 \leq \varphi(t)^{2s} \leq \varphi(u)^{2s}$ in T_1, we can estimate

$$T_1 \leq Cn\|\varphi^{2s}(\varphi^2 + n^{-1})^r g^{(2r)}\|_{\infty}$$
$$\times \sum_{k=0}^{\infty} p_{n,k+s}(x) \int_0^x p_{n,k+s}(t) \int_t^x (u-t)^{2r-1}(\varphi(u)^2 + n^{-1})^{-r}\,du\,dt. \tag{41}$$

Since $0 \leq t \leq u \leq x$ in T_1, we have $(u-t)(\varphi(x)^2 + n^{-1}) \leq (x-t)(\varphi(u)^2 + n^{-1})$, and hence

$$(u-t)^{2r-1}(\varphi(x)^2 + n^{-1})^r \leq (x-t)^{2r-1}(\varphi(u)^2 + n^{-1})^r.$$

Together with (41), this leads to

$$T_1 \leq Cn\|\varphi^{2s}(\varphi^2 + n^{-1})^r g^{(2r)}\|_{\infty}(\varphi(x)^2 + n^{-1})^{-r}$$
$$\times \sum_{k=0}^{\infty} p_{n,k+s}(x) \int_0^x p_{n,k+s}(t)(x-t)^{2r}\,dt$$
$$\leq C\|\varphi^{2s}(\varphi^2 + n^{-1})^r g^{(2r)}\|_{\infty}(\varphi(x)^2 + n^{-1})^{-r}[S_n(x-t)^{2r}](x) \tag{42}$$
$$\leq Cn^{-r}\|\varphi^{2s}(\varphi^2 + n^{-1})^r g^{(2r)}\|_{\infty},$$

where the last estimate holds true by [11, Korollar 5.8].

For T_2 we take into account that $0 \leq \varphi(x)^{2s} \leq \varphi(u)^{2s}$ to derive

$$T_2 \leq Cn\|\varphi^{2s}(\varphi^2 + n^{-1})^r g^{(2r)}\|_{\infty}$$
$$\times \sum_{k=0}^{\infty} p_{nk}(x) \int_x^{\infty} p_{n,k+2s}(t) \int_x^t (t-u)^{2r-1}(\varphi(u)^2 + n^{-1})^{-r}\,du\,dt.$$

We have $(t-u)^{2r-1}(\varphi(x)^2 + n^{-1})^r \leq (t-x)^{2r-1}(\varphi(u)^2 + n^{-1})^r$ as $0 \leq x \leq u \leq t$. Hence

$$T_2 \leq Cn\|\varphi^{2s}(\varphi^2 + n^{-1})^r g^{(2r)}\|_\infty (\varphi(x)^2 + n^{-1})^{-r}$$

$$\times \sum_{k=0}^\infty p_{nk}(x) \int_x^\infty p_{n,k+2s}(t)(t-x)^{2r}\,dt \tag{43}$$

$$\leq C\|\varphi^{2s}(\varphi^2 + n^{-1})^r g^{(2r)}\|_\infty (\varphi(x)^2 + n^{-1})^{-r}[S_{n,2s}(t-x)^{2r}](x)$$

$$\leq Cn^{-r}\|\varphi^{2s}(\varphi^2 + n^{-1})^r g^{(2r)}\|_\infty,$$

where the last inequality follows from $|S_{n,s}^{r*} f_\mu(x)| \leq Cn^{-[\frac{\mu+1}{2}]} x^{[\frac{\mu}{2}]}$ if $x \in [1/n, \infty)$, $2r - 1 \leq \mu$, which we proved in [14, Korollar 4.9]. With (42) and (43) we proved (39) for the case $p = \infty$.

Case 2: $p = 1$. With (14) for $\varphi(x)^{2s}[S_{n,2s}R_{2r}(g,t,x)]$ we get by applying Fubini's theorem twice

$$\|\varphi^{2s} S_{n,2s} R_{2r}(g,t,\cdot)\|_1$$

$$\leq Cn\left\{ \int_I \sum_{k=0}^\infty p_{n,k+s}(x) \int_0^x p_{n,k+s}(t)\varphi(t)^{2s} \int_t^x (u-t)^{2r-1}|g^{(2r)}(u)|\,du\,dt\,dx \right.$$

$$\left. + \int_I \sum_{k=0}^\infty p_{n,k+s}(x) \int_x^\infty p_{n,k+s}(t)\varphi(t)^{2s} \int_x^t (t-u)^{2r-1}|g^{(2r)}(u)|\,du\,dt\,dx \right\}$$

$$= Cn\left\{ \int_I |g^{(2r)}(u)| \int_u^\infty \int_0^u \varphi(t)^{2s}(u-t)^{2r-1} \sum_{k=0}^\infty p_{n,k+s}(x)p_{n,k+s}(t)\,dt\,dx\,du \right.$$

$$\left. + \int_I |g^{(2r)}(u)| \int_0^u \int_u^\infty \varphi(t)^{2s}(t-u)^{2r-1} \sum_{k=0}^\infty p_{n,k+s}(x)p_{n,k+s}(t)\,dt\,dx\,du \right\}.$$

Splitting the second term into two summands and using (9), we get

$$\|\varphi^{2s} S_{n,2s} R_{2r}(g,t,\cdot)\|_1$$

$$\leq Cn\left\{ \int_I |g^{(2r)}(u)| \int_u^\infty \int_0^u \varphi(t)^{2s}(u-t)^{2r-1} \sum_{k=0}^\infty p_{n,k+s}(x)p_{n,k+s}(t)\,dt\,dx\,du \right.$$

$$+ \int_0^{1/n} |g^{(2r)}(u)| \int_0^u \int_u^\infty (t-u)^{2r-1} \sum_{k=0}^\infty \varphi(x)^{2s} p_{nk}(x)p_{n,k+2s}(t)\,dt\,dx\,du$$

$$\left. + \int_{1/n}^\infty |g^{(2r)}(u)| \int_0^u \int_u^\infty \varphi(t)^{2s}(t-u)^{2r-1} \sum_{k=0}^\infty p_{nk}(x)p_{nk}(t)\,dt\,dx\,du \right\}$$

$$=: Cn\{T_1 + T_2 + T_3\}.$$

$$\tag{44}$$

Now we look at the terms T_i separately. In T_1 we have $0 \leq \varphi(t)^{2s} \leq \varphi(u)^{2s}$. Thus

$$T_1 \leq \int_I |\varphi(u)^{2s} g^{(2r)}(u)| \int_u^\infty \int_0^u (u-t)^{2r-1} \sum_{k=0}^\infty p_{nk}(x) p_{nk}(t) \, dt \, dx \, du. \quad (45)$$

Adding

$$\int_I |\varphi(u)^{2s} g^{(2r)}(u)| \int_0^u \int_u^\infty (t-u)^{2r-1} \sum_{k=0}^\infty p_{nk}(x) p_{nk}(t) \, dt \, dx \, du \geq 0,$$

using the definition of $H_{n,2r,0}$ (cf. (17)) along with the fact that $H_{n,2r,0}(u) \leq Cn^{-r}(\varphi(u)^2 + n^{-1})^r$ (cf. [13, Korollar 4.11]), we get

$$\begin{aligned} T_1 &\leq \int_I |\varphi(u)^{2s} g^{(2r)}(u)| n^{-1} H_{n,2r,0}(u) \, du \\ &\leq Cn^{-r-1} \|\varphi^{2s}(\varphi^2 + n^{-1})^r g^{(2r)}(u)\|_1. \end{aligned} \quad (46)$$

Now we consider the second term in (44). Taking into account that $0 \leq \varphi(x)^{2s} \leq \varphi(u)^{2s}$ we can estimate

$$T_2 \leq \int_0^{1/n} |\varphi(u)^{2s} g^{(2r)}(u)| \int_0^u \int_u^\infty (t-u)^{2r-1} \sum_{k=0}^\infty p_{nk}(x) p_{n,k+2s}(t) \, dt \, dx \, du. \quad (47)$$

Adding

$$\int_0^{1/n} |\varphi(u)^{2s} g^{(2r)}(u)| \int_u^\infty \int_0^u (u-t)^{2r-1} \sum_{k=0}^\infty p_{nk}(x) p_{n,k+2s}(t) \, dt \, dx \, du \geq 0,$$

and using the definition of $\widetilde{H}_{n,2r,2s}$ (cf. (18)) and the corresponding estimate in Corollary 5, we get

$$\begin{aligned} T_2 &\leq Cn^{-1} \int_0^{1/n} |\varphi(u)^{2s} g^{(2r)}(u)| \widetilde{H}_{n,2r,2s}(u) \, du \\ &\leq Cn^{-2r-1} \|\varphi^{2s} g^{(2r)}\|_1. \end{aligned} \quad (48)$$

Adding $\int_{1/n}^\infty |g^{(2r)}(u)| \int_u^\infty \int_0^u \varphi(t)^{2s} (u-t)^{2r-1} \sum_{k=0}^\infty p_{nk}(x) p_{nk}(t) \, dt \, dx \, du \geq 0$ and applying the inequality for $H_{n,2r,2s}(u)$, which was proved in Corollary 3, we estimate T_3 by

$$\begin{aligned} T_3 &\leq Cn^{-1} \int_{1/n}^\infty |g^{(2r)}(u)| H_{n,2r,2s}(u) \, du \\ &\leq Cn^{-r-1} \|\varphi^{2s+2r} g^{(2r)}\|_1. \end{aligned} \quad (49)$$

Now we can derive (39) for the case $p = 1$ by putting in the estimates (46), (48), and (49) in (44). Thus we proved the inequality (33).

Together with the estimates (32) and (33), we have (31), and by putting (31) in (30) we get

$$\|\varphi^{2s}(S_n^{r*}f - f)^{(2s)}\|_p \leq C\left\{\|\varphi^{2s}(f^{(2s)} - g)\|_p + n^{-r}\|\varphi^{2s}(\varphi^2 + n^{-1})^r g^{(2r)}\|_p\right\}.$$

Taking the infimum over all g on the right-hand side and using the equivalence of $\bar{K}_\varphi^{2r}(f^{(2s)}, n^{-r})_{\varphi^{2s},p}$ to $\omega_\varphi^{2r}(f^{(2s)}, n^{-1/2})_{\varphi^{2s},p}$ proves the theorem. ∎

§6. The Inverse Result

Now we prove the corresponding inverse of Theorem 6.

Theorem 7. *Let* $\varphi(x) = \sqrt{x}$, $s \in \mathbb{N}_0$, f, $\varphi^{2s}f^{(2s)} \in L_p(I)$, $1 \leq p < \infty$, $0 < \alpha - s < r$, *and suppose* $\|\varphi^{2s}(S_n^r f - f)^{(2s)}\|_p = \mathcal{O}(n^{s-\alpha})$. *Then*
$$\omega_\varphi^{2r+2s}(f, t)_p = \mathcal{O}(t^{2\alpha}).$$

Proof: By the definition of the weighted K–functional $K_\varphi^{2r}(f^{(2s)}, h^{2r})_{\varphi^{2s},p}$ and the relation (16) for the derivatives of $S_n^r f$ and the operators $S_{n,2s}^r$, under the assumptions of the theorem, we have

$$K_\varphi^{2r}(f^{(2s)}, h^{2r})_{\varphi^{2s},p}$$
$$\leq C\left\{\|\varphi^{2s}(S_n^r f - f)^{(2s)}\|_p + h^{2r}\|\varphi^{2s}\varphi^{2r}(S_{n,2s}^r f^{(2s)})^{(2r)}\|_p\right\} \qquad (50)$$
$$\leq C\left\{n^{s-\alpha} + h^{2r}\|\varphi^{2s}\varphi^{2r}(S_{n,2s}^r f^{(2s)})^{(2r)}\|_p\right\}.$$

Since $(S_{n,2s}^r f^{(2s)})^{(2r)} = \sum_{i=0}^{r-1} \alpha_i(n)(S_{n_i,2s} f^{(2s)})^{(2r)}$, it follows from (50) and (3) that

$$K_\varphi^{2r}(f^{(2s)}, h^{2r})_{\varphi^{2s},p} \leq C\left\{n^{s-\alpha} + h^{2r}\|\varphi^{2s}\varphi^{2r}(S_{n,2s} f^{(2s)})^{(2r)}\|_p\right\}$$
$$\leq C\left\{n^{s-\alpha} + h^{2r}n^r K_\varphi^{2r}(f^{(2s)}, n^{-r})_{\varphi^{2s},p}\right\}. \qquad (51)$$

Here we have used $\|\varphi^{2s}\varphi^{2r}(S_{n,2s} f^{(2s)})^{(2r)}\|_p \leq Cn^r \omega_\varphi^{2r}(f^{(2s)}, n^{-1/2})_{\varphi^{2s},p}$ (cf. [15, Lemma 3.3]) along with the fact that $\omega_\varphi^{2r}(f^{(2s)}, n^{-1/2})_{\varphi^{2s},p}$ and $K_\varphi^{2r}(f^{(2s)}, n^{-r})_{\varphi^{2s},p}$ are equivalent.

With the Berens-Lorentz lemma (cf. [2, p.696, 1, p.100]), we get
$$K_\varphi^{2r}(f^{(2s)}, t^{2r})_{\varphi^{2s},p} = \mathcal{O}(t^{2\alpha-2s}).$$

By the equivalence of this K–functional to the corresponding weighted modulus of smoothness we have $\omega_\varphi^{2r}(f^{(2s)}, t)_{\varphi^{2s},p} = \mathcal{O}(t^{2\alpha-2s})$, which is equivalent to $\omega_\varphi^{2r+2s}(f, t)_p = \mathcal{O}(t^{2\alpha})$ by [8, Corollary 6.3.2]. ∎

§7. Equivalence Results

The following equivalence result for the linear combinations S_n^{r*} follows partly as a corollary from Theorems 6 and 7.

Theorem 8. Let $\varphi(x) = \sqrt{x}$, $s \in \mathbb{N}_0$, $f \in L_p(I)$, $1 \leq p < \infty$ and $0 < \alpha - s < r$. Then the following statements are equivalent.

$$(i) \quad \omega_\varphi^{2r+2s}(f,t)_p = \mathcal{O}(t^{2\alpha})$$

$$(ii) \quad \varphi^{2s} f^{(2s)} \in L_p(I) \quad \text{and} \quad \|\varphi^{2s}(S_n^{r*}f - f)^{(2s)}\|_p = \mathcal{O}(n^{s-\alpha}).$$

Proof: Since $\omega_\varphi^{2r+2s}(f,t)_p = \mathcal{O}(t^{2\alpha})$ and $\omega_\varphi^{2r}(f^{(2s)},t)_{\varphi^{2s},p} = \mathcal{O}(t^{2\alpha-2s})$ are equivalent for $\varphi^{2s} f^{(2s)} \in L_p(I)$ (cf. [8, Corollary 6.3.2]), by Theorems 6 and 7 we only have to prove the implication

$$\omega_\varphi^{2r+2s}(f,t)_p = \mathcal{O}(t^{2\alpha}) \quad \Longrightarrow \quad \varphi^{2s} f^{(2s)} \in L_p(I).$$

For fixed n we consider for $k \in \mathbb{N}$, $U_k = S_{2^k n}^{r+s*}f - S_{2^{k-1}n}^{r+s*}f$. Since $\sum_{k=1}^{N} U_k = S_{2^N n}^{r+s*}f - S_n^{r+s*}f$ and $\|S_{2^N n}^{r+s*}f - f\|_p \longrightarrow 0$ $(N \longrightarrow \infty)$ we have the representation $f - S_n^{r+s*}f = \sum_{k=1}^{\infty} U_k$ in the sense of the L_p–metric. Thus we have convergence in the sense of distributions, i.e.

$$\sum_{k=1}^{\infty} \langle U_k, \psi \rangle = \langle f - S_n^{r+s*}, \psi \rangle, \quad \forall \psi \in C_0^\infty(I).$$

With the theory of distributions we get

$$D^{2s}\left\{\sum_{k=1}^{\infty} \langle U_k, \psi \rangle\right\} = \sum_{k=1}^{\infty} \left\langle U_k^{(2s)}, \psi \right\rangle = \sum_{k=1}^{\infty} \left\langle U_k, \psi^{(2s)} \right\rangle$$

$$= \left\langle f - S_n^{r+s*}f, \psi^{(2s)} \right\rangle = \left\langle (f - S_n^{r+s*}f)^{(2s)}, \psi \right\rangle.$$

Since $\varphi^{2s} \in C^\infty(I)$ we can multiply the last equation by φ^{2s} which gives

$$\sum_{k=1}^{\infty} \left\langle \varphi^{2s} U_k^{(2s)}, \psi \right\rangle = \left\langle \varphi^{2s}(f - S_n^{r+s*}f)^{(2s)}, \psi \right\rangle. \tag{52}$$

Now we prove that $\left\|\sum_{k=1}^{\infty} \varphi^{2s} U_k^{(2s)}\right\|_p < \infty$. By the commutativity of the operators which implies the commutativity of the linear combinations as well, we get for every $m, n \in \mathbb{N}$

$$\|\varphi^{2s}(S_n^{r+s*}f - S_m^{r+s*}f)^{(2s)}\|_p$$

$$\leq \|\varphi^{2s}(S_n^{r+s*}(f - S_m^{r+s*}f))^{(2s)}\|_p + \|\varphi^{2s}(S_m^{r+s*}(f - S_n^{r+s*}f))^{(2s)}\|_p.$$

By the definition of the linear combinations and the Bernstein–type inequality $\|\varphi^{2s}(S_n f)^{(2s)}\|_p \leq C n^s \|f\|_p$ (cf. [13, Lemma 4.5, (3.6) in Theorem 3.2]) we derive

$$\|\varphi^{2s}(S_n^{r+s*} f - S_m^{r+s*} f)^{(2s)}\|_p$$
$$\leq C[\max\{n,m\}]^s \left\{ \|f - S_m^{r+s*} f\|_p + \|f - S_n^{r+s*} f\|_p \right\}. \tag{53}$$

By Theorem 6 and the fact that $\omega_\varphi^{2r+2s}(f,t)_p = \mathcal{O}(t^{2\alpha})$, we get

$$\|\varphi^{2s}(S_n^{r+s*} f - S_m^{r+s*} f)^{(2s)}\|_p \leq C[\max\{n,m\}]^s [\min\{n,m\}]^{-\alpha}.$$

Since $s - \alpha < 0$ we get from the above mentioned estimate

$$\left\| \sum_{k=1}^{\infty} \varphi^{2s} U_k^{(2s)} \right\|_p \leq \sum_{k=1}^{\infty} \left\| \varphi^{2s}(S_{2^k n}^{r+s*} f - S_{2^{k-1} n}^{r+s*} f)^{(2s)} \right\|_p$$
$$\leq C \sum_{k=1}^{\infty} (2^k n)^{s-\alpha} \leq C n^{s-\alpha} < \infty.$$

Hence $\sum_{k=1}^{\infty} \varphi^{2s} U_k^{(2s)} \in L_p(I)$ and $\sum_{k=1}^{\infty} \langle \varphi^{2s} U_k^{(2s)}, \psi \rangle$ is a regular distribution. So by (52) $\langle \varphi^{2s}(f - S_n^{r+s*} f)^{(2s)}, \psi \rangle$ is also a regular distribution. Thus

$$\sum_{k=1}^{\infty} \varphi^{2s} U_k^{(2s)} = \varphi^{2s}(f - S_n^{r+s*} f)^{(2s)}$$

a. e. in I. Therefore $\|\varphi^{2s}(f - S_n^{r+s*} f)^{(2s)}\|_p < \infty$, i.e., we proved that $\|\varphi^{2s} f^{(2s)}\|_p < \infty$ holds true, and the theorem is proved. $\blacksquare$

Several direct and inverse theorems can be combined now to give a nice equivalence result which corresponds to theorems of Butzer and Scherer e.g. for the case of trigonometric best approximation [5], and theorems of Butzer, Jansche and Stens e.g., for the case of weighted and nonweighted polynomial best approximation [4].

Theorem 9. *Let $f \in L_p(I)$, $1 \leq p < \infty$, $s \in \mathbb{N}_0$ and $\alpha \in \mathbb{R}$ with $0 < \alpha - s < r$. Then the following statements are equivalent:*

$$\omega_\varphi^{2r+2s}(f,t)_p = \mathcal{O}(t^{2\alpha}) \tag{54}$$

$$\varphi^{2s} f^{(2s)} \in L_p(I) \quad \text{and} \quad \|\varphi^{2s}(S_n^{r*} f - f)^{(2s)}\|_p = \mathcal{O}(n^{s-\alpha}) \tag{55}$$

$$\|\varphi^{2s+2r}(S_n f)^{(2s+2r)}\|_p = \mathcal{O}(n^{s+r-\alpha}) \tag{56}$$

$$\|S_n^{r+s*} f - f\|_p = \mathcal{O}(n^{-\alpha}). \tag{57}$$

Proof: For (54) $\Longleftrightarrow$ (55), see Theorem 8. For (54) $\Longleftrightarrow$ (56), see [11, Satz 7.1]. Finally, for (54) $\Longleftrightarrow$ (57), see Theorem 8 for the special case $s = 0$ and replace r by $r + s$. ∎

References

1. M. Becker, R. J. Nessel, An elementary approach to inverse approximation theorems, J. Approx. Th. **23** (1978), 99–103.
2. H. Berens, G. G. Lorentz, Inverse theorems for Bernstein polynomials, Indiana Univ. Math. J. **21** (1972), 693–708.
3. J. Bergh, J. Löfström, *Interpolation Spaces, An Introduction*, Grundlehren der mathematischen Wissenschaften 223, Springer-Verlag, Berlin, Heidelberg, New-York, 1976.
4. P. L. Butzer, S. Jansche, R. L. Stens, Functional analytic methods in the solution of the fundamental theorems on best weighted algebraic approximation, preprint.
5. P. L. Butzer, K. Scherer, Über die Fundamentalsätze der klassischen Approximationstheorie in abstrakten Räumen, in *Abstract Spaces and Approximation*, P. L. Butzer, B. Szökefalvi-Nagy (eds.), Birkhäuser Verlag, Basel, 1969, 113–125.
6. M. M. Derriennic, Sur l'approximation de fonctions intégrables sur $[0, 1]$ par des polynômes de Bernstein modifiés, J. Approx. Th. **31** (1981), 325–343.
7. Z. Ditzian, K. Ivanov, Bernstein-type operators and their derivatives, J. Approx. Th. **56** (1989), 72–90.
8. Z. Ditzian, V. Totik, *Moduli of Smoothness*, Springer Series in Computational Mathematics, Springer Verlag, New York, 1987.
9. J. L. Durrmeyer, Une formule d'inversion de la transformée de Laplace: applications à la théorie des moments, Thèse de 3e cycle, Faculté des Sciences de l'Université de Paris, 1967.
10. H. W. Gould, *Combinatorial Identities*, Morgantown Printing and Binding Co., Morgantown, W. Va., 1972.
11. M. Heilmann, Approximation auf $[0, \infty)$ durch das Verfahren der Operatoren vom Baskakov-Durrmeyer Typ, Dissertation, Universität Dortmund, 1987.
12. M. Heilmann, Commutativity of operators from Baskakov-Durrmeyer type, in *Constructive Theory of Functions*, B. Sendov, P. Petrushev, K.

Ivanov, R. Maleev (eds.), Publishing House of the Bulgarian Academy of Sciences, Sofia, 1988, 197–206.

13. M. Heilmann, Direct and converse results for operators of Baskakov-Durrmeyer type, Approx. Th. and its Appl. **5:1** (1989), 105–127.
14. M. Heilmann, Erhöhung der Konvergenzgeschwindigkeit bei der Approximation von Funktionen mit Hilfe von Linearkombinationen spezieller positiver linearer Operatoren, Habilitationsschrift, 1991.
15. M. Heilmann, M. W. Müller, Direct and converse results on weighted simultaneous approximation by the method of operators of Baskakov-Durrmeyer type, Results in Mathematics **16** (1989), 228–242.
16. M. Heilmann, M. W. Müller, Equivalence of a weighted modulus of smoothness and a modified weighted K-functional, in *Progress in Approximation Theory*, P. Nevai, A. Pinkus (eds.), Academic Press, 1991, 467–473.
17. S. M. Mazhar, V. Totik, Approximation by modified Szász operators, Acta Sc. Math. **49** (1985), 257–269.
18. E. M. Stein, G. Weiss, *Introduction to Fourier Analysis on Euclidian Spaces*, Princeton University Press, Princeton, New Jersey, 1971.
19. O. Szász, Generalization of S. Bernstein's polynomials to the infinite interval, J. Res. Nat. Bureau Stand. **45** (1950), 239–245.

Margareta Heilmann
University of Dortmund
Vogelpothsweg 87
Dortmund, Germany
uma019@ddohrz11.bitnet

Numerical Methods of Approximation Theory, Vol. 9
Dietrich Braess and Larry L. Schumaker (eds.), pp. 117–134.
International Series of Numerical Mathematics, Vol. 105

Approximation by Multivariate Splines: an Application of Boolean Methods

Rong-Qing Jia

Dedicated to the memory of Lothar Collatz

Abstract. Multivariate smooth splines on nonuniform rectangular grids are investigated. A general theory of Boolean methods is developed in such a way that it can be applied to noncommutative operators. Based on this theory, an explicit quasi-interpolant is constructed so that it gives rise to an efficient scheme of approximation by multivariate smooth splines. This scheme is shown to achieve the optimal rate of approximation.

§1. Introduction

In this paper we investigate multivariate splines on nonuniform rectangular grids. Using Boolean methods, we construct explicitly a quasi-interpolant and establish an efficient scheme of approximation by multivariate splines which achieves the optimal approximation order.

As usual, $\mathbb{N}$, $\mathbb{Z}$ and $\mathbb{R}$ denote the set of nonnegative integers, integers, and real numbers, respectively. We denote by $C^\rho(\mathbb{R}^d)$ the linear space of all ρ-times continuously differentiable functions on $\mathbb{R}^d$. When $\rho = 0$ we simply write $C(\mathbb{R}^d)$ instead of $C^0(\mathbb{R}^d)$.

Suppose we are given d strictly increasing sequences $(t_i^{(m)})_{i \in \mathbb{Z}}$ of real numbers $(m = 1, \ldots, d)$ with

$$|\Delta| := \max_{1 \le m \le d} \sup_{i \in \mathbb{Z}} (t_{i+1}^{(m)} - t_i^{(m)}) < \infty,$$

where Δ is the partition of $\mathbb{R}^d$ given by

$$\Delta = \left\{ [t_{i_1}^{(1)}, t_{i_1+1}^{(1)}] \times \cdots \times [t_{i_d}^{(d)}, t_{i_d+1}^{(d)}] : (i_1, \ldots, i_d) \in \mathbb{Z}^d \right\}, \qquad (1.1)$$

and $|\Delta|$ is called the *mesh-size* of Δ.

Given two nonnegative integers k and ρ, let $S_k^\rho(\Delta)$ denote the linear space of those ρ-times continuously differentiable functions on $\mathbb{R}^d$ which agree with some polynomial (in d variables) of (total) degree $\leq k$ on each rectangle $(t_{i_1}^{(1)}, t_{i_1+1}^{(1)}) \times \cdots \times (t_{i_d}^{(d)}, t_{i_d+1}^{(d)})$, $(i_1, \ldots, i_d) \in \mathbb{Z}^d$. We are interested in the approximation power of $S_k^\rho(\Delta)$.

In the univariate case $(d = 1)$, the approximation properties of the spline space $S_k^\rho(\Delta)$ has been extensively studied. The optimal approximation order of $S_k^\rho(\Delta)$ has been shown to be $k + 1$, regardless of what the smoothness ρ is, as long as $\rho < k$. There are essentially two approaches to obtaining spline approximants. The first approach (see [14] and [8]) uses induction on k, the degree of the splines. Suppose f is a sufficiently smooth function to be approximated. One can approximate the derivative of f by splines of lower degree and smoothness, and then integrate and make local corrections for the approximation to f. The second approach (see [3] and [13]) employs a quasi-interpolation scheme. A quasi-interpolant Q is a linear projector onto the spline space which preserves certain polynomials. Once a quasi-interpolant is constructed, one can simply use Qf as an approximant to f. The second approach, in my opinion, is preferable to the first, because the quasi-interpolant Q is local in the sense that $Qf(x)$ depends only on values of f in a small neighborhood of x, and the quasi-interpolation scheme is easy to implement.

When Δ is a uniform rectangular grid in $\mathbb{R}^2$, de Boor and DeVore in [4] have shown that the approximation power of $S_k^\rho(\Delta)$ is adversely affected by the smoothness requirement. Specifically, if $k < 2\rho+2$, then $S_k^\rho(\Delta)$ is not effective for approximation at all; if $k \geq 2\rho+2$, then the approximation order of $S_k^\rho(\Delta)$ is $k-\rho$. These results are in sharp contrast to the univariate situation. On the other hand, their results also demonstrate that approximation using bivariate splines of a certain total degree is more efficient than using tensor-products of univariate splines with the same total degree and smoothness. However, the approximation scheme used in [4] is not local and difficult to carry out.

The purpose of this paper is to construct explicitly a linear operator (a quasi-interpolant) from $C(\mathbb{R}^d)$ into $S_k^\rho(\Delta)$ such that Q is local and Qf approximates smooth functions with the optimal order of accuracy. This will be accomplished in §4. In §2 we review the quasi-interpolation schemes in [3], [13], and [15]. In §3 we develop a general theory of Boolean methods which can be applied to noncommutative operators. Based on this theory, we construct the desired quasi-interpolants in §4 and establish the direct theorems of approximation for both continuous functions and functions whose pth powers are integrable. Finally, in §5, we point out that the space $S_k^\rho(\Delta)$ is effective for approximation only if $k \geq d(\rho + 1)$. In such a case we prove that the optimal order of approximation is $k + 1 - (d - 1)(\rho + 1)$. This demonstrates that the quasi-interpolation scheme described in §4 achieves the optimal order of approximation.

§2. Quasi-interpolation Schemes

An important role in the construction of spline approximants is played by the B-splines. Let $(t_i)_{i \in \mathbb{Z}}$ be a (strictly) increasing sequence of real numbers such that

$$\sup_i (t_{i+1} - t_i) < \infty.$$

The ith (normalized) B-spline of order r for the above knot sequence is denoted by $N_{i,r}$ and is defined by the rule

$$N_{i,r}(x) := (t_{i+r} - t_r)[t_i, \ldots, t_{i+r}](\cdot - x)_+^{r-1}, \quad x \in \mathbb{R},$$

where $[t_i, \ldots, t_{i+r}]$ denotes the rth order divided difference at $t_i, \ldots, t_{i+r}$, and $a_+ := \max\{a, 0\}$ for $a \in \mathbb{R}$. The B-spline $N_{i,r}$ is supported on $[t_i, t_{i+r}]$ and is positive on (t_i, t_{i+r}). It belongs to $C^{r-2}(\mathbb{R})$ and agrees with some polynomial of degree $< r$ on each interval (t_j, t_{j+1}), $j \in \mathbb{Z}$. Moreover,

$$\sum_{i \in \mathbb{Z}} N_{i,r} = 1.$$

We denote by Δ the partition of $\mathbb{R}$ given by

$$\Delta - \{[t_i, t_{i+1}] : i \in \mathbb{Z}\}$$

and by $|\Delta|$ its mesh-size. Let $\mathcal{S}_r(\Delta)$ be the linear space of all (possibly infinite) linear combinations of the B-splines $N_{i,r}$, $i \in \mathbb{Z}$. Then a function f lies in $\mathcal{S}_r(\Delta)$ if and only if it belongs to $C^{r-2}(\mathbb{R})$ and agrees with some polynomial of degree $< r$ on each interval (t_j, t_{j+1}), $j \in \mathbb{Z}$. In particular, $\mathcal{S}_r(\Delta)$ contains $\Pi_{r-1}(\mathbb{R})$, the linear space of all polynomials of degree $< r$. These facts are well known and can be found in [2].

It was first proved by de Boor in [3] that there exist linear functionals $\lambda_{i,r}$ $(i \in \mathbb{Z})$ such that

$$\lambda_{i,r} N_{j,r} = \delta_{ij} \quad \text{for all } i, j \in \mathbb{Z}, \tag{2.1}$$

where δ_{ij} is the Kronecker symbol. Moreover, for $f \in C(\mathbb{R})$,

$$|\lambda_{i,r}(f)| \leq \mathrm{const}_r \|f\|_\infty([t_{i+1}, t_{i+r-1}]), \quad i \in \mathbb{Z}. \tag{2.2}$$

Here and henceforth $\|f\|_\infty(I)$ denotes the essential supremum of f over I. Lyche and Schumaker in [13] constructed such functionals explicitly based on point evaluations. See also [15, p. 224]. Now let Q_r be the operator on $C(\mathbb{R})$ given by

$$Q_r(f) := \sum_{i \in \mathbb{Z}} \lambda_{i,r}(f) \, N_{i,r}.$$

Then Q_r is a projector on $\mathcal{S}_r(\Delta)$ because of (2.1). In particular, since $\Pi_{r-1}(\mathbb{R})$ is contained in $\mathcal{S}_r(\Delta)$, we have

$$Q_r(q) = q \quad \text{for all } q \in \Pi_{r-1}(\mathbb{R}). \tag{2.3}$$

Moreover, Q_r is local in the sense that

$$|(Q_r f)(x)| \leq \mathrm{const}_r \|f\|_\infty([x - r|\Delta|, x + r|\Delta|]), \quad x \in \mathbb{R}.$$

It was proved in [3] and [13] that for any $f \in C^r(\mathbb{R})$

$$\|Q_r f - f\|_\infty \leq \mathrm{const}_r |\Delta|^r \|f^{(r)}\|_\infty.$$

In other words, the quasi-interpolant Q_r produces an approximation scheme which achieves the optimal order of approximation. Note that Q_r and Q_s do *not* commute if $r \neq s$.

§3. Boolean Methods

The theory of distributive lattices was first applied by Gordon in [9] to multivariate interpolation. This method is commonly referred to as *the blending method*. Here, following Delvos and Schempp [6], we call such a method *the Boolean method*, since Boolean algebras play a key role in its development. A successful application of Boolean methods usually requires that the operators involved be commutative projectors. "In several important situations, however", observed Gordon and Cheney in [10], "the basic approximation operators do not commute with each other". The quasi-interpolation operators Q_r ($r \geq 1$) described in §2 provide such an example. In this section we develop a general theory of Boolean methods in such a way that it can be applied to noncommutative operators. It is expected that this theory will have applications to a wide range of problems in multivariate approximation.

We begin with a study of the lattice $\mathbb{N}^d$, where d is a positive integer. The ith component of an element $\alpha \in \mathbb{N}^d$ is often written α_i. The length of $\alpha = (\alpha_1, \ldots, \alpha_d) \in \mathbb{N}^d$ is defined to be

$$|\alpha| := \alpha_1 + \cdots + \alpha_d.$$

For two elements $\alpha = (\alpha_1, \ldots, \alpha_d)$ and $\beta = (\beta_1, \ldots, \beta_d)$ in $\mathbb{N}^d$, $\alpha \leq \beta$ means

$$\alpha_j \leq \beta_j \quad \text{for all } j = 1, \ldots, d.$$

This partial order makes $\mathbb{N}^d$ a *lattice*. Indeed, the greatest lower bound of α and β, denoted by $\alpha \wedge \beta$, is the element in $\mathbb{N}^d$ whose jth component is $\min\{\alpha_j, \beta_j\}$, $j = 1, \ldots, d$; the least upper bound of α and β, denoted by $\alpha \vee \beta$, is the element in $\mathbb{N}^d$ whose jth component is $\max\{\alpha_j, \beta_j\}$, $j = 1, \ldots, d$. Evidently, $\mathbb{N}^d$ with $\wedge$ and $\vee$ becomes a distributive lattice. The infimum of a finite subset A of $\mathbb{N}^d$ is denoted by $\inf A$.

Now let G be a free abelian group with a basis $\{u_\alpha : \alpha \in \mathbb{N}^d\}$ (see [11, pp. 70–74]). Any element f of G has a unique representation of the form

$$f = \sum_{\alpha \in \mathbb{N}^d} a_\alpha u_\alpha,$$

where a_α are integers and $a_\alpha = 0$ except for finitely many α. Given two elements $f, g \in G$, $f = \sum_{\alpha \in \mathbb{N}^d} a_\alpha u_\alpha$ and $g = \sum_{\beta \in \mathbb{N}^d} b_\beta u_\beta$, the product of f and g, denoted by $f \circ g$, is defined by the rule

$$f \circ g := \sum_{\alpha, \beta \in \mathbb{N}^d} a_\alpha b_\beta u_{\alpha \wedge \beta}.$$

It is easily seen that the multiplication so defined is commutative and associative:

$$f \circ g = g \circ f, \qquad (f \circ g) \circ h = f \circ (g \circ h).$$

Moreover, the multiplication satisfies the distributive law with respect to addition in G:

$$f \circ (g + h) = f \circ g + f \circ h.$$

The *Boolean sum* of two elements f and g in G is defined by the rule

$$f \oplus g := f + g - f \circ g.$$

Note that G with $\circ$ and $\oplus$ does *not* constitute a lattice. But the Boolean addition is commutative and associative. The commutativity follows immediately from the definition. To see the associativity, we let $f, g, h \in G$ and observe that both $(f \oplus g) \oplus h$ and $f \oplus (g \oplus h)$ equal

$$f + g + h - f \circ g - g \circ h - h \circ f + f \circ g \circ h.$$

Thus, given finitely many elements $f_1, \ldots, f_n$ of G, the Boolean sum

$$f_1 \oplus \cdots \oplus f_n$$

is well defined. When $n = 1$, this sum is interpreted as f_1. We caution the reader that the multiplication in G does *not* satisfy the distributive law with respect to the Boolean addition.

Let A be a nonempty finite subset of $\mathbb{N}^d$. Consider the Boolean sum

$$\bigoplus_{\alpha \in A} u_\alpha =: \sum_{\beta \in \mathbb{N}^d} b_\beta u_\beta. \tag{3.1}$$

We want to find the coefficients b_β, $\beta \in \mathbb{N}^d$. From the definition of the Boolean addition, one can easily deduce that

$$\bigoplus_{\alpha \in A} u_\alpha = \sum_{\emptyset \neq B \subseteq A} (-1)^{\#B - 1} u_{\inf B},$$

where $\#B$ denotes the cardinality of B. It follows that

$$b_\beta = \sum_{\substack{\inf B = \beta \\ B \subseteq A}} (-1)^{\#B - 1}, \quad \beta \in \mathbb{N}^d. \tag{3.2}$$

Here we adopt the convention that the empty sums mean 0. We want to derive another formula for b_β, which is often easier to calculate than (3.2). To this end, we denote by R_α the set $\{\beta \in \mathbb{N}^d : \beta \leq \alpha\}$, and by R_A the union $\cup_{\alpha \in A} R_\alpha$. In particular

$$E_d := \{(\varepsilon_1, \ldots, \varepsilon_d) : \varepsilon_j = 0 \text{ or } 1 \text{ for } j = 1, \ldots, d\} = R_e,$$

where e is the d-vector $(1, \ldots, 1)$. It is easily seen that

$$\sum_{\varepsilon \in R_\alpha} (-1)^\varepsilon = \prod_{j=1}^d \Big(\sum_{0 \leq \varepsilon_j \leq \alpha_j} (-1)^{\varepsilon_j} \Big) = \delta_{\alpha 0} \quad \text{for all } \alpha \leq e. \tag{3.3}$$

The set R_A is a *lower set* in the sense that

$$\beta \in R_A \text{ and } \gamma \leq \beta \implies \gamma \in R_A.$$

Let i_A denote the characteristic function of the set R_A.

Theorem 3.1. *The coefficients b_β in (3.1) are given by*

$$b_\beta = \sum_{\varepsilon \in E_d} (-1)^{|\varepsilon|} i_A(\beta + \varepsilon), \quad \beta \in \mathbb{N}^d. \tag{3.4}$$

In particular, $b_\beta = 0$ if $\beta \notin R_A$ or $\beta + e \in R_A$.

Proof: We first prove that

$$\sum_{\beta \geq \gamma} b_\beta = i_A(\gamma) \quad \text{for all } \gamma \in \mathbb{N}^d. \tag{3.5}$$

This will be done by induction on $\#A$. If $\#A = 1$, then (3.5) holds trivially. Suppose (3.5) has been established for a nonempty finite subset A of $\mathbb{N}^d$. We wish to prove that it is also true for $A' = A \cup \alpha'$, where α' is an arbitrary element of $\mathbb{N}^d$. It follows from (3.1) that

$$\bigoplus_{\alpha \in A'} u_\alpha = \sum_{\beta \in \mathbb{N}^d} b_\beta u_\beta + u_{\alpha'} - \sum_{\beta \in \mathbb{N}^d} b_\beta u_{\beta \wedge \alpha'} =: \sum_{\beta \in \mathbb{N}^d} b'_\beta u_\beta.$$

There are two possible cases: either $\gamma \in R_{\alpha'}$ or $\gamma \notin R_{\alpha'}$. In the former case, we have

$$\sum_{\beta \geq \gamma} b'_\beta = \sum_{\beta \geq \gamma} b_\beta + 1 - \sum_{\beta \geq \gamma} b_\beta = 1 = i_{A'}(\gamma).$$

In the latter case, we have

$$\sum_{\beta \geq \gamma} b'_\beta = \sum_{\beta \geq \gamma} b_\beta = i_A(\gamma) = i_{A'}(\gamma).$$

This completes the induction procedure.

Next, we show that (3.4) is a consequence of (3.5). This can be done using Möbius functions (see [16, pp. 118–119]). Here, for the reader's convenience, we give a direct proof. We deduce from (3.5) that for $\beta \in \mathbb{N}^d$

$$\sum_{\varepsilon \in E_d} (-1)^{|\varepsilon|} i_A(\beta + \varepsilon) = \sum_{\varepsilon \in E_d} (-1)^{|\varepsilon|} \sum_{\gamma \geq \beta + \varepsilon} b_\gamma = \sum_{\gamma \geq \beta} b_\gamma \sum_{\varepsilon \in E_d \cap R_{\gamma - \beta}} (-1)^{|\varepsilon|}.$$

It follows from (3.3) that

$$\sum_{\gamma \geq \beta} b_\gamma \sum_{\varepsilon \in E_d \cap R_{\gamma - \beta}} (-1)^{|\varepsilon|} = \sum_{\gamma \geq \beta} b_\gamma \delta_{\gamma \beta} = b_\beta.$$

This proves the desired formula (3.4).

If $\beta \notin R_A$, then $b_\beta = 0$ because of (3.4). Finally, if $\beta + e \in R_A$, then $i_A(\beta + \varepsilon) = 1$ for all $\varepsilon \in E_d$. Hence, (3.3) and (3.4) imply $b_\beta = 0$. $\blacksquare$

Of particular interest is the case when $A = \{\alpha \in \mathbb{N}^d : |\alpha| = k\}$ for some integer $k \geq 0$. In this case $i_A(\gamma) = 1$ if and only if $|\gamma| \leq k$. By Theorem 3.1 we have $b_\beta = 0$ if $|\beta| > k$ or $|\beta| \leq k - d$. The latter is true because $|\beta| \leq k - d$ implies

$$|\beta + e| = |\beta| + |e| \leq k - d + d = k.$$

Suppose now $|\beta| = k - \ell$, where $0 \leq \ell \leq d - 1$. Observe that the quantity $i_A(\beta + \varepsilon)$ in (3.4) is 1 if and only if $|\beta + \varepsilon| \leq k$, i.e., $|\varepsilon| \leq k - |\beta| = \ell$. But, for $j = 0, \ldots, \ell$, the cardinality of the set $\{\varepsilon \in E_d : |\varepsilon| = j\}$ is $\binom{d}{j}$. Thus, by Theorem 3.1 we have

$$b_\beta = \sum_{j=0}^{\ell} (-1)^j \binom{d}{j}.$$

The above sum equals $(-1)^\ell \binom{d-1}{\ell}$. This can be proved by induction on ℓ. The case $\ell = 0$ is obvious. Suppose $1 \leq \ell \leq d - 1$ and

$$\sum_{j=0}^{\ell-1} (-1)^j \binom{d}{j} = (-1)^{\ell-1} \binom{d-1}{\ell-1}.$$

Then

$$\sum_{j=0}^{\ell} (-1)^j \binom{d}{j} = \sum_{j=0}^{\ell-1} (-1)^j \binom{d}{j} + (-1)^\ell \binom{d}{\ell}$$

$$= (-1)^{\ell-1} \binom{d-1}{\ell-1} + (-1)^\ell \binom{d}{\ell} = (-1)^\ell \binom{d-1}{\ell}.$$

This completes the induction procedure. The above discussion is summarized in the following.

Theorem 3.2. *The identity*

$$\bigoplus_{|\alpha|=k} u_\alpha = \sum_{\ell=0}^{d-1} (-1)^\ell \binom{d-1}{\ell} \sum_{|\beta|=k-\ell} u_\beta$$

holds for every $k \in \mathbb{N}$.

When u_α $(\alpha \in \mathbb{N}^d)$ are commutative interpolating operators, this theorem was first proved by Delvos in [5] (also see [1]).

One essential idea of Boolean methods is that the Boolean sum of the elements u_α ($\alpha \in A$) often combines certain properties of every individual member. The precise meaning of this idea is formulated as follows. Let H be another abelian group and V a nonempty set of homomorphisms from G to H. Given $v \in V$ and $g \in G$, the image of g under the mapping v is written as $g(v)$. For $\alpha \in \mathbb{N}^d$ and $i \in \{1, \ldots, d\}$, we denote by $\alpha(\hat{i})$ the d-tuple obtained from α by replacing its ith component with 0.

Theorem 3.3. *Let f be a mapping from V to H. Suppose there is a sequence of elements v_α in V ($\alpha \in \mathbb{N}^d$) satisfying the following two conditions.*
   (i) For any $\alpha \in \mathbb{N}^d$, $u_\alpha(v_\gamma) = f(v_\gamma)$ for all $\gamma \leq \alpha$.
   (ii) If for some $i \in \{1, \ldots, d\}$, $\alpha(\hat{i}) = \beta(\hat{i})$ and $\gamma_i \leq \min\{\alpha_i, \beta_i\}$, then
      $u_\alpha(v_\gamma) = u_\beta(v_\gamma).$

Then for any finite nonempty subset A of $\mathbb{N}^d$,

$$\left(\bigoplus_{\alpha \in A} u_\alpha \right)(v_\gamma) = f(v_\gamma) \quad \text{for all } \gamma \in R_A. \tag{3.6}$$

Proof: Let $\alpha, \beta, \gamma \in \mathbb{N}^d$ satisfy the condition that for $j = 1, \ldots, d$, either $\alpha_j = \beta_j$ or $(\alpha \wedge \beta)_j \geq \gamma_j$. We shall show $u_\alpha(v_\gamma) = u_\beta(v_\gamma)$. This is certainly true if $\alpha = \beta$. Consider the set $\{j : (\alpha \wedge \beta)_j \geq \gamma_j\}$. Without loss of generality we may assume that this set has the form $\{1, \ldots, m\}$, where $m \in \{1, \ldots, d\}$. It follows that $\alpha_j = \beta_j$ for $j = m+1, \ldots, d$. For $n = 0, 1, \ldots, m$, let $\alpha^{(n)} \in \mathbb{N}^d$ be defined as follows:

$$\alpha_j^{(n)} = \begin{cases} \beta_j, & \text{if } j = 1, \ldots, n; \\ \alpha_j, & \text{if } j = n+1, \ldots, d. \end{cases}$$

Then $\alpha^{(0)} = \alpha$ and $\alpha^{(m)} = \beta$. Note that $\alpha^{(n-1)}$ and $\alpha^{(n)}$ ($1 \leq n \leq m$) differ by only one component. We have $\alpha^{(n-1)}(\hat{n}) = \alpha^{(n)}(\hat{n})$ and $\gamma_n \leq \min\{\alpha_n^{(n-1)}, \alpha_n^{(n)}\}$. Thus the condition (ii) implies $u_{\alpha^{(n-1)}}(v_\gamma) = u_{\alpha^{(n)}}(v_\gamma)$. This is true for $n = 1, \ldots, m$; hence $u_\alpha(v_\gamma) = u_\beta(v_\gamma)$.

Let us prove (3.6). The case $\#A = 1$ is evidently true. So we assume that $\#A \geq 2$. Given $\gamma \in R_A$, one can find an element $\alpha' \in A$ such that $\gamma \leq \alpha'$. Express the Boolean sum $\oplus_{\alpha \in A \setminus \{\alpha'\}} u_\alpha$ as $\sum_{\beta \in \mathbb{N}^d} c_\beta u_\beta$. Then

$$\bigoplus_{\alpha \in A} u_\alpha = u_{\alpha'} \bigoplus \left(\sum_{\beta \in \mathbb{N}^d} c_\beta u_\beta \right) = u_{\alpha'} + \sum_{\beta \in \mathbb{N}^d} c_\beta (u_\beta - u_{\alpha' \wedge \beta}).$$

By the condition (i), $u_{\alpha'}(v_\gamma) = f(v_\gamma)$. Thus, in order to prove (3.6) we only need to verify $u_\beta(v_\gamma) = u_{\alpha' \wedge \beta}(v_\gamma)$ for $\beta \in \mathbb{N}^d$. But $\alpha' \geq \gamma$ implies that for

$j = 1, \ldots, d$, either $(\alpha' \wedge \beta)_j = \beta_j$ or $(\alpha' \wedge \beta)_j \geq \gamma_j$. Hence by what has been proved before, we have $u_\beta(v_\gamma) = u_{\alpha' \wedge \beta}(v_\gamma)$. ∎

Combining Theorem 3.1 with Theorem 3.3 yields the following result.

Theorem 3.4. *Let G and H be two abelian groups, V a nonempty set of homomorphisms from G to H, and f a mapping from V to H. Suppose there are two sequences $\{g_\alpha : \alpha \in \mathbb{N}^d\} \subseteq G$ and $\{v_\alpha : \alpha \in \mathbb{N}^d\} \subseteq V$ satisfying the conditions (i) and (ii) of Theorem 3.3 with u_α replaced by g_α $(\alpha \in \mathbb{N}^d)$. Then for any nonempty finite subset A of $\mathbb{N}^d$,*

$$\Big(\sum_{\beta \in \mathbb{N}^d} b_\beta g_\beta \Big)(v_\gamma) = f(v_\gamma) \quad \text{for all } \gamma \in R_A,$$

where the coefficients b_β are given by (3.2) or (3.4).

Proof: Note that G is not required to be free. Let U be a free abelian group with a basis $\{u_\alpha : \alpha \in \mathbb{N}^d\}$. There is a homomorphism σ from U to G such that $\sigma(u_\alpha) = g_\alpha$ for all $\alpha \in \mathbb{N}^d$. Define

$$u(v) := (\sigma(u))(v) \quad \text{for } u \in U \text{ and } v \in V.$$

Then

$$g_\beta(v_\gamma) = u_\beta(v_\gamma) \quad \text{for all } \beta, \gamma \in \mathbb{N}^d.$$

The two sequences $\{u_\alpha : \alpha \in \mathbb{N}^d\}$ and $\{v_\alpha : \alpha \in \mathbb{N}^d\}$ satisfy the conditions (i) and (ii) of Theorem 3.3. By Theorems 3.1 and 3.3 we have

$$\Big(\sum_{\beta \in \mathbb{N}^d} b_\beta u_\beta \Big)(v_\gamma) = f(v_\gamma) \quad \text{for all } \gamma \in R_A.$$

The desired result follows from the above two identities. ∎

A remark on Theorem 3.4 is in order. In most applications, G is a linear space spanned by some linear projectors from a linear space onto its subspaces. In order to apply Boolean methods it is usually required that two projectors P and Q commute: $PQ = QP$. However, as we mentioned before, two quasi-interpolation operators do not commute with each other in general. In Theorem 3.4, it is only required that G be an abelian group, hence only one operation — addition — is defined on G. Whether two elements in G can be multiplied or not is irrelevant to our theory. We believe that this development of Boolean methods will greatly enhance their applicability. In particular, many problems in multivariate spline approximation, to which the previous theory of Boolean methods is not accessible, can be solved by using Theorem 3.4.

§4. Approximation by Smooth Splines

In this section we apply the Boolean method developed in the previous section to construct a quasi-interpolant which gives rise to a desirable approximation scheme. Also, we discuss L_p approximation $(1 \leq p \leq \infty)$ by multivariate smooth splines on nonuniform rectangular grids.

For $\alpha \in \mathbb{N}^d$, we denote by q_α the monomial

$$x \mapsto x^\alpha := x_1^{\alpha_1} \cdots x_d^{\alpha_d}, \quad x = (x_1, \ldots, x_d) \in \mathbb{R}^d.$$

For $k \in \mathbb{N}$, the linear span of q_α, $|\alpha| \leq k$, is denoted by $\Pi_k = \Pi_k(\mathbb{R}^d)$. In other words, Π_k consists of all polynomials of (total) degree $\leq k$.

Let Δ be the partition of $\mathbb{R}^d$ given in (1.1). We are concerned with the spline space $S_k^\rho(\Delta)$, where $k, \rho \in \mathbb{N}$. For each $m = 1, \ldots, d$, let $N_{i,r}^{(m)}$ be the ith (normalized) B-spline of order r for the sequence $(t_i^{(m)})_{i \in \mathbb{Z}}$. From the discussion in §2 (see (2.1) and (2.2)) one can construct linear functionals $\lambda_{i,r}^{(m)}$ $(i \in \mathbb{Z})$ such that

$$\lambda_{i,r}^{(m)} N_{j,r}^{(m)} = \delta_{ij}$$

and

$$|\lambda_{i,r}^{(m)} g| \leq \mathrm{const}_r \|g\|_\infty ([t_i^{(m)}, t_{i+r}^{(m)}]) \quad \text{for all } g \in C(\mathbb{R}).$$

Consider the parametrically extended quasi-interpolation operator $Q_r^{(m)}$ on $C(\mathbb{R}^d)$ given by

$$Q_r^{(m)} f(x) := \sum_{i \in \mathbb{Z}} \lambda_{i,r}^{(m)} \big(f(x_1, \ldots, x_{m-1}, \cdot, x_{m+1}, \ldots, x_d) \big) N_{i,r}^{(m)}(x_m)$$

where $x = (x_1, \ldots, x_d) \in \mathbb{R}^d$. For $\alpha = (\alpha_1, \ldots, \alpha_d)$, let Q_α be the product of the operators $Q_{\alpha_1}^{(1)}, \ldots, Q_{\alpha_d}^{(d)}$. Then the operator Q_α can be expressed as

$$Q_\alpha f(x) = \sum_{(i_1, \ldots, i_d) \in \mathbb{Z}^d} (\lambda_{i_1, \alpha_1}^{(1)} \otimes \cdots \otimes \lambda_{i_d, \alpha_d}^{(d)} f) N_{i_1, \alpha_1}^{(1)}(x_1) \cdots N_{i_d, \alpha_d}^{(d)}(x_d) \quad (4.1)$$

Let $P_\alpha := Q_{\alpha+e}$ for $\alpha \in \mathbb{N}^d$, where e is the d-vector $(1, \ldots, 1)$. First, we point out that

$$P_\alpha(q_\gamma) = q_\gamma \quad \text{for all } \gamma \leq \alpha. \tag{4.2}$$

Indeed, if $\gamma \leq \alpha$, then $\gamma_i \leq \alpha_i$ for $i = 1, \ldots, d$, and hence a repeated use of (2.3) gives (4.2). Next, let α and β be two elements of $\mathbb{N}^d$ satisfying $\alpha(\hat{i}) = \beta(\hat{i})$ for some $i \in \{1, \ldots, d\}$. We claim that

$$P_\alpha(q_\gamma) = P_\beta(q_\gamma) \quad \text{for all } \gamma \in \mathbb{N}^d \text{ with } \gamma_i \leq \min\{\alpha_i, \beta_i\}. \tag{4.3}$$

To justify this claim we may assume that $i = d$ without loss of generality. Note that for fixed $x_1, \ldots, x_{d-1}$, $q_\gamma(x_1, \ldots, x_{d-1}, \cdot)$ is a monomial of degree γ_d. Since $\gamma_d \leq \min\{\alpha_d, \beta_d\}$, by (2.3) we have

$$(Q^{(d)}_{\alpha_d+1} - Q^{(d)}_{\beta_d+1})(q_\gamma(x_1, \ldots, x_{d-1}, \cdot)) = 0.$$

But $\alpha(\hat{d}) = \beta(\hat{d})$, hence

$$P_\alpha - P_\beta = Q^{(1)}_{\alpha_1+1} \cdots Q^{(d-1)}_{\alpha_{d-1}+1}(Q^{(d)}_{\alpha_d+1} - Q^{(d)}_{\beta_d+1}).$$

Combining the above two equalities gives (4.3), as desired.

Given $\alpha \in \mathbb{N}^d$, the range of the operator P_α is contained in $S_k^\rho(\Delta)$ if and only if $|\alpha| \leq k$ and $\alpha \geq (\rho + 1)e$. Let

$$A := \{\alpha \in \mathbb{N}^d : |\alpha| = k \text{ and } \alpha \geq (\rho + 1)e\}.$$

The set A is nonempty if and only if $k \geq d(\rho + 1)$. In what follows we assume that this is the case.

Let us compute the coefficients b_β in (3.1) for the set A as given in the above. From (3.2) we see that $b_\beta \neq 0$ only if $|\beta| \leq k$ and $\beta \geq (\rho + 1)e$. Moreover, $b_\beta = 0$ for $|\beta| \leq k - d$ by Theorem 3.1. Now let $\beta \geq (\rho + 1)e$ and $|\beta| = k - \ell$ for some $\ell = 0, \ldots, d - 1$. Let $A' := \{\alpha \in \mathbb{N}^d : |\alpha| = k\}$. Since $\beta \geq (\rho + 1)e$, $\beta + \varepsilon \in R_A$ if and only if $\beta + \varepsilon \in R_{A'}$. Thus, by (3.4) and the results in §3 we have

$$b_\beta = \sum_{\varepsilon \in E_d} (-1)^{|\varepsilon|} i_A(\beta + \varepsilon) = \sum_{\varepsilon \in E_d} (-1)^{|\varepsilon|} i_{A'}(\beta + \varepsilon) = (-1)^\ell \binom{d-1}{\ell}.$$

Theorem 4.1. *If $k \geq d(\rho + 1)$, then the operator*

$$P := \sum_{\ell=0}^{d-1} (-1)^\ell \binom{d-1}{\ell} \sum_{\substack{|\beta|=k-\ell \\ \beta \geq (\rho+1)e}} P_\beta \tag{4.4}$$

satisfies

$$Pq = q \quad \text{for all } q \in \Pi_{k-(d-1)(\rho+1)}.$$

Proof: By virtue of (4.2) and (4.3), Theorem 3.4 can be applied to the present situation, so we have

$$P(q_\gamma) = q_\gamma \quad \text{for all } \gamma \in R_A.$$

Thus, it suffices to show $\gamma \in R_A$ for $|\gamma| \leq k - (d-1)(\rho+1)$. Let $\alpha_i = \max\{\gamma_i, \rho+1\}$ for $i = 1, \ldots, d$. If all $\gamma_i \leq \rho+1$, then

$$|\alpha| = \alpha_1 + \cdots + \alpha_d = d(\rho+1) \leq k;$$

otherwise $\gamma_i > \rho+1$ for some i, hence

$$|\alpha| \leq |\gamma| + (d-1)(\rho+1) \leq k.$$

This shows $\alpha \in R_A$. It follows that $\gamma \in R_\alpha \subseteq R_A$. $\blacksquare$

In the following, we denote by D_j the partial derivative operator with respect to the jth coordinate, and by D^α the partial differential operator $D_1^{\alpha_1} \cdots D_d^{\alpha_d}$, where $\alpha = (\alpha_1, \ldots, \alpha_d)$. Let I be the cube $[-1,1]^d$. Given a measurable subset Ω of $\mathbb{R}^d$, let $\|f\|_\infty(\Omega)$ be the essential supremum of f over Ω, and

$$|f|_{s,\infty}(\Omega) := \sum_{|\alpha|=s} \|D^\alpha f\|_\infty(\Omega).$$

Theorem 4.2. *The operator P given in (4.4) maps $C(\mathbb{R}^d)$ to $S_k^\rho(\Delta)$. Moreover, if $k \geq d(\rho+1)$, then for any $s \leq k+1-(d-1)(\rho+1)$, any function $f \in C^s(\mathbb{R}^d)$, and any measurable subset Ω of $\mathbb{R}^d$,*

$$\|f - Pf\|_\infty(\Omega) \leq \mathrm{const}_k |\Delta|^s |f|_{s,\infty}(\Omega + (k+1)|\Delta|I). \tag{4.5}$$

Proof: We have proved before that P maps $C(\mathbb{R}^d)$ to $S_k^\rho(\Delta)$. Let $x = (x_1, \ldots, x_d) \in \mathbb{R}^d$ be fixed for the time being. Consider the expression (4.1) of the quasi-interpolant Q_α, where $\alpha \geq (\rho+2)e$ and $|\alpha| \leq k+d$. Since $\max\{\alpha_1, \ldots, \alpha_d\} \leq k+1$, it follows from (2.2) that

$$|\lambda_{i_1,\alpha_1}^{(1)} \otimes \cdots \otimes \lambda_{i_d,\alpha_d}^{(d)} f| \leq \mathrm{const}_k \|f\|_\infty(E_{i,\alpha}) \quad \text{for } f \in C(\mathbb{R}^d),$$

where $E_{i,\alpha}$ is the rectangle $[t_{i_1}^{(1)}, t_{i_1+\alpha_1}^{(1)}] \times \cdots \times [t_{i_d}^{(d)}, t_{i_d+\alpha_d}^{(d)}]$. But

$$N_{i_1,\alpha_1}^{(1)}(x_1) \cdots N_{i_d,\alpha_d}^{(d)}(x_d) \neq 0 \Longrightarrow x \in E_{i,\alpha} \Longrightarrow E_{i,\alpha} \subseteq x + (k+1)|\Delta|I.$$

Also, note that

$$\sum_{(i_1,\ldots,i_d)\in\mathbb{Z}^d} N_{i_1,\alpha_1}^{(1)}(x_1) \cdots N_{i_d,\alpha_d}^{(d)}(x_d) = 1.$$

Hence we can deduce from (4.1) that

$$|Q_\alpha f(x)| \le \mathrm{const}_k \|f\|_\infty (x + (k+1)|\Delta|I).$$

This, together with the expression (4.4) of P, gives the following estimate

$$|Pf(x)| \le \mathrm{const}_k \|f\|_\infty (x + (k+1)|\Delta|I). \tag{4.6}$$

Suppose now that $k \ge d(\rho+1)$, $1 \le s \le k + 1 - (d-1)(\rho+1)$ and $f \in C^s(\mathbb{R}^d)$. Let q be the Taylor polynomial of f of degree $s-1$ about x. By Theorem 4.1, $Pq = q$. Consequently,

$$f(x) - Pf(x) = q(x) - Pf(x) = P(q-f)(x).$$

By (4.6) we have

$$|f(x) - Pf(x)| = |P(q-f)(x)| \le \mathrm{const}_k \|q-f\|_\infty (x + (k+1)|\Delta|I).$$

But q is the Taylor polynomial of degree $s-1$ of f about x, hence

$$\|q-f\|_\infty (x + (k+1)|\Delta|I) \le \mathrm{const}_k |\Delta|^s |f|_{s,\infty} (x + (k+1)|\Delta|I).$$

Putting the above estimates together, we obtain

$$|f(x) - Pf(x)| \le \mathrm{const}_k |\Delta|^s |f|_{s,\infty} (x + (k+1)|\Delta|I).$$

If Ω is a measurable subset of $\mathbb{R}^d$, then the above estimate is valid for every point $x \in \Omega$, and therefore (4.5) follows immediately. ∎

In the rest of this section we discuss L_p approximation $(1 \le p \le \infty)$ from the spline space $S_k^\rho(\Delta)$. Given a measurable subset Ω of $\mathbb{R}^d$, the space $L_p(\Omega)$ consists of all functions f on Ω such that $\|f\|_p(\Omega) := (\int_\Omega |f|^p)^{1/p}$ is finite. The Sobolev space $W_p^s(\Omega)$ consists of those functions f on Ω for which $\|f\|_{s,p}(\Omega) := \sum_{j=0}^s |f|_{j,p}(\Omega)$ is finite, where $|f|_{j,p}(\Omega) := \sum_{|\alpha|=j} \|D^\alpha f\|_p(\Omega)$. When $\Omega = \mathbb{R}^d$, the reference to Ω is often omitted.

To deal with L_p approximation we shall apply the smoothing technique as employed in [12]. Choose a function $\chi \in C^{k+1}(\mathbb{R}^d)$ such that χ is supported in the cube $[-1,1]^d$, $\chi \ge 0$ and $\int \chi = 1$. For $h > 0$, set $\chi_h := \chi(\cdot/h)/h^d$. Given a function $f \in L_p(\mathbb{R}^d)$ and $h > 0$, let

$$f_h(x) := \int_{\mathbb{R}^d} (f - \nabla_u^k f)(x)\chi_h(u)\, du, \quad x \in \mathbb{R}^d,$$

where ∇_u denotes the difference operator given by

$$\nabla_u f = f - f(\cdot - u).$$

The proof of the following theorem concerning L_p approximation is similar to that of Theorem 4.1 in [12].

Theorem 4.3. *Let P be the operator from $C(\mathbb{R}^d)$ to $S_k^\rho(\Delta)$ as given in (4.4). If $k \geq d(\rho+1)$, then for any $s \leq k+1-(d-1)(\rho+1)$, any function $f \in W_p^s(\mathbb{R}^d)$ $(1 \leq p \leq \infty)$, any measurable subset Ω of $\mathbb{R}^d$,*

$$\|f - Pf_h\|_p(\Omega) \leq \mathrm{const}_k |\Delta|^s |f|_{s,p}(\Omega + (3k+2)|\Delta|I),$$

where $h = |\Delta|$.

§5. The Approximation Order

In this section we show that the approximation scheme given in the previous section achieves the optimal order of approximation.

Let $\mathcal{S}$ be a subspace of $L_p(\mathbb{R}^d)$ $(1 \leq p \leq \infty)$. We denote by $\mathrm{dist}_p(f, \mathcal{S})$ the distance from $\mathcal{S}$ to f in the L_p norm:

$$\mathrm{dist}_p(f, \mathcal{S}) := \inf_{g \in \mathcal{S}} \|f - g\|_p.$$

For $h > 0$, let σ_h be the h-dilation operator:

$$\sigma_h f(x) := f(x/h) \quad \text{for all } f \text{ and } x.$$

The L_p approximation order of $\mathcal{S}$, by definition, is the largest real number n such that

$$\mathrm{dist}_p(f, \sigma_h(\mathcal{S})) \leq \mathrm{const}_f h^n, \quad h > 0,$$

holds for all sufficiently smooth functions f, where const_f is a constant depending on f but independent of h.

Now let Δ be the partition of $\mathbb{R}^d$ given in (1.1) and let $\mathcal{S} = S_k^\rho(\Delta)$ for some integers $k, \rho \geq 0$. When $d = 2$ and Δ is a uniform rectangular grid, it was shown by de Boor and DeVore in [4] that the L_∞ approximation order of $S_k^\rho(\Delta)$ is $k-\rho$ if $k \geq 2\rho+2$. When $k < 2\rho+2$, the space $S_k^\rho(\Delta)$ is not effective for approximation. Our goal is to extend their results to higher dimensions and quasi-uniform rectangular grids. We say that Δ is *quasi-uniform* if

$$\max_{1 \leq m \leq d} \sup_{i,j \in \mathbb{Z}} (t_{i+1}^{(m)} - t_i^{(m)})/(t_{j+1}^{(m)} - t_j^{(m)}) < \infty.$$

Let $\mathcal{B}$ be the set of the functions g in $\mathcal{S}$ of the form

$$g(x) = g_1(x_1) \cdots g_d(x_d), \quad x = (x_1, \ldots, x_d) \in \mathbb{R}^d, \tag{5.1}$$

where each g_m $(m = 1, \ldots, d)$ is either a monomial $x_m \mapsto x_m^{s_m}$ for some $s_m \leq k$, or a truncated power $x_m \mapsto (x_m - t_i^{(m)})_+^{s_m}$ or $x_m \mapsto (t_i^{(m)} - x_m)_+^{s_m}$ for some s_m between $\rho + 1$ and k. Evidently, $g \in \mathcal{S}$ implies $s_1 + \cdots + s_d \leq k$. It is also easily seen that $\mathcal{B}$ spans $\mathcal{S}$, that is, any element of $\mathcal{S}$ is an (infinite) linear combination of functions in $\mathcal{B}$.

Let g be a function in $\mathcal{B}$ of the form (5.1). If $k < d(\rho + 1)$, then one of the factors $g_1, \ldots, g_d$ is a monomial of degree $\leq k$. Hence $\nabla_m^{(k+1)} g = 0$ for some $m \in \{1, \ldots, d\}$, where $\nabla_m = \nabla_{e_m}$ with e_m being the mth coordinate vector, i.e., the vector whose mth component is 1 and other components are 0. Let ∇^α denote the difference operator $\nabla_1^{\alpha_1} \cdots \nabla_d^{\alpha_d}$ for $\alpha = (\alpha_1, \ldots, \alpha_d) \in \mathbb{N}^d$. Thus, we have proved that

$$k < d(\rho + 1) \quad \Longrightarrow \quad \nabla^{(k+1)e} g = 0, \tag{5.2}$$

where e is the d-vector $(1, \ldots, 1)$. Since $\mathcal{B}$ spans $\mathcal{S}$, (5.2) is valid for all $g \in \mathcal{S}$. Next, consider the case $k \geq d(\rho + 1)$. In this case, we claim that for $g \in \mathcal{B}$,

$$D_1^n (\nabla_2 \cdots \nabla_d)^{k+1} g(x) = 0 \quad \text{for } t_i^{(1)} < x_1 < t_{i+1}^{(1)}, \ i \in \mathbb{Z}, \tag{5.3}$$

where

$$n := (k + 1) - (d - 1)(\rho + 1).$$

Indeed, if one of the factors $g_2, \ldots, g_d$ in (5.1) is a monomial of degree $\leq k$, then (5.3) is certainly true. If all the factors $g_2, \ldots, g_d$ are truncated powers, then $s_2, \ldots, s_d \geq \rho + 1$. It follows that

$$s_1 \leq k - (d - 1)(\rho + 1) < n.$$

Hence (5.3) is valid for $g \in \mathcal{B}$. Since $\mathcal{B}$ spans $\mathcal{S}$, (5.3) holds for all $g \in \mathcal{S}$.

Theorem 5.1. *Let Δ be a quasi-uniform rectangular grid. The approximation order of $S_k^\rho(\Delta)$ is 0 if $k < d(\rho + 1)$, and is $k + 1 - (d - 1)(\rho + 1)$ if $k \geq d(\rho + 1)$.*

Proof: Let $f \in L_p(\mathbb{R}^d)$ for some $p \in [1, \infty]$ and $g \in \sigma_h(\mathcal{S})$, where $\mathcal{S} = S_k^\rho(\Delta)$ and $h > 0$. Then

$$\|\nabla^{(k+1)e} f - \nabla^{(k+1)e} g\|_p \leq \mathrm{const}_k \|f - g\|_p. \tag{5.4}$$

Consider the case $k < d(\rho + 1)$ first. Note that $\sigma_h(\mathcal{S}) = S_k^\rho(h\Delta)$. Hence $\nabla^{(k+1)e} g = 0$ by (5.2). This in connection with (5.4) gives

$$\mathrm{dist}_p(f, \sigma_h(\mathcal{S})) \geq \mathrm{const}_k \|\nabla^{(k+1)e} f\|_p.$$

In other words, the approximation order of $S_k^\rho(\Delta)$ is 0 when $k < d(\rho + 1)$.

Second, consider the case $k \geq d(\rho + 1)$. In this case we have proved that the approximation order of $S_k^\rho(\Delta)$ is at least n. Thus, it remains to prove that n is optimal. Choose f to be a C^∞ function with compact support such that

$$(\nabla_2 \cdots \nabla_d)^{k+1} f(x) = x_1^n/n! \quad \text{for } x = (x_1, \ldots, x_d) \in J := (0,1)^d.$$

Let $f_1 := (\nabla_2 \cdots \nabla_d)^{k+1} f$ and $g_1 := (\nabla_2 \cdots \nabla_d)^{k+1} g$. Since the measure of J is 1, we have

$$\text{const}_k \|f - g\|_p \geq \|f_1 - g_1\|_p(J) \geq \|f_1 - g_1\|_1(J). \tag{5.5}$$

Hence it suffices to deal with the case $p = 1$. For $i \in \mathbb{Z}$, let

$$K_i := \{(x_1, \ldots, x_d) \in (0,1)^d : ht_i^{(1)} < x_1 < ht_{i+1}^{(1)}\}.$$

Applying (5.3) to the mesh $h\Delta$, we see that $D_1^n g_1(x) = 0$ for $x \in K_i$, any $i \in \mathbb{Z}$. Since Δ is quasi-uniform, we can invoke Markov's inequality and obtain

$$\|f_1 - g_1\|_1(K_i) \geq \text{const}_k h^n \|D_1^n(f_1 - g_1)\|_1(K_i) = \text{const}_k h^n \|D_1^n f_1\|_1(K_i).$$

But $D_1^n f_1 = 1$ on K_i provided $K_i \subseteq J$, hence $\|D_1^n f_1\|_1(K_i)$ equals the measure of K_i. Note that

$$\|f_1 - g_1\|_1(J) \geq \sum_i \|f_1 \dot- g_1\|_1(K_i)$$

where i runs over those integers for which the interval $(ht_i^{(1)}, ht_{i+1}^{(1)})$ is contained in $(0,1)$. Combining the above estimates, we see that for sufficiently small $h > 0$,

$$\|f_1 - g_1\|_1(J) \geq \text{const}_k h^n. \tag{5.6}$$

Finally, we conclude from (5.5) and (5.6) that

$$\text{dist}_p(f, \sigma_h(\mathcal{S})) \geq \text{const}_k h^n.$$

This shows that $n = k + 1 - (d-1)(\rho + 1)$ is the optimal approximation order of $S_k^\rho(\Delta)$. $\blacksquare$

References

1. Baszenski, G., and F.-J. Delvos, Boolean algebra and multivariate interpolation, in *Approximation and Function Spaces*, Banach Center Publications, Vol. 22, PWN-Polish Scientific Publishers, Warsaw, 1989, 25–44.

2. de Boor, C., *A Practical Guide to Splines*, Springer-Verlag, New York, 1978.

3. de Boor, C., Uniform approximation by splines, J. Approx. Th. **1** (1968), 219–235.

4. de Boor, C., and R. DeVore, Approximation by smooth multivariate splines, Trans. Amer. Math. Soc. **276** (1983), 775–788.

5. Delvos, F.-J., d-variate Boolean interpolation, J. Approx. Th. **34** (1982), 99–114.

6. Delvos, F.-J., and W. Schempp, *Boolean Methods in Interpolation and Approximation*, Longman Scientific & Technical, Harlow, Essex, UK, 1989.

7. Dahmen, W., R. DeVore, and K. Scherer, Multi-dimensional spline approximation, SIAM J. Numer. Anal. **17** (1980), 380–402.

8. DeVore, R., Degree of approximation, in *Approximation Theory II*, G. G. Lorentz, C. K. Chui and L. L. Schumaker (eds.), Academic Press, New York, 1976, 117–161.

9. Gordon, W. J., Distributive lattices and approximation of multivariate functions, in *Approximation with Special Emphasis on Spline Functions*, I. J. Schoenberg (ed.), Academic Press, New York, 1969, 223–277.

10. Gordon, W. J., and E. W. Cheney, Bivariate and multivariate interpolation with noncommutative projectors, in *Linear Spaces and Approximations*, P. L. Butzer and S. Sz. Nagy (eds.), ISNM Vol. 40, Birkhäuser, Basel, 1977, 381–387.

11. Hungerfold, T. W., *Algebra*, Springer-Verlag, New York, 1980.

12. Jia, R. Q., and J. J. Lei, Approximation by multiinteger translates of functions having global support, J. Approx. Th. , to appear.

13. Lyche, T., and L. L. Schumaker, Local spline approximation methods, J. Approx. Th. **15** (1975), 294–325.

14. Popov, V. A., and B. K. Sendov, The classes that are characterized by the best approximation by spline functions, Math. Notes **8** (1970), 550–557.

15. Schumaker, L. L., *Spline Functions: Basic Theory*, John Wiley & Sons, New York, 1981.

16. Stanley, R. P., *Enumerative Combinatorics, Vol. 1*, Wadsworth & Brooks, Monterey, California, 1986.

Numerical Methods of Approximation Theory, Vol. 9
Dietrich Braess and Larry L. Schumaker (eds.), pp. 135–154.
International Series of Numerical Mathematics, Vol. 105
Copyright © 1992 by Birkhäuser Verlag, Basel
ISBN 3-7643-2746-4.

$L^{m,\ell,s}$–splines in $\mathbb{R}^d$

A. Le Méhauté and A. Bouhamidi

Dedicated to the memory of Lothar Collatz

Abstract. In this paper, we introduce $L^{m,\ell,s}$–splines, which are a natural generalization of the thin plate splines of Duchon [5]. Here we investigate some of the properties of these functions and the space $D_\ell^{-m}(\widetilde{H}^s)$ in which they are defined. This enables us to characterize thin plate splines under tension. The $L^{m,\ell,s}$–splines are explicitly given in terms of the Hilbert kernel.

§0. Notations and Preliminaries

Let s be a real number, d the dimension of $\mathbb{R}^d$, $m > 0$ and $\ell \geq 0$ be two integers. For any $\xi = (\xi_1, \cdots, \xi_d) \in \mathbb{R}^d$, $|\xi|$ denotes the Euclidian norm, and $d\xi$ denotes the Lebesgue measure on $\mathbb{R}^d$. We use the usual multi–index notation, namely we write for any $\alpha = (\alpha_1, \ldots, \alpha_d) \in \mathbb{N}^d$,

$$|\alpha| = \alpha_1 + \cdots + \alpha_d, \qquad \xi^\alpha = \xi_1^{\alpha_1} \cdots \xi_d^{\alpha_d},$$

$$D^\alpha = \frac{\partial^{|\alpha|}}{\partial \xi_1^{\alpha_1} \cdots \partial \xi_d^{\alpha_d}}, \qquad \alpha! = \alpha_1! \cdots \alpha_d!.$$

Let us recall some notations for the usual functional spaces. Let $\mathcal{F}(\mathbb{C}, \mathbb{R}^d)$ be the space of complex valued functions on $\mathbb{R}^d$. For k an integer or $k = \infty$,

$C^k(\mathbb{R}^d)$ denotes the subspace of $\mathcal{F}(\mathbb{C}, \mathbb{R}^d)$ of functions of class k on $\mathbb{R}^d$; $L^2(\mathbb{R}^d)$ is the space of (classes of) measurable complex valued functions on $\mathbb{R}^d$; $\mathcal{D}(\mathbb{R}^d)$ denotes the subspace of $C^\infty(\mathbb{R}^d)$ of functions which are compactly supported and infinitely differentiable, and $\mathcal{D}'(\mathbb{R}^d)$ its topologic dual, the space of distributions on $\mathbb{R}^d$; $\mathcal{S}(\mathbb{R}^d)$ is the subspace of $C^\infty(\mathbb{R}^d)$ of rapidly decreasing functions, and $\mathcal{S}'(\mathbb{R}^d)$ denotes its topologic dual, the space of tempered distributions.

If E is a Fréchet space and $\mathcal{H}$ a subspace of E, then $\mathcal{H}$ is a Hilbert subspace of E if and only if $\mathcal{H}$ is a Hilbert space continuously imbedded in E. If $\mathcal{H}$ is a Hilbert subspace of E, the Hilbert kernel of $\mathcal{H}$ relative to E (Schwartz [14]) is the unique mapping H from $\overline{E'}$ to $\mathcal{H}$, such that, for any element $\overline{e'}$ in $\overline{E'}$ and for any h in $\mathcal{H}$, the relation $(h|H\overline{e'})_{\mathcal{H}} = \langle h, \overline{e'} \rangle$ holds, where $(\,.\,|\,.\,)_{\mathcal{H}}$ is the scalar product on $\mathcal{H}$, and $\langle .,. \rangle$ is for duality.

§1. Space of the $\mathbf{L}^{m,\ell,s}$–splines

Let $\widetilde{L}^2_s(\mathbb{R}^d)$ be the space of (classes of) measurable functions u in $\mathbb{R}^d$ such that

$$\int_{\mathbb{R}^d} |2\pi\xi|^{2s} |u(\xi)|^2 \, d\xi < +\infty, \tag{1.1}$$

and let us define the scalar product

$$(u, v)_s = \int_{\mathbb{R}^d} |2\pi\xi|^{2s} \, u(\xi) \, \overline{v(\xi)} \, d\xi \tag{1.2}$$

and the (pre-Hilbert) associated norm

$$\|u\|_s = (u, u)_s^{1/2}. \tag{1.3}$$

For any real number $s < \frac{d}{2}$, the space $\widetilde{L}^2_s(\mathbb{R}^d)$ is a Hilbert subspace of $\mathcal{S}'(\mathbb{R}^d)$, and for any real number $s > -\frac{d}{2}$, the space $\mathcal{S}(\mathbb{R}^d)$ is continuously imbedded in the space $\widetilde{L}^2_s(\mathbb{R}^d)$.

Duchon [5] defined $\widetilde{H}^s(\mathbb{R}^d)$ as the space of tempered distributions whose Fourier transforms are in $\widetilde{L}^2_s(\mathbb{R}^d)$, namely

$$\widetilde{H}^s(\mathbb{R}^d) = \left\{ u \in \mathcal{S}'(\mathbb{R}^d) \quad ; \quad \mathcal{F}u \in \widetilde{L}^2_s(\mathbb{R}^d) \right\}. \tag{1.4}$$

On this space, we consider the following scalar product

$$(u, v)_{0,s} = (\mathcal{F}u, \mathcal{F}v)_s = \int_{\mathbb{R}^d} |2\pi\xi|^{2s} \mathcal{F}u(\xi) \; \overline{\mathcal{F}v(\xi)} \, d\xi \tag{1.5}$$

and the associated (pre-Hilbert) norm

$$|u|_{0,s} = (u,u)_{0,s}^{1/2} = \|\mathcal{F}u\|_s = \Big[\int_{\mathbb{R}^d} |2\pi\xi|^{2s}|\mathcal{F}u(\xi)|^2\, d\xi \Big]^{1/2}. \tag{1.6}$$

If $s < \frac{d}{2}$, the space $\widetilde{H}^s(\mathbb{R}^d)$ is a Hilbert subspace of $\mathcal{S}'(\mathbb{R}^d)$ (Duchon [5]). Hereafter, we assume that

$$s < \frac{d}{2}. \tag{1.7}$$

For any integers $m \in \mathbb{N}^*$ and $\ell \in \mathbb{N}$, we introduce the Beppo–Levi type space $D_\ell^{-m}(\widetilde{H}^s)$ of distributions, all of whose derivatives of order k, with $m \le k \le m+\ell$, are in $\widetilde{H}^s(\mathbb{R}^d)$, namely

$$D_\ell^{-m}(\widetilde{H}^s) = \Big\{ v \in \mathcal{D}'(\mathbb{R}^d) : \ \forall \alpha \in \mathbb{N}^d, m \le |\alpha| \le m+\ell, \ \ D^\alpha v \in \widetilde{H}^s(\mathbb{R}^d) \Big\}. \tag{1.8}$$

In the space $D_\ell^{-m}(\widetilde{H}^s)$ a semi–inner product is defined by

$$(u,v)_{m,\ell,s} = \sum_{m \le |\alpha| \le m+\ell} C_\alpha \int_{\mathbb{R}^d} |2\pi\xi|^{2s}\mathcal{F}D^\alpha u(\xi)\ \overline{\mathcal{F}D^\alpha v(\xi)}\, d\xi, \tag{1.9}$$

where the real positive constants C_α are specified by

$$\begin{cases} \forall k \in \mathbb{N} \ ; \ m \le k \le m+\ell \ ; \ \exists\, C_k, C'_k > 0 \quad \text{such that:} \\[2mm] C'_k\,|\xi|^{2k} \ge \displaystyle\sum_{|\alpha|=k} C_\alpha \xi^{2\alpha} \ge C_k\,|\xi|^{2k} \quad \forall \xi \in \mathbb{R}^d. \end{cases} \tag{1.10}$$

The associated semi–norm is

$$|u|_{m,\ell,s} = (u,u)_{m,\ell,s}^{1/2}. \tag{1.11}$$

Proposition 1. *With the semi–scalar product $(\,.\,,\,.\,)_{m,\ell,s}$ and the associated semi–norm $|\cdot|_{m,\ell,s}$, the space $D_\ell^{-m}(\widetilde{H}^s)$ is a semi–Hilbert space, algebraically enclosed in $\mathcal{S}'(\mathbb{R}^d)$ and continuously imbedded in $X^{m,s}$.*

Proof: First, we show that $D_\ell^{-m}(\widetilde{H}^s)$ is a complete space. It is obvious that this space is enclosed in $X^{m,s} = D_0^{-m}(\widetilde{H}^s)$, which is itself a subspace of $\mathcal{S}'(\mathbb{R}^d)$ (Duchon [5]). From (1.10) and

$$\mathcal{F}(D^\alpha T) = (2i\pi\xi)^\alpha \mathcal{F}T, \tag{1.12}$$

which holds for any tempered distribution T and for any multi–index α of $\mathbb{N}^d$, it's clear that for any element u of $D_\ell^{-m}(\widetilde{H}^s)$, we have

$$|u|_{m,0,s} \leq \frac{1}{\sqrt{C_m}}|u|_{m,\ell,s}. \qquad (1.13)$$

which establishes the result. ∎

This immediately implies

Corollary 1. *Let r be a positive integer; if $-m + r + \frac{d}{2} < s < \frac{d}{2}$, the space $D_\ell^{-m}(\widetilde{H}^s)$ is continuously imbedded in $C^r(\mathbb{R}^d)$.*

Proof: If $-m + r + \frac{d}{2} < s < \frac{d}{2}$, the space $X^{m,s}$ is continuously imbedded in $C^r(\mathbb{R}^d)$ (Duchon [5]). ∎

Proposition 2. *If $-m - \frac{d}{2} < s < \frac{d}{2}$, then $\mathcal{S}(\mathbb{R}^d)$ and $\mathcal{D}(\mathbb{R}^d)$ are algebraically enclosed in $D_\ell^{-m}(\widetilde{H}^s)$, and $D_\ell^{-m}(\widetilde{H}^s)$ separates points in $\mathbb{R}^d$.*

Proof: Let u be an element of $\mathcal{S}(\mathbb{R}^d)$, and let $\alpha \in \mathbb{N}^d$ with $k = |\alpha|$ be such that $m \leq k \leq m + \ell$. Then from (1.12) we obtain the following inequality

$$|2\pi\xi|^{2s}|\mathcal{F}(D^\alpha u(\xi))|^2 \leq |2\pi\xi|^{2s+2k}|\mathcal{F}(u(\xi))|^2. \qquad (1.14)$$

But $\mathcal{F}u$ is an element of $\mathcal{S}(\mathbb{R}^d)$, which is a subspace of $\widetilde{L}^2_{s+k}(\mathbb{R}^d)$ for $s + k > -\frac{d}{2}$. Thus for $k \geq m > -\frac{d}{2} - s$, we see that u is also an element of $D_\ell^{-m}(\widetilde{H}^s)$. Finally, $\mathcal{D}(\mathbb{R}^d)$ is a subspace of $\mathcal{S}(\mathbb{R}^d)$ and with Urysohn's lemma (Khoan [9]), we see that $D_\ell^{-m}(\widetilde{H}^s)$ separates points in $\mathbb{R}^d$. ∎

Henceforth, we assume that m is a positive integer and s a real number satisfying

$$-m - \frac{d}{2} < s < \frac{d}{2}. \qquad (1.15)$$

It is well known that for any distribution T on $\mathbb{R}^d$ and any positive integer k, the Fourier transform of the k-th iterate of the usual Laplace operator in $\mathbb{R}^d$ is given by

$$\mathcal{F}(\Delta^k T) = (-1)^k \, |2\pi\xi|^{2k} \, \mathcal{F}T. \qquad (1.16)$$

For a real number r, Rabut [12] introduced the following definition, which is a natural extension of formula (1.16) to Δ^r:

Definition 1. *Let r be a real number and T a tempered distribution on $\mathbb{R}^d$. Then $\Delta^r T$ is the tempered distribution whose Fourier transform is given by*

$$\mathcal{F}(\Delta^r T) = e^{i\pi r}\,|2\,\pi\,\xi\,|^{2r}\,\mathcal{F}T\,. \tag{1.17}$$

Let us now introduce a linear differential operator $\Delta_{m,\ell,s}$, defined from $\mathcal{S}'(\mathbb{R}^d)$ onto itself.

Definition 2. *For any tempered distribution T on $\mathbb{R}^d$, let*

$$\Delta_{m,\ell,s}T = e^{-i\pi s}\,\Delta^s\big[\,\Delta_{m,\ell}\,T\,\big]\ with\ \Delta_{m,\ell}T = \sum_{m\le|\alpha|\le m+\ell}(-1)^{|\alpha|}\,C_\alpha\,D^{2\alpha}\,T,\tag{1.18}$$

where the constants C_α are specified by (1.10).

From Definition 1 and $(1.10)-(1.12)$, we obtain that for any tempered distribution T on $\mathbb{R}^d$, the Fourier transform of $\Delta_{m,\ell,s}\,T$ satisfies the inequality

$$\mathcal{F}(\Delta_{m,\ell,s}\,T) \ge \Big\{\,\sum_{k=m}^{m+\ell} C_k\,|2\pi\xi|^{2(k+s)}\,\Big\}\,\mathcal{F}T. \tag{1.19}$$

Proposition 3. *For any element u of $D_\ell^{-m}(\widetilde{H}^s)$ and for any function φ of $\mathcal{S}(\mathbb{R}^d)$, we have the following equalities, in the distributional sense:*

$$(\,u,\varphi\,)_{m,\ell,s} \;=\; \langle\Delta_{m,\ell,s}\,u,\overline{\varphi}\rangle \;=\; \langle u,\overline{\Delta_{m,\ell,s}\,\varphi}\rangle\,.$$

Proof: We have

$$
\begin{aligned}
(u,\varphi)_{m,\ell,s} \;&=\; \sum_{m\le|\alpha|\le m+\ell} C_\alpha \int_{\mathbb{R}^d} |2\pi\xi|^{2s}\,(2i\pi\xi)^\alpha\,\mathcal{F}u(\xi)\;\overline{(2i\pi\xi)^\alpha\,\mathcal{F}\varphi(\xi)}\,d\xi\\[2mm]
&=\; \sum_{m\le|\alpha|\le m+\ell} (-1)^{|\alpha|}\,C_\alpha \int_{\mathbb{R}^d} |2\pi\xi|^{2s}\,(2i\pi\xi)^{2\alpha}\,\mathcal{F}u(\xi)\overline{\mathcal{F}\varphi(\xi)}\,d\xi,
\end{aligned}\tag{1.20}
$$

or equivalently,

$$
\begin{aligned}
(u,\varphi)_{m,\ell,s} \;&=\; \int_{\mathbb{R}^d} |2\pi\xi|^{2s}\,\mathcal{F}\big[\Delta_{m,\ell}\,u\,\big](\xi)\;\overline{\mathcal{F}\varphi(\xi)}\,d\xi\\[2mm]
&=\; \int_{\mathbb{R}^d} |2\pi\xi|^{2s}\,\mathcal{F}u(\xi)\;\overline{\mathcal{F}\big[\Delta_{m,\ell}\,\varphi\,\big](\xi)}\,d\xi.
\end{aligned}\tag{1.21}
$$

Thus,

$$(u, \varphi)_{m,\ell,s} = \langle \mathcal{F}(\Delta_{m,\ell,s}\, u), \overline{\mathcal{F}\varphi} \rangle = \langle \mathcal{F}u, \overline{\mathcal{F}(\Delta_{m,\ell,s}\, \varphi)} \rangle.$$

Since for any distribution T of $\mathcal{S}'(\mathbb{R}^d)$ and for any function φ of $\mathcal{S}(\mathbb{R}^d)$ we have $\langle T, \overline{\mathcal{F}\varphi} \rangle = \langle \overline{\mathcal{F}T}, \overline{\varphi} \rangle$, our result follows. $\blacksquare$

Proposition 4. *Let u be an element of $D_\ell^{-m}(\widetilde{H}^s)$ such that $\Delta_{m,\ell,s}u = 0$ in the distributional sense. Then u belongs to $\mathbb{P}_{m-1}(\mathbb{R}^d)$.*

Proof: Since u is a tempered distribution, its Fourier transform $\mathcal{F}u$ satisfies

$$|2\pi\xi|^{2(m+s)} \left\{ \sum_{k=0}^{\ell} C_{m+k}\, |2\pi\xi|^{2k} \right\} \mathcal{F}u = 0. \tag{1.23}$$

But $|2\pi\xi|^{2(m+s)}$ vanishes only at the origin. It follows that $\mathcal{F}u$ is a distribution supported at the origin. From Schwartz's theorem [13], page 100, and because the only polynomial in $\widetilde{H}^s(\mathbb{R}^d)$ is the polynomial 0, u belongs to $\mathbb{P}_{m-1}(\mathbb{R}^d)$. $\blacksquare$

Corollary 2. *The null space $\mathcal{N}$ of $(D_\ell^{-m}(\widetilde{H}^s), |\cdot|_{m,\ell,s})$ is $\mathbb{P}_{m-1}(\mathbb{R}^d)$.*

Proof: It is obvious that $\mathbb{P}_{m-1}(\mathbb{R}^d)$ is a subset of $\mathcal{N}$. I that that that thatf u is an element of $\mathcal{N}$, for any element φ of $\mathcal{S}(\mathbb{R}^d)$, we have $(u, \varphi)_{m,\ell,s} = 0$; thus from Proposition 3, $\Delta_{m,\ell,s}\, u = 0$. By Proposition 4, we obtain that u is in $\mathbb{P}_{m-1}(\mathbb{R}^d)$. $\blacksquare$

It is usual that for a positive integer s, and two distributions S and T such that the convolution $S \star T$ exists, we have

$$\Delta^s(S \star T) = \Delta^s S \star T = S \star \Delta^s T. \tag{1.24}$$

But, when s is a real positive number, this is not always valid. We still have the following:

Proposition 5. *For a real number s, a rapidly decreasing distribution S, and a tempered distribution T on $\mathbb{R}^d$, we have:*

$$\Delta^s(S \star T) = S \star \Delta^s T.$$

Proof: For a rapidly decreasing distribution S and a tempered distribution T on $\mathbb{R}^d$, we have

$$\mathcal{F}(S \star T) = (\mathcal{F}S)(\mathcal{F}T). \tag{1.25}$$

The result follows from Definition 1 and the fact that $\mathcal{F}$ and $\overline{\mathcal{F}}$ are two reciprocal isomorphisms. $\blacksquare$

We obtain immediately the following corollary:

Corollary 3. *Let s be a real number, S a compactly supported distribution, and T a tempered distribution on $\mathbb{R}^d$. Then*

(i) $\Delta^s(S \star T) = S \star \Delta^s T$.

(ii) $\Delta_{m,\ell,s}(S \star T) = S \star (\Delta_{m,\ell,s} T)$.

Definition 3. *A fundamental solution for the differential operator $\Delta_{m,\ell,s}$ on $\mathbb{R}^d$ is a distribution T on $\mathbb{R}^d$ such that*

$$\Delta_{m,\ell,s} T = \delta. \tag{1.26}$$

Later on, we will assume the existence of a fundamental solution of $\Delta_{m,\ell,s}$ which is a tempered distribution $K_{m,\ell,s}$ associated with a function defined on $\mathbb{R}^d$ and still denoted by $K_{m,\ell,s}$. In the particular case when s is a positive integer, it is not necessary to assume that $K_{m,\ell,s}$ is a tempered distribution.

§2. Quotient Spaces and Reproducing Kernels

The space $\mathcal{D}'(\mathbb{R}^d)$ of the distributions on $\mathbb{R}^d$ is a Fréchet space; the differentiation operator $D^\alpha, |\alpha| = m$, is linear and continuous, so $\mathbb{P}_{m-1}(\mathbb{R}^d)$ is a closed subspace of $\mathcal{D}'(\mathbb{R}^d)$.

The quotient space of $\mathcal{D}'(\mathbb{R}^d)$ by $\mathbb{P}_{m-1}(\mathbb{R}^d)$, denoted by

$$E_m = \mathcal{D}'(\mathbb{R}^d)/\mathbb{P}_{m-1}(\mathbb{R}^d), \tag{2.1}$$

is also a Fréchet space. Its antidual space is

$$\overline{E'}_m = \mathcal{D}(\mathbb{R}^d)/\mathbb{P}_{m-1}(\mathbb{R}^d), \tag{2.2}$$

which is isomorphic to the space of compactly supported and $\mathcal{C}^\infty$ functions orthogonal to $\mathbb{P}_{m-1}(\mathbb{R}^d)$, *i.e.*, to

$$\overline{E'}_m \simeq \mathcal{D}(\mathbb{R}^d) \bigcap \mathbb{P}_{m-1}^{\perp}(\mathbb{R}^d), \tag{2.3}$$

where $\mathbb{P}_{m-1}^{\perp}(\mathbb{R}^d)$ is the subspace orthogonal to $\mathbb{P}_{m-1}(\mathbb{R}^d)$ in $\mathcal{D}'(\mathbb{R}^d)$. We can write

$$\overline{E'}_m = \left\{ \varphi \in \mathcal{D}(\mathbb{R}^d) : \langle \varphi, \nu \rangle = 0 \quad , \quad \forall \nu \in \mathbb{P}_{m-1}(\mathbb{R}^d) \right\}. \tag{2.4}$$

Let r be a positive integer, and let

$$G_{m,r} = \mathcal{C}^r(\mathbb{R}^d)/\mathbb{P}_{m-1}(\mathbb{R}^d) \tag{2.5}$$

be the quotient space of $\mathcal{C}^r(\mathbb{R}^d)$ by $\mathbb{P}_{m-1}(\mathbb{R}^d)$. Its antidual space is

$$\overline{G'}_{m,r} = \mathcal{C}'^r(\mathbb{R}^d)/\mathbb{P}_{m-1}(\mathbb{R}^d). \tag{2.6}$$

$\overline{G'}_{m,r}$ is the quotient of the space of compactly supported distributions of order $\leq r$, by the subspace of polynomials $\mathbb{P}_{m-1}(\mathbb{R}^d)$ and is isomorphic to the space of compactly supported distributions of order $\leq r$ orthogonal to $\mathbb{P}_{m-1}(\mathbb{R}^d)$:

$$\overline{G'}_{m,r} \simeq \mathcal{C}'^r(\mathbb{R}^d) \bigcap \mathbb{P}_{m-1}^{\perp}(\mathbb{R}^d), \tag{2.7}$$

or equivalently:

$$\overline{G'}_{m,r} = \left\{ T \in \mathcal{C}'^r(\mathbb{R}^d) : \langle T, \nu \rangle = 0 \quad ; \quad \forall \nu \in \mathbb{P}_{m-1}(\mathbb{R}^d) \right\}. \tag{2.8}$$

Finally, let us consider

$$\overset{\bullet}{D}_{\ell}{}^{-m}(\widetilde{H}^s) = D_{\ell}^{-m}(\widetilde{H}^s)/\mathbb{P}_{m-1}(\mathbb{R}^d). \tag{2.9}$$

As $D_{\ell}^{-m}(\widetilde{H}^s)$ has a semi–norm $|\,.\,|_{m,\ell,s}$, it is natural to define a norm in $\overset{\bullet}{D}_{\ell}{}^{-m}(\widetilde{H}^s)$ by

$$\|\,\overset{\bullet}{u}\,\|_{m,\ell,s} = \inf_{v \in \overset{\bullet}{u}} |v|_{m,\ell,s} \quad \forall \, \overset{\bullet}{u} \in \overset{\bullet}{D}_{\ell}{}^{-m}(\widetilde{H}^s). \tag{2.10}$$

In fact, the inf is superfluous, since

$$\|\,\overset{\bullet}{u}\,\|_{m,\ell,s} = |u|_{m,\ell,s} \tag{2.11}$$

is well defined and does not depend on the representative of the class $\overset{\bullet}{u}$.

We can define a scalar product in $\overset{\bullet}{D}_{\ell}{}^{-m}(\widetilde{H}^s)$ with

$$[\,\overset{\bullet}{u}\,,\,\overset{\bullet}{v}\,]_{m,\ell,s} = (u,v)_{m,\ell,s}, \qquad \forall \, \overset{\bullet}{u}, \overset{\bullet}{v} \in \overset{\bullet}{D}_{\ell}{}^{-m}(\widetilde{H}^s) \tag{2.12}$$

Proposition 6. *With the norm* $\|\,.\,\|_{m,\ell,s}$, *the space* $\overset{\bullet}{D}_{\ell}{}^{-m}(\widetilde{H}^s)$ *is a Hilbert subspace of* E_m.

Proof: Because $\mathbb{P}_{m-1}(\mathbb{R}^d)$ is the null space of the semi-Hilbert space $D_\ell^{-m}(\widetilde{H}^s)$, it is obvious that $\overset{\bullet}{D_\ell}{}^{-m}(\widetilde{H}^s)$ is a Hilbert space. Moreover, from its topological properties, and with the closed graph theorem, we obtain that $\overset{\bullet}{D_\ell}{}^{-m}(\widetilde{H}^s)$ is a Hilbert subspace of E_m. $\blacksquare$

Thus, from Schwartz [14], page 142, $\overset{\bullet}{D_\ell}{}^{-m}(\widetilde{H}^s)$ has a reproducing kernel relative to E_m which is the unique mapping from $\overline{E'}_m$ in $\overset{\bullet}{D_\ell}{}^{-m}(\widetilde{H}^s)$ such that $\varphi \in \overline{E'}_m$ and

$$[\overset{\bullet}{u}, H\varphi]_{m,\ell,m} = \langle u, \overline{\varphi}\rangle, \qquad \forall\, \overset{\bullet}{u}\in\overset{\bullet}{D_\ell}{}^{-m}(\widetilde{H}^s), \qquad \forall \varphi \in \overline{E'}_m. \qquad (2.13)$$

Proposition 7. *Assuming that* $-m + \frac{d}{2} < s < \frac{d}{2}$, *the Hilbert kernel of* $\overset{\bullet}{D_\ell}{}^{-m}(\widetilde{H}^s)$ *relative to* E_m *is the mapping* H *defined from* $\overline{E'}_m$ *to* $\overset{\bullet}{D_\ell}{}^{-m}(\widetilde{H}^s)$ *by*

$$H \,:\, \varphi \longmapsto \varphi \star K_{m,\ell,s} + \mathbb{P}_{m-1}(\mathbb{R}^d)\,.$$

Proof: First, H is well defined from $\overline{E'}_m$ to $\overset{\bullet}{D_\ell}{}^{-m}(\widetilde{H}^s)$ because, for any real s such that $-m + \frac{d}{2} < s < \frac{d}{2}$ and for any $\varphi \in \overline{E'}_m$, the convolution product $\varphi \star K_{m,\ell,s}$ is in $D_\ell^{-m}(\widetilde{H}^s)$. Then, H is characterized by (2.13) as a reproducing kernel. $\blacksquare$

Hereafter, we assume that

$$-m + \frac{d}{2} < s < \frac{d}{2}. \qquad (2.14).$$

Proposition 8. *If the (distribution) function* $K_{m,\ell,s}$ *is in* $C^r(\mathbb{R}^d)$, *the space* $\overset{\bullet}{D_\ell}{}^{-m}(\widetilde{H}^s)$ *is a Hilbert subspace of* $G_{m,r}$ *whose reproducing kernel is the mapping* $\widehat{H} \,:\, \overline{G'}_{m,r} \mapsto \overset{\bullet}{D_\ell}{}^{-m}(\widetilde{H}^s)$ *such that*

$$\widehat{H} \,:\, T \longmapsto T \star K_{m,\ell,s} + \mathbb{P}_{m-1}(\mathbb{R}^d), \qquad \forall T \in \overline{G'}_{m,r}.$$

Proof: The proof is a consequence of a corollary of Schwartz [14], pp 179–180. In fact we prove first that the kernel H of $\overset{\bullet}{D_\ell}{}^{-m}(\widetilde{H}^s)$ relative to E_m can be extended to $\widehat{H}$, then that $\widehat{H}$ is weakly continuous from $\overline{G'}_{m,r}$ to $G_{m,r}$. $\blacksquare$

Corollary 4. *With the same assumptions,* $D_\ell^{-m}(\widetilde{H}^s)$ *is continously imbedded in* $C^r(\mathbb{R}^d)$.

From now on, we assume the hypothesis of Proposition 8. Let us remark that one or the other of the two Corollaries 1 and 4 provides a sufficient condition for $D_\ell^{-m}(\widetilde{H}^s)$ to be continously imbedded in $C^r(\mathbb{R}^d)$.

§3. $L^{m,\ell,s}$–splines

3.1. Definition, Existence, Unicity

Let $d(m)$ be the dimension of the space $\mathbb{P}_{m-1}(\mathbb{R}^d)$:

$$d(m) = d_m = \dim(\mathbb{P}_{m-1}(\mathbb{R}^d)) = \binom{m+d-1}{d}. \tag{3.1}$$

Definition 4. *An interpolation scheme is a set of $p \geq d(m)$ compactly supported distributions of order $\leq r$, i.e., a set of p linear and continuous functionals defined on $C^r(\mathbb{R}^d)$.*

Let $L = (\ell_1, \cdots, \ell_p)$ be such a scheme and let r be a positive integer such that $-m + r + \frac{d}{2} < s < \frac{d}{2}$, which ensures us that the ℓ_i are also continuous on $D_\ell^{-m}(\widetilde{H}^s)$ (Corollary 1). For any $z = (z_1, \cdots, z_p) \in \mathbb{C}^p$, let us denote by $G(z)$ the subset of $D_\ell^{-m}(\widetilde{H}^s)$ such that

$$G(z) = \left\{ u \in D_\ell^{-m}(\widetilde{H}^s);\ \langle \ell_i, u \rangle = z_i\ ;\ i = 1, \cdots, p \right\}. \tag{3.2}$$

We assume the following hypotheses:

- H1: The interpolation scheme L is topologically independent (Bourbaki [4]), *i.e.*, there exists a family $(\varphi_j)_{1 \leq j \leq p}$ of functions in $\mathcal{D}(\mathbb{R}^d)$ such that

$$\langle \ell_i, \varphi_j \rangle = \delta_{ij}, \qquad i, j = 1, \cdots, p.$$

- H2: the interpolation scheme L contains a subfamily $\widehat{L}$ which is $\mathbb{P}_{m-1}-$ unisolvent (Le Méhauté [11]). For convenience, we assume that $\widehat{L}$ is the set of the first $d(m)$-th distributions in L. This implies the existence of a dual basis $(\nu_j)_{1 \leq j \leq d(m)}$ in $\mathbb{P}_{m-1}(\mathbb{R}^d)$ associated with $\widehat{L}$ in such a way that

$$\langle \ell_i, \nu_j \rangle = \delta_{ij}, \qquad i, j = 1, \cdots, d(m).$$

Definition 5. *Let $z \in \mathbb{C}^p$. A $L^{m,\ell,s}$–spline (or interpolating spline) is an element σ of $D_\ell^{-m}(\widetilde{H}^s)$ which is a solution of*

$$(\mathcal{P}) \quad \begin{cases} \sigma \in G(z) \\ \qquad \text{and} \\ |\sigma|_{m,\ell,s} = \inf_{u \in G(z)} |u|_{m,\ell,s}. \end{cases} \tag{3.3}$$

Let us remark that $H2$ implies that $G(z) \cap \mathbb{P}_{m-1}(\mathbb{R}^d)$ is either empty or is the singleton $\{\sigma*\}$. In the latter case, it is obvious that $\sigma*$ is the unique solution of $(\mathcal{P})$, so we will assume now that $G(z) \cap \mathbb{P}_{m-1}(\mathbb{R}^d)$ is empty; this and the fact that $G(z)$ is a closed subset of $D_\ell^{-m}(\widetilde{H}^s)$ is equivalent to

$$\inf_{u \in G(z)} |u|_{m,\ell,s} > 0, \tag{3.4}$$

which we'll assume now.

The set $G(z)$ is a closed convex nonempty subset of $\mathcal{D}'(\mathbb{R}^d)$. Its asymptotic cone (Attéia [2], Laurent [10]) is the set

$$G_\infty(z) = \bigcap_{\lambda > 0} \lambda[G(z) - u], \tag{3.5}$$

where $u \in G(z)$. $G_\infty(z)$ does not depend on the choice of u, and

$$G_\infty(z) = G(0). \tag{3.6}$$

Let Π_m be the canonical surjection from $\mathcal{D}'(\mathbb{R}^d)$ onto E_m and let

$$\widetilde{G}(z) = \Pi_m[G(z)]. \tag{3.7}$$

We obtain the following result :

Proposition 9. The set $\widetilde{G}(z)$ is a closed subset of E_m.

Proof: The proof is an application of Theorem 1 in (Attéia [2]), using the following:

    h1) $G(z) \cap \mathbb{P}_{m-1}(\mathbb{R}^d) = \emptyset$.

    h2) $G_\infty(z) \cap \mathbb{P}_{m-1}(\mathbb{R}^d) = \{0\}$.

    h3) $\dim \mathbb{P}_{m-1}(\mathbb{R}^d) = d(m) < +\infty$. ∎

Proposition 10. $\widetilde{G}(z)$ is a closed convex nonempy subset of $\overset{\bullet}{D_\ell}{}^{-m}(\widetilde{H}^s)$.

Proof: Since $G(z)$ is closed and nonempty, so is $\widetilde{G}(z)$. Also $\widetilde{G}(z)$ is closed in $\overset{\bullet}{D_\ell}{}^{-m}(\widetilde{H}^s)$ for the Hilbert topology because it is closed for the topology induced by E_m and $\overset{\bullet}{D_\ell}{}^{-m}(\widetilde{H}^s)$, while the topology of its Hilbert structure is continuously imbedded in E_m. ∎

Proposition 11. For any $z \in C^p$, there exists one and only one $L^{m,\ell,s}$–spline σ interpolating z.

Proof: As $G(z) \bigcap \mathbb{P}_{m-1}(\mathbb{R}^d) = \emptyset$, we know that $\overset{\bullet}{0} \notin \widetilde{G}(z)$. Then we use the projection theorem on the closed convex nonempty subset $\widetilde{G}(z)$ of the Hilbert space $\overset{\bullet}{D}_\ell{}^{-m}(\widetilde{H}^s)$ to define the unique element $\overset{\bullet}{\sigma}$, projection of $\overset{\bullet}{0}$ on $\widetilde{G}(z)$, and the result follows because $\overset{\bullet}{\sigma} \bigcap G(z)$ is the singleton $\{\sigma\}$. ∎

3.2. The $\mathbf{L}^{m,\ell,s}$–spline

Proposition 12. $\overline{\Delta_{m,\ell,s}\,\sigma}$ *is orthogonal to* $\mathbb{P}_{m-1}(\mathbb{R}^d)$ *and is a linear combination of the distributions of the family* $L = (\ell_1, \cdots, \ell_p)$.

Proof: As $\overset{\bullet}{\sigma}$ is the best approximation of $\overset{\bullet}{0}$ from $\overset{\bullet}{D}_\ell{}^{-m}(\widetilde{H}^s)$ in the convex set $\widetilde{G}(z)$, it is obvious that $(\sigma, v)_{m,\ell,s} = 0$ for any $v \in G(0)$. The interpolation scheme L being topologically independent, there exists a family $\phi = (\varphi_1, \cdots, \varphi_p)$ in $\mathcal{D}(\mathbb{R}^d)$ such that for $i,j = 1, \cdots, p$, $\langle \ell_i, \varphi_j \rangle = \delta_{ij}$. For any $v \in D_\ell^{-m}(\widetilde{H}^s)$, $v - \sum_{i=1}^{p} \langle \ell_i, v \rangle \varphi_i$ is an element of $G(0)$. Thus

$$(v, \sigma)_{m,\ell,s} = \langle \sum_{i=1}^{p} \lambda_i \, \ell_i, v \rangle, \tag{3.8}$$

where the coefficients λ_i are defined by

$$\lambda_i = (\varphi_i, \sigma)_{m,\ell,s}. \tag{3.9}$$

Consequently, for any function $\varphi \in \mathcal{D}(\mathbb{R}^d)$, we obtain from Prop. 4 that

$$(\varphi, \sigma)_{m,\ell,s} = \langle \overline{\Delta_{m,\ell,s}\,\sigma}, \varphi \rangle, \tag{3.10}$$

and the result follows from (3.8)-(3.10).

Moreover, we have the orthogonality property

$$\langle \sum_{i=1}^{p} \lambda_i \, \ell_i, \nu \rangle = (\nu, \sigma)_{m,\ell,s} = 0, \quad \forall \nu \in \mathbb{P}_{m-1}(\mathbb{R}^d). \quad ∎$$

Let us denote by S the linear combination

$$S = \sum_{i=1}^{p} \lambda_i \, \ell_i, \tag{3.12}$$

where the λ_i are defined in (3.9). S is a compactly supported distribution of order $\leq r$ and is orthogonal to $\mathbb{P}_{m-1}(\mathbb{R}^d)$.

Proposition 13. *The Hilbert kernel* $\widehat{H}$ *of* $\overset{\bullet}{D}_\ell{}^{-m}(\widetilde{H}^s)$ *relative to* $G_{m,r}$ *maps* S *into* $\overset{\bullet}{\sigma}$, *i.e.,*

$$\overset{\bullet}{\sigma} = \widehat{H}\, S \;=\; S \star K_{m,\ell,s} + \mathbb{P}_{m-1}(\mathbb{R}^d)\,.$$

Proof: From the definition of $\widehat{H}$ as a kernel,

$$[\overset{\bullet}{v}, \widehat{H} S]_{m,\ell,s} \;=\; \langle S, v \rangle, \quad \forall\, \overset{\bullet}{v} \in \overset{\bullet}{D}_\ell{}^{-m}(\widetilde{H}^s). \tag{3.13}$$

But from (3.8),

$$\langle S, v \rangle \;=\; [\overset{\bullet}{v}, \overset{\bullet}{\sigma}]_{m,\ell,s}, \quad \forall\, \overset{\bullet}{v} \in \overset{\bullet}{D}_\ell{}^{-m}(\widetilde{H}^s). \tag{3.14}$$

Thus,

$$[\overset{\bullet}{v}, \widehat{H} S - \overset{\bullet}{\sigma}]_{m,\ell,s} \;=\; 0, \tag{3.15}$$

which means that $\widehat{H} S = \overset{\bullet}{\sigma}$. ∎

Corollary 5. *The* $L^{m,\ell,s}$-*spline* σ *solution of* $(\mathcal{P})$ *is the unique element of* $\overset{\bullet}{\sigma} \bigcap G(z)$ *and is explicitly given by*

$$\sigma = \sum_{i=1}^{p} \lambda_i\, \ell_i \star K_{m,\ell,s} + \sum_{j=1}^{d(m)} \alpha_j \nu_j,$$

where $(\nu_j)_{j=1,\cdots,d(m)}$ *is a basis of* $\mathbb{P}_{m-1}(\mathbb{R}^d)$ *and the coefficients* λ_i, $i = 1,\cdots,p$ *and* α_j, $j = 1,\cdots,d(m)$ *are solutions of the linear system*

$$\left\{ \begin{array}{l} \text{interpolation conditions :} \\[4pt] \quad \left\{ \begin{array}{l} \langle \ell_k, \sigma \rangle - z_k \\ k = 1,\ldots,p \end{array} \right. \\[14pt] \text{orthogonality conditions :} \\[4pt] \quad \left\{ \begin{array}{l} \displaystyle\sum_{i=1}^{p} \lambda_i \langle \ell_i, \nu_j \rangle = 0 \\ j = 1,\ldots,d(m). \end{array} \right. \end{array} \right. \tag{S}$$

Because $G(0) \bigcap \mathbb{P}_{m-1}(\mathbb{R}^d) = \emptyset$, it is obvious that the semi-scalar product $(.\,,.)_{m,\ell,s}$ and the associated semi-norm $|\,.\,|_{m,\ell,s}$ of $D_\ell^{-m}(\widetilde{H}^s)$ provide on $\mathcal{H} = G(0)$ a scalar product and the associated norm defined by

$$(u,v)_{\mathcal{H}} = (u,v)_{m,\ell,s} \quad \text{and} \quad \|u\|_{\mathcal{H}} = |u|_{m,\ell,s}. \tag{3.16}$$

Thus, $\mathcal{H}$ becomes a Hilbert subspace of $G_{m,r}$ whose reproducing kernel is the mapping, still denoted by H and defined from $\overline{G'}_{m,r}$ to $\mathcal{H}$ as

$$H : T \longmapsto T \star K_{m,\ell,s}. \tag{3.17}$$

An important fact is that the kernel of a Hilbert subspace is positive (Schwartz [14]). Thus, for any compactly supported distribution T of order $\leq r$ and orthogonal to $\mathbb{P}_{m-1}(\mathbb{R}^d)$, we immediately have

$$\langle T, \overline{T} \star K_{m,\ell,s} \rangle \geq 0. \tag{3.18}$$

Example. Let $r = 0$, and μ be a compactly supported measure orthogonal to $\mathbb{P}_{m-1}(\mathbb{R}^d)$. Then (3.18) implies

$$\int \int_{\mathbb{R}^d} K_{m,\ell,s}(x - \xi) \, d\mu(x) \, d\overline{\mu}(\xi) \geq 0. \tag{3.19}$$

More precisely, for a measure μ which is a linear combination of Dirac measures and which is orthogonal to $\mathbb{P}_{m-1}(\mathbb{R}^d)$, *i.e.*, for

$$\mu = \sum_{i=1}^{N} c_i \, \delta_{x_i} \quad \text{and} \quad \sum_{i=1}^{N} c_i \, x_i^\alpha = 0, \qquad \forall \alpha \in \mathbb{N}^d, \quad |\alpha| < m \tag{3.20}$$

we have

$$\sum_{i=1}^{N} \sum_{j=1}^{N} c_i \, \overline{c_j} \, K_{m,\ell,s}(x_i - x_j) \geq 0. \tag{3.21}$$

§4. Practical Examples

4.1. Spline Curves Under Tension

In [15], Schweikert introduced the idea of spline curves under tension in the following way. Let $a = x_1 < x_2 < \cdots < x_p = b$ be a collection of points in $[a, b]$. Given ${}^t V = (r_1, \cdots, r_p) \in \mathbb{R}^p$, the spline curve under tension σ_k is the unique element in the space

$$\mathcal{H} = \{ f : f' \text{ absolutely continuous on } [a, b] \text{ and } f'' \in L^2([a, b]) \} \tag{4.1}$$

such that $\sigma_k(x_i) = r_i$, $i = 1, \cdots, p$, and minimizing

$$\int_a^b [f''(x)]^2 \, dx + k^2 \int_a^b [f'(x)]^2 \, dx, \tag{4.2}$$

where k is the tension parameter. Then, the spline σ_k is defined locally on each subinterval. For $x \in [x_i, x_{i+1}]$, $i = 1, \cdots, p - 1$, we have

$$\sigma_k(x) = a_i + b_i x + c_i \, sh(kx) + d_i \, ch(kx), \tag{4.3}$$

where the coefficients a_i, b_i, c_i, and d_i are defined by the interpolation and continuity conditions.

Another way to consider splines under tension is to characterize σ_k as a $L^{m,\ell,s}$ spline. Let the interpolation scheme be $L = (\delta_{x_1}, \cdots, \delta_{x_p})$. As soon as we consider more than 2 distinct points x_i, L is topologically independent and $\mathbb{P}_0(\mathbb{R})$–unisolvent. Then, we choose $s = 0$ and $\ell = m = d = 1$. The spline under tension will be in the space

$$D_0^{-1}(L^2(\mathbb{R})) = \{\, u \in \mathcal{D}'(\mathbb{R}) \quad ; \quad u', u'' \in L^2(\mathbb{R}) \,\}, \tag{4.4}$$

equipped with the semi–norm

$$|u|_{1,0,0} = \int_{\mathbb{R}} \left[\frac{d^2 u(x)}{dx^2}\right]^2 dx + k^2 \int_{\mathbb{R}} \left[\frac{du(x)}{dx}\right]^2 dx \tag{4.5}$$

and the semi–scalar product

$$(u, v)_{1,0,0} = \int_{\mathbb{R}} \left(\frac{d^2 u(x)}{dx^2}\right)\left(\frac{d^2 v(x)}{dx^2}\right) dx + k^2 \int_{\mathbb{R}} \left(\frac{du(x)}{dx}\right)\left(\frac{dv(x)}{dx}\right) dx, \tag{4.6}$$

associated with the differential operator

$$\Delta_k u = \Delta^2 u - k^2 \, \Delta u = \frac{d^4 u}{dx^4} - k^2 \frac{d^2 u}{dx^2}, \tag{4.7}$$

of which a fundamental solution (Garnir [8]) is the function $E_k(x)$ of class $C^2(\mathbb{R})$,

$$E_k(x) = -\frac{1}{2k^3}(e^{-k|x|} + k|x|). \tag{4.8}$$

The spline under tension now has the following expression:

$$\sigma_k(x) = \sum_{i=1}^{p} \lambda_i \, E_k(x - x_i) + C, \tag{4.9}$$

where the coefficients λ_i, $i = 1, \cdots, p$ and the constant C are the solutions of

$$\begin{cases} \displaystyle\sum_{i=1}^{p} \lambda_i \, E_k(x_j - x_i) + C \; = \; r_j \;\; ; \;\; j = 1, \cdots, p \\[2mm] \displaystyle\sum_{i=1}^{p} \lambda_i = 0, \end{cases} \tag{4.10}$$

or equivalently

$$\begin{bmatrix} K & U \\ {}^tU & 0 \end{bmatrix} \begin{bmatrix} \Lambda \\ C \end{bmatrix} = \begin{bmatrix} V \\ 0 \end{bmatrix} \tag{4.11}$$

with ${}^t\Lambda = (\lambda_1, \cdots, \lambda_p)$, ${}^tU = (1, \cdots, 1)$, ${}^tV = (r_1, \cdots, r_p)$ and K is the symmetric matrix with a zero diagonal $K = (E_k(x_j - x_i))_{i,j=1,\cdots,p}$.

4.2. $\mathbf{L}^{2,1,0}$–splines in $\mathbb{R}^3$

Franke [7] defined thin plate splines under tension as surfaces minimizing

$$\sum_{|\alpha|=2} \frac{2!}{\alpha!} \int_{\mathbb{R}^2} |D^\alpha u(\xi)|^2 \, d\xi + k^2 \sum_{|\alpha|=1} \int_{\mathbb{R}^2} |D^\alpha u(\xi)|^2 \, d\xi \tag{4.12}$$

with scattered Lagrange interpolation constraints. The energy defined in (4.12) is comparable to the one in (4.5).

In $\mathbb{R}^2$, we obtain the following operator

$$\Delta^2 u - k^2 \, \Delta u. \tag{4.13}$$

A fundamental solution is defined involving the Bessel function $K_0(kr)$, where $r = \sqrt{x^2 + y^2}$. But in this case, because the assumption (2.14) is not fulfilled with $m = \ell = 1$, $d = 2$ and $s = 0$, we are not at all certain that

$$\varphi \star K \in D_1^{-1}(L^2(\mathbb{R}^2)), \quad \forall \varphi \in \overline{E'}_1. \tag{4.14}$$

In fact, Franke himself does not take care of (4.14).

Another problem is of a numerical nature, since the Bessel function needs to be approximately evaluated (Abramowitz and Stegun [1]), and each entry of the matrix is involved. Thus, we propose to consider $\mathbb{R}^2$ as a section of $\mathbb{R}^3$ and consider splines under tension for $d = 3$.

Let us consider the space

$$D_1^{-2}(L^2(\mathbb{R}^3)) = \left\{ u \in \mathcal{D}'(\mathbb{R}^3)\,;\, \forall\, \alpha\ \mathbb{N}^3\,;\, D^\alpha u \in L^2(\mathbb{R}^3)\,;\, |\alpha| = 2, 3 \right\} \quad (4.15)$$

with the scalar product $(u, v)_{1,1,0}$

$$= \sum_{|\alpha|=3} \frac{3!}{\alpha!} \int_{\mathbb{R}^3} D^\alpha u(\xi)\, D^\alpha v(\xi)\, d\xi + k^2 \sum_{|\alpha|=2} \frac{2!}{\alpha!} \int_{\mathbb{R}^3} D^\alpha u(\xi)\, D^\alpha v(\xi),\, d\xi$$

$$(4.16)$$

and the associated semi-norm

$$|u|_{1,10} = \sum_{|\alpha|=3} \frac{3!}{\alpha!} \int_{\mathbb{R}^3} |D^\alpha u(\xi)|^2\, d\xi + k^2 \sum_{|\alpha|=2} \frac{2!}{\alpha!} \int_{\mathbb{R}^3} |D^\alpha u(\xi)|^2\, d\xi\,. \quad (4.17)$$

The differential operator involved is

$$\Delta_{1,1,0}\, u = -\Delta^3\, u + k^2\, \Delta^2 u\,. \quad (4.18)$$

In order to find a fundamental solution, let $v = \Delta\, u$, so that

$$-\Delta^2\, v + k^2\, \Delta\, v = \delta\,. \quad (4.19)$$

A fundamental solution of (4.19) is given by (Garnir [8])

$$v_k(r) = \frac{1}{4\pi r\, k^2}\, (e^{-kr} - 1), \quad (4.20)$$

where $r = |(x, y, z)| = \sqrt{x^2 + y^2 + z^2}$. We need to integrate twice in order to have a solution of (4.18) because, in $\mathbb{R}^3$,

$$\Delta\, u = \frac{1}{r}\, \frac{\partial^2}{\partial r^2}\, (ru)\,. \quad (4.21)$$

Thus, a solution of (4.18) is

$$E_k(x, y, z) = \frac{1}{r} \int \left[\int s v_k(s)\, ds \right] dt\,, \quad (4.22)$$

which gives

$$E_k(x, y, z) = \frac{1}{8\pi k^4 r} \left\{ 2e^{-kr} - (1 - kr)^2 + C \right\}, \qquad (4.23)$$

where C is an arbitrary constant.

For Lagrange interpolation at scattered data points $A_i = (x_i, y_i, z_i)$, $i = 1, \cdots, p$ in $\mathbb{R}^3$, interpolating ${}^t V = (r_1, \cdots, r_p)$,the $L^{2,1,0}$–spline σ_k is given by

$$\sigma_k(X) = \sum_{i=1}^{p} \lambda_i E_k(X - A_i) + \alpha_1 x + \alpha_2 y + \alpha_3 z + \alpha_4, \qquad (4.24)$$

where the coefficients λ_i, $i = 1, \cdots, p$ and $\alpha_1, \alpha_2, \alpha_3, \alpha_4$, are solutions of

$$\begin{cases} \begin{cases} \displaystyle\sum_{i=1}^{p} \lambda_i E_k(A_j - A_i) + \alpha_1 x_j + \alpha_2 y_j + \alpha_3 z_j + \alpha_4 = r_j \\ \qquad\qquad \text{with} \ \ j = 1, \cdots, p \ \ \text{and} \end{cases} \\ \displaystyle\sum_{i=1}^{p} \lambda_i x_i = 0 \\ \displaystyle\sum_{i=1}^{p} \lambda_i y_i = 0 \\ \displaystyle\sum_{i=1}^{p} \lambda_i z_i = 0 \\ \displaystyle\sum_{i=1}^{p} \lambda_i = 0 \end{cases} \qquad (4.25)$$

or equivalently,

$$\begin{bmatrix} K & {}^t M \\ M & 0 \end{bmatrix} \begin{bmatrix} \Lambda \\ \alpha \end{bmatrix} = \begin{bmatrix} V \\ 0 \end{bmatrix} \qquad (4.26)$$

with ${}^t\Lambda = (\lambda_1, \cdots, \lambda_p)$, ${}^t V = (r_1, \cdots, r_p)$, ${}^t\alpha = (\alpha_1, \alpha_2, \alpha_3, \alpha_4)$, K is the symmetric matrix with a zero diagonal $K = (E_k(A_j - A_i))_{i,j=1,\cdots,p}$, and M is the $p \times 4$ matrix

$$M = \begin{bmatrix} x_1 & \cdots & x_p \\ y_1 & \cdots & y_p \\ z_1 & \cdots & z_p \\ 1 & \cdots & 1 \end{bmatrix}.$$

4.3. Hermite Thin plate Splines

We define Hermite thin plate splines to be those (m,s)–splines (Duchon [5]) which interpolate Lagrange data and also Hermite data at a set of scattered data points. For simplicity, we consider here only the simplest case where the interpolation of Lagrange data is for $(p-1)$ points $A_i = (x_i, y_i)$ in $\mathbb{R}^2$ and we have only one point for Hermite interpolation, namely $A_p = (x_p, y_p)$.

We choose $s = \frac{1}{2}, m = 2$ and $\ell = 0$. Thus, we obtain the pseudo-cubic splines of Duchon([5]) which are, up to a continuous extension, in $\mathcal{C}^2(\mathbb{R}^2)$. The associated kernel is defined by

$$K_0(X) = |X|^3 \qquad \text{and} \qquad (4.27)$$

$$\begin{cases} K_1(X) = D^{(1,0)}\, K_0(X) = 3\,x\,|X| \\ K_2(X) = D^{(0,1)}\, K_0(X) = 3\,y\,|X|, \end{cases} \qquad (4.28)$$

where $X = (x,y)$ and $|X| = \sqrt{x^2 + y^2}$. The spline is now given by

$$\sigma(X) = \sum_{i=1}^{p} \lambda_i\, K_0(X - A_i) + \alpha_1\, K_1(X - A_p) + \alpha_2\, K_2(X - A_p) + \alpha\,x + \beta\,y + \gamma,$$

$$(4.29)$$

where the coefficients λ_i, $i = 1, \cdots, p$, α_1, α_2, α, β and γ are solutions of a $(p+5) \times (p+5)$ linear system,

$$\begin{cases} \begin{cases} \displaystyle\sum_{i=1}^{p} \lambda_i K_0(A_j - A_i) + \sum_{q=1}^{2} \alpha_q K_q(A_j - A_p) + \alpha x_j + \beta y_j + \gamma = r_j \\ \qquad\qquad \text{for} \quad j = 1, \cdots, p \end{cases} \\ \displaystyle\sum_{i-1}^{p} \lambda_i K_1(A_p - A_i) + \alpha = d_1 \\ \displaystyle\sum_{i=1}^{p} \lambda_i K_2(A_p - A_i) + \beta = d_2 \\ \displaystyle\sum_{i=1}^{p} \lambda_i x_i + \alpha_1 = 0 \\ \displaystyle\sum_{i=1}^{p} \lambda_i y_i + \alpha_2 = 0 \\ \displaystyle\sum_{i=1}^{p} \lambda_i = 0 \end{cases}$$

$$(4.30)$$

References

1. Abramowitz. M. and Stegun. I., A. (eds.) *Handbook of mathematical functions, Applied Mathematics Series 55*, National Bureau of Standards, U. S. Government Printing Office, Washington, D. C. 20402, 1960, 370–380.

2. Attéia. M., Fonctions spline et noyaux reproduisant d'Aronszajn-Bergman, RAIRO, R–3,1970, 192–210.

3. Bouhamidi. A., *Interpolation et approximation par des fonctions splines radiales à plusieurs variables*, Thèse, Nantes, 1992 .

4. Bourbaki. N., *Espaces vectoriels topologiques, Chapitres 1 à 5. Eléments de mathématiques*, Masson, Paris, 1981.

5. Duchon. J., Fonctions–spline à énergie invariante par rotation, Rapport de Recherche $n°27$, Grenoble, 1976.

6. Duchon. J., Interpolation des fonctions de deux variables suivant le principe de la flexion des plaques minces, R.A.I.R.O, Analyse numérique **10**, $n°12$ (1975), 5–12.

7. Franke. R. Thin plate splines with tension, Computer Aided Geometric Design **2** (1985), 87–95.

8. Garnir. H. G., Sur les distributions résolvantes des opérateurs de la physique mathématique, Bull. Société Royale des Sciences de Liège (1951), 174–195, 271–296.

9. Khoan. V. K., *Distributions, Analyse de Fourier, Opérateurs aux dérivées partielles*, Vuibert, Paris, 1972, Vol. 1 et 2.

10. Laurent. P. J., *Approximation et optimisation.* Hermann, Paris, 1972.

11. Le Méhauté. A., *Interpolation et approximation par des fonctions polynomiales par morceaux dans $\mathbb{R}^n$*, Thèse d'Etat, Rennes, 1984 .

12. Rabut, C., B-splines polyharmoniques cardinales: Interpolation, Quasi-interpolation, Filtrage, Thèse d'Etat, Toulouse, 1990.

13. Schwartz. L., *Théorie des distributions.* Hermann, Paris, 1966.

14. Schwartz. L., Sous–espaces hilbertiens d'espaces vectoriels topologiques et noyaux associés (noyaux reproduisants), J. Anal. Math., Paris, 1964.

15. Schweikert. D. G., An interpolation curve using a spline in tension. J. Math. and Phys **45** (1966), 312–317.

Département de Mathématiques et d'Informatique
Faculté des Sciences. Université de Nantes
2 rue de la Houssinière
44072 NANTES Cedex03 FRANCE
alm @cicb.fr

Numerical Methods of Approximation Theory, Vol. 9
Dietrich Braess and Larry L. Schumaker (eds.), pp. 155–175.
International Series of Numerical Mathematics, Vol. 105
Copyright © 1992 by Birkhäuser Verlag, Basel
ISBN 3-7643-2746-4.

Constructive Multivariate Approximation with Sigmoidal Functions and Applications to Neural Networks

Burkhard Lenze

Dedicated to the memory of Lothar Collatz

Abstract. In this paper, we show how to use sigmoidal functions in order to generate approximation operators for multivariate functions of bounded variation. We start with Lebesgue-Stieltjes type convolution operators, then – via numerical quadrature – we pass over to point-evaluation operators and give local and global approximation results for them. In the following, we discuss an important application of our results to neural networks with one hidden layer consisting of so-called sigma-pi units. In this context, our results should be seen as an explicit constructive contribution to Kolmogorov's well-known mapping existence theorem for three-layer feedforward neural networks. At the end, we apply our operators, resp. networks, to a special test function in order to get some concrete idea of their behaviour.

§1. Introduction

One of the main ideas for constructing multivariate approximation procedures is to use "simple" one-dimensional functions together with appropriately chosen multivariate arguments. Without regard to completeness, we mention convolution operators with radial-symmetric distribution kernels (cf. [1,7,33,2]), radial basis functions (cf. [18,32,34,13,14,6,8]), and, finally, approximation by ridge functions (cf. [12,15,11,29,9,10,3,30]). In this paper, we take a look at multivariate approximation induced by sigmoidal functions which have proved to be of some interest in connection with neural networks (cf. [15,11,9,10]). In our context, however, the multivariate functions are not obtained by *ridge-type* arguments (connected with usual sigmoidal units),

$$x \longmapsto \sum_{k=1}^{n} x_k v_k^{(i)}, \quad i = 1, 2, 3, \ldots, \tag{1.1}$$

but by *hyperbolic-type* ones (connected with so-called sigma-pi units, see below),

$$x \longmapsto \prod_{k=1}^{n} \left(x_k - v_k^{(i)} \right), \quad i = 1, 2, 3, \ldots, \tag{1.2}$$

with $v^{(i)} \in \mathbb{R}^n$, $i = 1, 2, 3, \ldots$, some fixed vectors. Therefore, from a mathematical point of view, the paper should be seen in the tradition of [26] (cf. also [25,27,28] for the corresponding one-sided convolutional approach).

§2. Notation

For $a, b \in \mathbb{R}^n$ with $a \leq b$ (i.e. $a_i \leq b_i$, $1 \leq i \leq n$) we define

$$[a, b] := \{x \in \mathbb{R}^n : a_i \leq x_i \leq b_i, \ 1 \leq i \leq n\}, \tag{2.1}$$

$$\mathrm{Cor}[a, b] := \{x \in \mathbb{R}^n : x_i = a_i \lor x_i = b_i, \ 1 \leq i \leq n\}, \tag{2.2}$$

$$\gamma(x, a) := \left| \{i \in \{1, \ldots, n\} : x_i = a_i\} \right|, \quad x \in \mathrm{Cor}[a, b]. \tag{2.3}$$

In (2.3), $|\cdot|$ denotes the number of distinct elements of the set under consideration, and $n - \gamma(x, a)$ is nothing else but the well-known Hamming distance of x and a (cf. [20], p. 43). Now, for a given function $f : \mathbb{R}^n \to \mathbb{R}$ the so-called corresponding interval function Δ_f of f is defined for all bounded intervals $[a, b] \subset \mathbb{R}^n$ by

$$\Delta_f[a, b] := \sum_{x \in \mathrm{Cor}[a,b]} (-1)^{\gamma(x,a)} f(x). \tag{2.4}$$

Definition 2.1. *A function* $f : \mathbb{R}^n \to \mathbb{R}$ *is said to be of (uniform) bounded variation on* $\mathbb{R}^n$ *(for short:* $f \in BV(\mathbb{R}^n)$*), if there exists a finite constant*

$K \geq 0$ *such that the interval function* $\bar{\Delta}_f$ *defined for all* $[a, b] \subset \mathbb{R}^n$ *by*

$$\bar{\Delta}_f[a, b] := \sup \left\{ \sum_{i=1}^{r} \left| \Delta_f[a^{(i)}, b^{(i)}] \right| : \ \left([a^{(i)}, b^{(i)}] \subset [a, b], \right. \right.$$
$$\left. \left. (a^{(i)}, b^{(i)}) \cap (a^{(j)}, b^{(j)}) = \emptyset, \ i \neq j \right), \ 1 \leq i, j \leq r, \ r \in \mathbb{N} \right\} \tag{2.5}$$

satisfies

$$\sup \left\{ \bar{\Delta}_f[a, b] : \ [a, b] \subset \mathbb{R}^n \right\} = K. \tag{2.6}$$

As is well-known, a function $f \in BV(\mathbb{R}^n)$ induces a signed Borel measure m_f, the so-called Lebesgue-Stieltjes measure associated with f. This measure determines the Lebesgue-Stieltjes integral with respect to f, and all integrals which will appear in the following have to be interpreted in the Lebesgue-Stieltjes sense (for details compare the classical books of KAMKE [22], MC-SHANE [31], and SAKS [36], or the compact introduction in [25] or even [26]).

We finish this part of the paper by introducing the notation of a so-called sigmoidal function.

Definition 2.2. *A continuous function* $\sigma : \mathbb{R} \to \mathbb{R}$ *is called a sigmoidal function if*

$$\lim_{\xi \to -\infty} \sigma(\xi) = 0 \quad \text{and} \quad \lim_{\xi \to \infty} \sigma(\xi) = 1. \tag{2.7}$$

Remark. Let us note that all results of this paper remain valid if instead of σ, we take an arbitrary parameter-dependent continuous function $\sigma_r : \mathbb{R} \to \mathbb{R}$, with $r \in \mathbb{R}$ fixed, satisfying

$$\lim_{\xi \to -\infty} \sigma_r(\xi) = r \quad \text{and} \quad \lim_{\xi \to \infty} \sigma_r(\xi) = r + 1. \tag{2.8}$$

§3. Approximation Operators Based on Sigmoidal Functions

We start with Lebesgue-Stieltjes type convolution operators induced by sigmoidal functions.

Theorem 3.1. *Let* $\sigma : \mathbb{R} \to \mathbb{R}$ *be a sigmoidal function. For* $f \in BV(\mathbb{R}^n)$ *with* $\lim_{|t| \to \infty} f(t) = 0$, *the operators* Ω_ϱ, $\varrho > 0$,

$$\Omega_\varrho(f)(x) := (-1)^n 2^{1-n} \int_{\mathbb{R}^n} \sigma\left(\varrho \prod_{k=1}^{n} (t_k - x_k) \right) df(t), \quad x \in \mathbb{R}^n, \tag{3.1}$$

are well-defined and map into the space $C(\mathbb{R}^n)$ of continuous functions on $\mathbb{R}^n$. Moreover, they are linear, and with $K \geq 0$ given by (2.6), they are bounded by

$$\sup_{x \in \mathbb{R}^n} \left| \Omega_\varrho(f)(x) \right| \leq 2^{1-n} K \sup_{\xi \in \mathbb{R}} \left| \sigma(\xi) \right|. \tag{3.2}$$

Finally, in case f is continuous at a point $x \in \mathbb{R}^n$, we have the local approximation result

$$\lim_{\varrho \to \infty} \Omega_\varrho(f)(x) = f(x). \tag{3.3}$$

Proof: See the proofs of Theorems 3.1 and 3.2 in [26], and make some simple modifications. ∎

Since we are interested in getting practical numerical schemes in order to approximate f, we have to evaluate the convolution integral appearing in (3.1), at least approximately. As a first step, we use a simple Riemann-type midpoint quadrature rule and get operators $\Omega_\varrho^{(h)}$, $\varrho, h > 0$,

$$\Omega_\varrho^{(h)}(f)(x) := (-1)^n 2^{1-n} \sum_{j \in \mathbb{Z}^n} \sigma\left(\varrho \prod_{k=1}^{n} \left(h(j_k + \tfrac{1}{2}) - x_k\right)\right) \Delta_f \left[hj, h(j+e)\right],$$

$$\tag{3.4}$$

with $e := (1, 1, \ldots, 1) \in \mathbb{Z}^n$. As a second step, we set $\varrho := h^{-n}$ and obtain the operators $\Omega^{(h)}$ depending only on $h > 0$,

$$\Omega^{(h)}(f)(x) := (-1)^n 2^{1-n} \sum_{j \in \mathbb{Z}^n} \sigma\left(\prod_{k=1}^{n} \left(j_k + \tfrac{1}{2} - \tfrac{x_k}{h}\right)\right) \Delta_f \left[hj, h(j+e)\right].$$

$$\tag{3.5}$$

In the following, we will examine the approximation properties of the hyperbolic cardinal translation-type point-evaluation operators $\Omega^{(h)}$ induced by σ.

Theorem 3.2. Let $\sigma : \mathbb{R} \to \mathbb{R}$ be a sigmoidal function and $\Omega^{(h)}$, $h > 0$, the family of operators defined for all $f \in BV(\mathbb{R}^n)$ with $\lim_{|t| \to \infty} f(t) = 0$ via (3.5). The operators are well-defined and map into $C(\mathbb{R}^n)$. Moreover, they are linear, and with $K \geq 0$ given by (2.6), they are bounded by

$$\sup_{x \in \mathbb{R}^n} \left| \Omega^{(h)}(f)(x) \right| \leq 2^{1-n} K \sup_{\xi \in \mathbb{R}} \left| \sigma(\xi) \right|. \tag{3.6}$$

Proof: Let $h > 0$, $x \in \mathbb{R}^n$, and $j^{(1)}, j^{(2)} \in \mathbb{Z}^n$, $j^{(1)} \le j^{(2)}$, be given arbitrarily. Since by means of (2.6) we have

$$
\begin{aligned}
\sum_{\substack{j \in \mathbb{Z}^n \\ j^{(1)} \le j \le j^{(2)}}} & \left| \sigma\Big(\prod_{k=1}^{n} \Big(j_k + \frac{1}{2} - \frac{x_k}{h} \Big) \Big) \Delta_f \left[hj, h(j+e) \right] \right| \\
& \le \sup_{\xi \in \mathbb{R}} |\sigma(\xi)| \sum_{\substack{j \in \mathbb{Z}^n \\ j^{(1)} \le j \le j^{(2)}}} \left| \Delta_f \left[hj, h(j+e) \right] \right| \\
& \le \sup_{\xi \in \mathbb{R}} |\sigma(\xi)| \cdot K \, ,
\end{aligned}
\tag{3.7}
$$

the infinite sum defining $\Omega^{(h)}(f)(x)$ is absolutely and uniformly convergent. Therefore, $\Omega^{(h)}(f)$ is always well-defined for $f \in BV(\mathbb{R}^n)$, $\lim_{|t| \to \infty} f(t) = 0$, and $\Omega^{(h)}(f) \in C(\mathbb{R}^n)$. Moreover, (3.6) is obvious and the linearity of $\Omega^{(h)}$ immediately follows from the linearity of Δ_f . $\blacksquare$

We now pass over to the local approximation properties of the operators $\Omega^{(h)}$, $h > 0$. We first need some additional notation which may already be found in [25] or [26], for example. Let $x \in \mathbb{R}^n$ be given arbitrarily, and let

$$
H(x) := \bigcup_{k=1}^{n} \{ t \in \mathbb{R}^n : \; t_k = x_k \}
\tag{3.8}
$$

be the so-called *hyperstar* associated with x. Let

$$
Q_+^0(x) := \Big\{ t \in \mathbb{R}^n : \; \prod_{k=1}^{n} (t_k - x_k) > 0 \Big\},
\tag{3.9}
$$

$$
Q_-^0(x) := \Big\{ t \in \mathbb{R}^n : \; \prod_{k=1}^{n} (t_k - x_k) < 0 \Big\},
\tag{3.10}
$$

denote the positive (resp. negative) open sets of quadrants with respect to x. Obviously, for each fixed $x \in \mathbb{R}^n$, the whole space $\mathbb{R}^n$ may be written as the disjoint union

$$
\mathbb{R}^n = H(x) \cup Q_+^0(x) \cup Q_-^0(x).
\tag{3.11}
$$

To complete the introduction of additional notation, we define the space $BVC_0(\mathbb{R}^n)$ of continuous multivariate functions of bounded variation vanishing at infinity,

$$
BVC_0(\mathbb{R}^n) := \Big\{ f \in BV(\mathbb{R}^n) \cap C(\mathbb{R}^n) : \; \lim_{|t| \to \infty} f(t) = 0 \Big\},
\tag{3.12}
$$

and agree to denote by m_f the well-known signed Lebesgue-Stieltjes measure induced by $f \in BVC_0(\mathbb{R}^n)$ and by m_{V_f} its corresponding total variation measure (for details cf. [22,31,36]).

Lemma 3.1. *Let* $f \in BVC_0(\mathbb{R}^n)$, $x \in \mathbb{R}^n$, *and* $[a,b] \subset \mathbb{R}^n$ *an arbitrary interval. Then we have*

$$m_f\left([a,b)\right) = \Delta_f[a,b] = m_f\left([a,b]\right), \tag{3.13}$$

$$m_f\left(H(x)\right) = m_{V_f}\left(H(x)\right) = 0. \tag{3.14}$$

Proof: Compare [36], p. 68, Theorem (6.2), p. 67, Lemma (6.1), and p. 63, Theorems (4.3) and (4.8). ∎

We now return to our operators $\Omega^{(h)}$, $h > 0$, given by (3.5).

Theorem 3.3. *Let* $\sigma : \mathbb{R} \to \mathbb{R}$ *be a sigmoidal function and* $\Omega^{(h)}$, $h > 0$, *the family of operators defined in (3.5). Then for all* $x \in \mathbb{R}^n$ *and all* $f \in BVC_0(\mathbb{R}^n)$ *the operators satisfy*

$$\lim_{h \to 0+} \Omega^{(h)}(f)(x) = f(x). \tag{3.15}$$

Proof: Let $x \in \mathbb{R}^n$ be fixed and $h > 0$. We define the piecewise constant functions $\sigma_x^{(h)} : \mathbb{R}^n \to \mathbb{R}$ by

$$\sigma_x^{(h)}(t) := \sigma\left(\prod_{k=1}^{n}\left(j_k + \frac{1}{2} - \frac{x_k}{h}\right)\right), \quad t \in [hj, h(j+e)),\ j \in \mathbb{Z}^n. \tag{3.16}$$

With these functions and identity (3.13) it is easily seen that for each $f \in BVC_0(\mathbb{R}^n)$ the operators may be rewritten as

$$\Omega^{(h)}(f)(x) = (-1)^n 2^{1-n} \int_{\mathbb{R}^n} \sigma_x^{(h)}(t)\, df(t), \quad h > 0. \tag{3.17}$$

In the first step, we show that the functions $\sigma_x^{(h)}$ converge to the characteristic function $\chi_{Q_+^0(x)}$ for all $t \in \mathbb{R}^n \setminus H(x)$, i.e., we have

$$\lim_{h \to 0+} \sigma_x^{(h)}(t) = \chi_{Q_+^0(x)}(t), \quad t \in \mathbb{R}^n \setminus H(x), \tag{3.18}$$

with

$$\chi_{Q_+^0(x)}(t) := \begin{cases} 1, & t \in Q_+^0(x) \\ 0, & t \in \mathbb{R}^n \setminus Q_+^0(x). \end{cases} \tag{3.19}$$

To show (3.18), let $t \in \mathbb{R}^n \setminus H(x)$ be given arbitrarily and

$$\delta := \min_{1 \leq k \leq n} \{|t_k - x_k|\} > 0. \tag{3.20}$$

Because of (3.11) and the fact that $t \notin H(x)$, we have $t \in Q_+^0(x)$ or $t \in Q_-^0(x)$. First, assume $t \in Q_+^0(x)$, which by means of (3.9) implies

$$\prod_{k=1}^{n} (t_k - x_k) > 0. \tag{3.21}$$

Now for $\epsilon > 0$ given arbitrarily, choose $M > 0$ such that for all $\xi \geq M$ we have

$$|1 - \sigma(\xi)| < \epsilon, \tag{3.22}$$

which is possible because of (2.7). Moreover, fix $h_1 > 0$ such that for all $h \in (0, h_1]$ we simultaneously have

$$\prod_{k=1}^{n} (h(j_k + \tfrac{1}{2}) - x_k) > \left(\frac{\delta}{4}\right)^n, \tag{3.23}$$

(with $j \in \mathbb{Z}^n$ satisfying $t \in [hj, h(j+e))$ and δ defined in (3.20)) and

$$h^{-n} \left(\frac{\delta}{4}\right)^n > M. \tag{3.24}$$

Summing up, we have

$$h^{-n} \prod_{k=1}^{n} (h(j_k + \tfrac{1}{2}) - x_k) > M. \tag{3.25}$$

Therefore, by means of (3.22) we obtain

$$\left|1 - \sigma_x^{(h)}(t)\right| = \left|1 - \sigma\left(h^{-n} \prod_{k=1}^{n} (h(j_k + \tfrac{1}{2}) - x_k)\right)\right| < \epsilon \tag{3.26}$$

for all $h \in (0, h_1]$ with $j \in \mathbb{Z}^n$ satisfying $t \in [hj, h(j+e))$. Since $\epsilon > 0$ was given arbitrarily, we have shown

$$\lim_{h \to 0+} \sigma_x^{(h)}(t) = 1 = \chi_{Q_+^0(x)}(t). \tag{3.27}$$

In the case that $t \in Q^0_-(x)$, i.e., if we have

$$\prod_{k=1}^{n}(t_k - x_k) < 0, \tag{3.28}$$

the argumentation is similar; instead of (3.22) we simply use that for all $\epsilon > 0$ there exists a constant $M > 0$ such that for all $\xi \le -M$ we have

$$|\sigma(\xi)| < \epsilon \tag{3.29}$$

(compare again (2.7)), which in this case implies

$$\lim_{h \to 0+} \sigma_x^{(h)}(t) = 0 = \chi_{Q^0_+(x)}(t). \tag{3.30}$$

In conclusion, we have proved the validity of (3.18).

Now, in a second step, we return to (3.17) and remember that by virtue of (3.14), $H(x)$ is a set of m_{V_f}-measure zero. Therefore, by means of the dominated convergence theorem for Lebesgue-Stieltjes integrals (cf. [22], p. 146, and note that σ is bounded) we finally obtain

$$\begin{aligned}
\lim_{h \to 0+} \Omega^{(h)}(f)(x) &= \lim_{h \to 0+}(-1)^n 2^{1-n} \int_{\mathbb{R}^n} \sigma_x^{(h)}(t)\, df(t) \\
&= (-1)^n 2^{1-n} \int_{\mathbb{R}^n} \chi_{Q^0_+(x)}(t)\, df(t) \tag{3.31} \\
&= (-1)^n 2^{1-n} m_f\left(Q^0_+(x)\right) \\
&= f(x).
\end{aligned}$$

Here, the last identity follows from Lemma 3.2 in [26] and the fact that f is continuous. ∎

Remark. Theorem 3.3 may be sharpened substantially by allowing f to be even discontinuous in a regular sense (cf. [25] for a proper definition of regular discontinuities in this context; also [1], pp. 190ff., for the precise analog in case of local inversion of multivariate Fourier integrals). A paper with results comparable to those of Theorem 3.2 in [26] is in preparation.

In applications, one usually is interested in having not only local but also global uniform approximation results. In order to get these results for our operators, we have to take a more detailed look at the total variation function V_f induced by $f \in BV(\mathbb{R}^n)$ and its continuity properties (cf. [27] or [28] for the precise definition of V_f which is – in our sense – uniquely determined by f). First of all, in the context of multivariate functions of bounded variation the more natural way to consider continuity is the so-called Δ-continuity. In general, let $f \in BV(\mathbb{R}^n)$ be given and $[a,b] \subset \mathbb{R}^n$, $a < b$. Following Saks (cf. [36], p. 59), we call f (uniformly) Δ-continuous on $[a,b]$ if for each $\epsilon > 0$ there exists a $\delta > 0$ such that for all intervals $[c,d] \subset [a,b]$ we have

$$\Delta[c,d] := \prod_{k=1}^{n}(d_k - c_k) < \delta \implies \left|\Delta_f[c,d]\right| < \epsilon. \tag{3.32}$$

Now, in the special case that $f \in BV(\mathbb{R}^n)$ is continuous on $[a,b]$ in the usual sense, a simple uniform continuity argument together with definition (2.4) implies the Δ-continuity of f on $[a,b]$ and, implicitly, even the Δ-continuity of V_f on $[a,b]$ (cf. [36], p. 63, Theorem (4.8)). This argument together with (3.13) rewritten for V_f instead of f will be essential in the proof of the following theorem (comp. (3.37)). Before its proper formulation, we finally have to introduce the usual supremum norm of f over $\mathbb{R}^n$ as

$$\|f\|_\infty := \sup\{|f(x)| : \; x \in \mathbb{R}^n\}. \tag{3.33}$$

Theorem 3.4. *Let* $\sigma : \mathbb{R} \to \mathbb{R}$ *be a sigmoidal function and* $\Omega^{(h)}$, $h > 0$, *the family of operators defined in (3.5). Then for all functions* $f \in BVC_0(\mathbb{R}^n)$, *the operators satisfy*

$$\lim_{h \to 0+} \|f - \Omega^{(h)}(f)\|_\infty = 0. \tag{3.34}$$

Proof: Let $f \in BVC_0(\mathbb{R}^n)$ and $\epsilon > 0$ be given. First, choose $\alpha \in \mathbb{R}, \alpha > 0$, such that for $a := (\alpha, \alpha, \ldots, \alpha) \in \mathbb{R}^n$ we have

$$\int_{\mathbb{R}^n \setminus [-a,a]} |df(t)| = m_{V_f}(\mathbb{R}^n \setminus [-a,a]) < \epsilon. \tag{3.35}$$

Next, choose $h_1 \in (0,1]$ such that for all $h \in (0, h_1]$ we simultaneously obtain

$$|\sigma(-h^{-\frac{n}{2}})| < \epsilon \quad \text{and} \quad |1 - \sigma(h^{-\frac{n}{2}})| < \epsilon, \tag{3.36}$$

and for all intervals $[c,d] \subset [-a,a]$,

$$\Delta[c,d] < (2\alpha)^{n-1}9\sqrt{h} \implies \Delta_{V_f}[c,d] = m_{V_f}([c,d]) < \epsilon. \tag{3.37}$$

Now, since for each $x \in \mathbb{R}^n$ we may rewrite $f(x)$ as

$$f(x) = (-1)^n 2^{1-n} m_f(Q^0_+(x)) = (-1)^n 2^{1-n} \int_{\mathbb{R}^n} \chi_{Q^0_+(x)}(t) df(t), \qquad (3.38)$$

(cf. again [26], Lemma 3.2, together with the continuity of f), we can conclude using (3.16) and (3.17)

$$\left| f(x) - \Omega^{(h)}(f)(x) \right| = 2^{1-n} \left| \int_{\mathbb{R}^n} \left(\chi_{Q^0_+(x)}(t) - \sigma_x^{(h)}(t) \right) df(t) \right|$$

$$\leq 2^{1-n} \Bigg\{ \int_{\mathbb{R}^n \setminus [-a,a]} (1 + \sup_{\xi \in \mathbb{R}} |\sigma(\xi)|) |df(t)|$$

$$+ \int_{\substack{[-a,a] \\ \min_{1 \leq k \leq n} |t_k - x_k| < 4\sqrt{h}}} (1 + \sup_{\xi \in \mathbb{R}} |\sigma(\xi)|) |df(t)| \qquad (3.39)$$

$$+ \int_{\substack{[-a,a] \\ \min_{1 \leq k \leq n} |t_k - x_k| \geq 4\sqrt{h}}} |\chi_{Q^0_+(x)}(t) - \sigma_x^{(h)}(t)| |df(t)| \Bigg\}$$

$$\leq 2^{1-n} \Big\{ (1 + \sup_{\xi \in \mathbb{R}} |\sigma(\xi)|)(\epsilon + n\epsilon) + \epsilon \, m_{V_f}(\mathbb{R}^n) \Big\}.$$

In detail, the last inequality is obtained as follows: The first term is estimated from above using (3.35). For the estimation of the second term, we note that it is easily seen that for each $k \in \{1, 2, \ldots, n\}$

$$\Delta\left(\left\{ t \in [-a,a] : |t_k - x_k| \leq 4\sqrt{h} \right\} \right) < (2\alpha)^{n-1} 9\sqrt{h} \qquad (3.40)$$

independently of the location of x, and apply (3.37) and the monotonicity and subadditivity of m_{V_f} (resp. Δ_{V_f}). Finally, to treat the third term we have to use the fact that for each t with

$$\min_{1 \leq k \leq n} |t_k - x_k| \geq 4\sqrt{h} \qquad (3.41)$$

and $j \in \mathbb{Z}^n$, the fixed index with $t \in [hj, h(j+e))$, we get the estimate

$$h^{-n} \prod_{k=1}^{n} |h\left(j_k + \frac{1}{2}\right) - x_k| > h^{-n} \prod_{k=1}^{n} \sqrt{h} = h^{-\frac{n}{2}}. \tag{3.42}$$

The rest follows by means of (3.36) with similar arguments as already used in proving Theorem 3.3.

Summing up, the upper bound in (3.39) is independent of x and becomes arbitrarily small in terms of ϵ which gives the desired uniform convergence result. ∎

Remarks.

(1) A uniform approximation theorem of the above type may also be formulated for the convolution operators (3.1). We omit the even easier details.

(2) Note that the above proof doesn't give any hint on how the operators behave in case $f \in BV(\mathbb{R}^n)$ is only locally continuous. Perhaps, in this case we will have uniform convergence on appropriate compact subsets on which f is continuous, but this is only a conjecture. However, it is in some sense reasonable since numerical results show that even for discontinuous functions $f \in BV(\mathbb{R}^n)$ vanishing at infinity which are not too strange (regular discontinuities) the results of the operators are quite satisfactory (smoothing effect). This is simply caused by the convolutional background of the operators (cf. also Theorems 3.1 and 3.2 in [26]) and the usually very general approximation results obtained in this context. In any case, the weakest local restriction for local uniform convergence remains an open question, although this seems to be of less importance in view of the following application.

§4. Application

As already pointed out in the introduction, there is a nice application of our results in neural network theory. The theoretical starting-point is Kolmogorov's mapping existence theorem for three-layer feedforward neural networks which we don't want to discuss here, but which is highly recommended to be looked at for sake of a proper motivation (cf. [19]; also [20], pp. 122ff., for further background information). Instead, we immediately start with a network consisting of three different layers with usual sigmoidal units in the hidden layer. More precisely, the input layer consists of n fixed input units, the hidden layer of arbitrary N sigmoidal units, and the output layer of one single output unit. Each input unit is connected with all hidden units (weights

$w_j^{(i)}, 1 \leq j \leq n, 1 \leq i \leq N$) and each hidden unit is connected with the output unit (weights $\alpha_i, 1 \leq i \leq N$). At the output unit the network answers with

$$\mathrm{OUT}(x) = \sum_{i=1}^{N} \alpha_i \sigma \Big(\sum_{j=1}^{n} w_j^{(i)} x_j - \Theta^{(i)} \Big) \tag{4.1}$$

for any given input $x \in \mathbb{R}^n$, where σ is any fixed sigmoidal function and $\Theta^{(i)}, 1 \leq i \leq N$, are so-called threshold values (cf. [35,24,20,23,4,5] and the references given there for more detailed information; see also Figure 1).

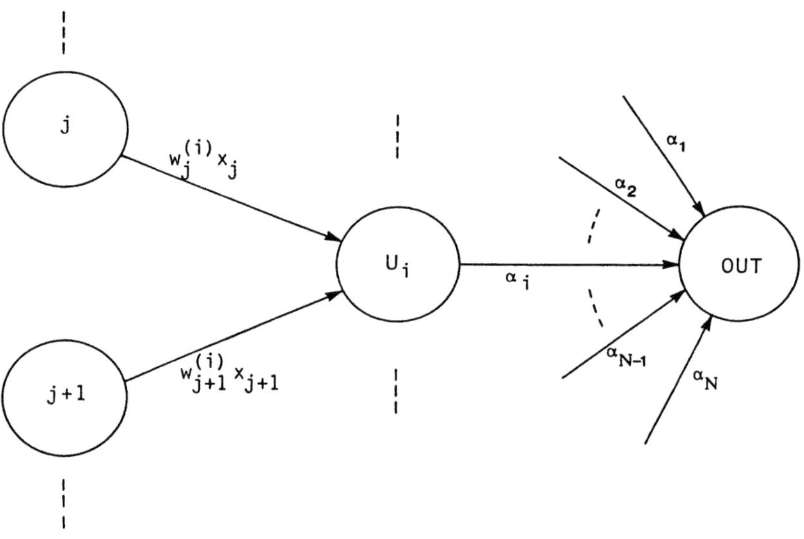

input layer　　　　　　hidden layer　　　　　　output layer

Figure 1.

One of the most interesting questions in connection with the above network is the following: Is it possible to design a network such that for a given continuous function f on a fixed compact set $K \subset \mathbb{R}^n$, $K \neq \emptyset$, and given $\epsilon > 0$,

$$|f(x) - \mathrm{OUT}(x)| < \epsilon \tag{4.2}$$

holds for all inputs $x \in K$? In other words, for f, K, and ϵ given as above we look for $N \in \mathbb{N}$ (number of hidden units), input weights $w_j^{(i)}, 1 \le j \le n, 1 \le i \le N$, thresholds $\Theta^{(i)}, 1 \le i \le N$, and output weights $\alpha_i, 1 \le i \le N$, such that (4.2) holds for all $x \in K$. The answer to this question is "yes" and, moreover, even an algorithm is known in order to determine the network parameters (depending on f, K, and ϵ), at least in case σ is of bounded variation (cf. [21,15,11,9] for existence proofs and related results, and [10] for the constructive design scheme).

We now generalize our network by allowing the input values to be multiplied by each other before they are weighted and handed over to the hidden units. In terms of network theory, we simply substitute the usual sigmoidal units by so-called sigma-pi units, also called higher-order units (cf. [35,4] for some more details and [16,17] for applications of these units; also [23] for a proper justification of using these units in view of real neural systems, German key-word: "präsynaptische Hemmung"); in terms of approximation theory we pass over from linear polynomials with total degree one in the σ-argument to n-linear polynomials (see also the footnote in [20], p. 39, and Figure 2).

In this situation, the analog of (4.1) is simply given by

$$\mathrm{OUT}(x) = \sum_{i=1}^{N} \alpha_i \, \sigma\Big(\sum_{\substack{R \neq \emptyset \\ R \subset \{1,2,\ldots,n\}}} w_R^{(i)} \prod_{k \in R} x_k - \Theta^{(i)} \Big). \tag{4.3}$$

Note, that since (4.1) is a special case of (4.3) (set all "proper multiplication weights" corresponding to subsets $R \subset \{1, 2, \ldots, n\}$ with at least two elements equal to zero), it is clear that any continuous function on a given compact set can be recovered approximately in the sense of (4.2) by networks of the above type with sigma-pi units in the hidden layer. The question is, however, is it possible to use the additional freedom of input weights in order to give an easy explicit realization of the approximating neural network without starting a sophisticated algorithm? By means of Theorem 3.4 the answer to this question is also "yes", but we have to assume some restrictions on f which fortunately only play a minor role in practice.

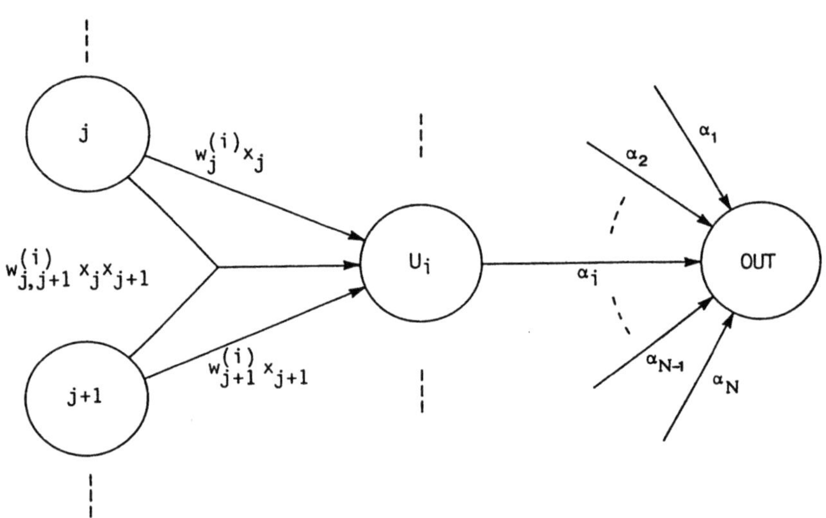

input layer   hidden layer   output layer

Figure 2.

Theorem 4.1. *Let $\sigma : \mathbb{R} \to \mathbb{R}$ be a sigmoidal function. Then for each function $f \in BVC_0(\mathbb{R}^n)$ which is not identically zero ($f \equiv 0$ is trivial) and each $\epsilon > 0$ we can construct a function OUT_f of type (4.3) such that for all $x \in \mathbb{R}^n$ we have*

$$|f(x) - OUT_f(x)| < \epsilon. \tag{4.4}$$

In detail, OUT_f is simply given by

$$OUT_f(x) := (-1)^n 2^{1-n} \sum_{-J \leq j < J} \sigma\left(\prod_{k=1}^{n}\left(j_k + \frac{1}{2} - \frac{x_k}{h}\right)\right) \Delta_f[hj, h(j+e)], \tag{4.5}$$

where $h \in (0,1]$ and $J \in \mathbb{Z}^n, J > 0$, have to be chosen as follows:

First, choose $\alpha \in \mathbb{R}, \alpha > 0$, such that with $a := (\alpha, \alpha, \ldots, \alpha) \in \mathbb{R}^n$ we have

$$\int_{\mathbb{R}^n \setminus [-a,a]} |df(t)| = m_{V_f}\left(\mathbb{R}^n \setminus [-a,a]\right) < \left(2^{1-n}(1 + \sup_{\xi \in \mathbb{R}} |\sigma(\xi)|)\right)^{-1} \frac{\epsilon}{4}. \quad (4.6)$$

Then, choose $h \in (0,1]$ such that we simultaneously have

$$|\sigma(-h^{-\frac{n}{2}})| < (2^{1-n} m_{V_f}(\mathbb{R}^n))^{-1} \frac{\epsilon}{4}, \quad (4.7)$$

$$|1 - \sigma(h^{-\frac{n}{2}})| < (2^{1-n} m_{V_f}(\mathbb{R}^n))^{-1} \frac{\epsilon}{4}, \quad (4.8)$$

and for all intervals $[c,d] \subset [-a,a]$,

$$\Delta[c,d] < (2\alpha)^{n-1} 9\sqrt{h} \implies \Delta_{V_f}[c,d] < \left(n 2^{1-n}(1 + \sup_{\xi \in \mathbb{R}} |\sigma(\xi)|)\right)^{-1} \frac{\epsilon}{4}. \quad (4.9)$$

Finally, choose $J \in \mathbb{Z}^n, J > 0$, such that

$$[-a,a] \subset (-Jh, Jh). \quad (4.10)$$

Proof: Let all parameters be given as in the statement of the theorem. Then for all $x \in \mathbb{R}^n$ we have

$$|f(x) - \mathrm{OUT}_f(x)| \le |f(x) - \Omega^{(h)}(f)(x)| + |\Omega^{(h)}(f)(x) - \mathrm{OUT}_f(x)|. \quad (4.11)$$

By means of our special choice of a and h, the first term of (4.11) is bounded by $\frac{3}{4}\epsilon$ as essentially shown in (3.39). Moreover, since $J \in \mathbb{Z}^n, J > 0$, satisfies (4.10), the second term may be estimated from above by

$$|\Omega^{(h)}(f)(x) - \mathrm{OUT}_f(x)| \le 2^{1-n} \sup_{\xi \in \mathbb{R}} |\sigma(\xi)| \sum_{\substack{j \in \mathbb{Z}^n \\ j \notin [-J,J)}} |\Delta_f[hj, h(j+e)]|$$

$$\le 2^{1-n} \sup_{\xi \in \mathbb{R}} |\sigma(\xi)| \, m_{V_f}(\mathbb{R}^n \setminus (-Jh, Jh)) \quad (4.12)$$

$$\le 2^{1-n} \sup_{\xi \in \mathbb{R}} |\sigma(\xi)| \, m_{V_f}(\mathbb{R}^n \setminus [-a,a]) \quad < \quad \frac{\epsilon}{4},$$

where the last inequality follows from (4.6). Summing up, we get (4.4). $\blacksquare$

Remarks.

(1) In order to see that (4.5) is really of type (4.3), we would have to evaluate the product in the argument of σ and reorganize it in terms of products of $x_k, 1 \leq k \leq n$; more precisely,

$$\prod_{k=1}^{n} \left(j_k + \frac{1}{2} - \frac{x_k}{h} \right) = \sum_{\substack{R \cap S = \emptyset \\ R \cup S = \{1,2,\ldots,n\}}} (-h)^{-|R|} \prod_{k \in S} (j_k + \frac{1}{2}) \prod_{k \in R} x_k \qquad (4.13)$$

with

$$\Theta^{(j)} := -\prod_{k=1}^{n} \left(j_k + \frac{1}{2} \right) \qquad (4.14)$$

the threshold given by the summation term corresponding to $R = \emptyset$, except for sign. Moreover, we would have to enumerate our hidden units not in terms of multiindices but in terms of natural numbers, i.e., we would have to identify the set $\{j \in \mathbb{Z}^n : -J \leq j < J\}$ with an appropriate set $\{i \in \mathbb{N} : 1 \leq i \leq N\}$. We omit these technical details.

(2) Note that the result of the above theorem may be used as follows: For given $\epsilon > 0, \alpha > 0, a := (\alpha, \alpha, \ldots, \alpha)$, and $h \in (0,1]$ choose $J \in \mathbb{Z}^n, J > 0$, such that (4.10) holds. Then, for all functions $f \in BVC_0(\mathbb{R}^n)$ which do not vanish identically and which satisfy (4.6), (4.7), (4.8), and (4.9), the output function OUT_f defined by (4.5) satisfies

$$|f(x) - \mathrm{OUT}_f(x)| < \epsilon, \quad x \in \mathbb{R}^n. \qquad (4.15)$$

This means that for the whole space of functions defined above you can almost work with the same network without update of thresholds and input weights (this is the memory aspect of the network, and there is no need to start a learning algorithm). The only weights you have to modify are the output weights and this can be done massively parallel by means of sampling: Sample f on the cardinal h-grid hj for $-J \leq j \leq J$; then, calculate $\Delta_f[hj, h(j+e)], -J \leq j < J$, simply by means of (2.4) and use these values (multiplied by $(-1)^n 2^{1-n}$) as output weights for the new function f to be approximated. Especially, in case $f : \mathbb{R}^n \to \mathbb{R}^s$ is a vector-valued function $f = (f_1, f_2, \ldots, f_s)$ with all coordinate functions $f_r : \mathbb{R}^n \to \mathbb{R}, 1 \leq r \leq s$, of type specified above, then simply fan out from each hidden unit to the s output units $\mathrm{OUT}_{f_1}, \mathrm{OUT}_{f_2}, \ldots, \mathrm{OUT}_{f_s}$ by using the same thresholds and input weights. Again, the only weights you have to modify are the output weights. However, in this case the number of or weights is $s \cdot N$ instead of

N . This easy real-time modification of the network when changing the function(s) to be simulated by means of a pure parallel sampling update of the output weights should be of some interest in practice. The usefulness of our network from a practical point of view may be confirmed even with respect to reliability: Note that, because of the continuity of f and the boundedness of σ, the contribution of each hidden unit to the output becomes arbitrarily small for h sufficiently close to zero. In other words, a failure of a few hidden units doesn't have a crucial effect on the output behaviour of the network.

§5. Example

In order to get some visual idea of the behaviour of our operators (and, therefore, implicitly also for our network realizations) we now consider as an example the two-dimensional case ($n = 2$) and take a look at the operators (3.5) generated by the special sigmoidal function

$$\sigma(\xi) := \frac{1}{2}\left(1 + \tanh(\xi)\right), \quad \xi \in \mathbb{R}. \tag{5.1}$$

We get

$$\Omega^{(h)}(f)(x) = \frac{1}{2}\sum_{j \in \mathbb{Z}^2} \sigma\left(\prod_{k=1}^{2}\left(j_k + \frac{1}{2} - \frac{x_k}{h}\right)\right) \Delta_f\left[hj, h(j+e)\right], \tag{5.2}$$

resp. in full-length

$$\Omega^{(h)}(f)(x_1, x_2) =$$

$$\frac{1}{2}\sum_{j_1=-\infty}^{\infty}\sum_{j_2=-\infty}^{\infty} \sigma\left(\left(j_1 + \frac{1}{2} - \frac{x_1}{h}\right)\left(j_2 + \frac{1}{2} - \frac{x_2}{h}\right)\right) \bullet \tag{5.3}$$

$$\left(f\left(h(j_1 + 1, j_2 + 1)\right) - f\left(h(j_1, j_2 + 1)\right) - f\left(h(j_1 + 1, j_2)\right) + f\left(h(j_1, j_2)\right)\right)$$

for $h > 0$ and $x = (x_1, x_2) \in \mathbb{R}^2$. Moreover, we apply (5.2) resp. (5.3) to the test function f,

$$f(x) = f(x_1, x_2) := 5e^{-5\left(x_1^2 + (x_2-1)^2\right)} - e^{-5\left((x_1-1)^2 + (x_2+1)^2\right)}, \tag{5.4}$$

and assume that f is numerically zero outside $[-2, 2] \times [-2, 2]$ in order to get finite sums in (5.3) or, in other words, to get proper networks with a

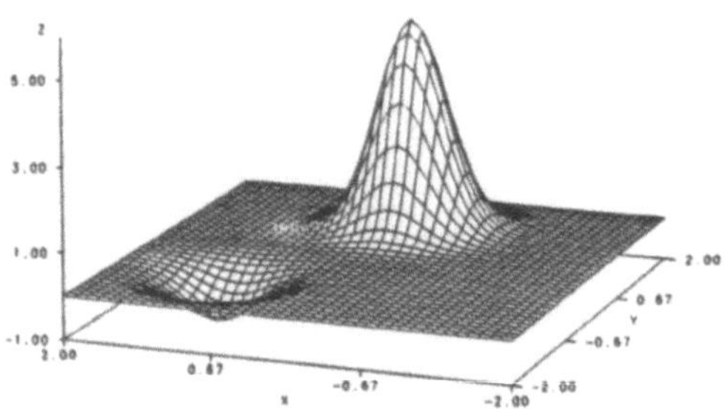

test function f

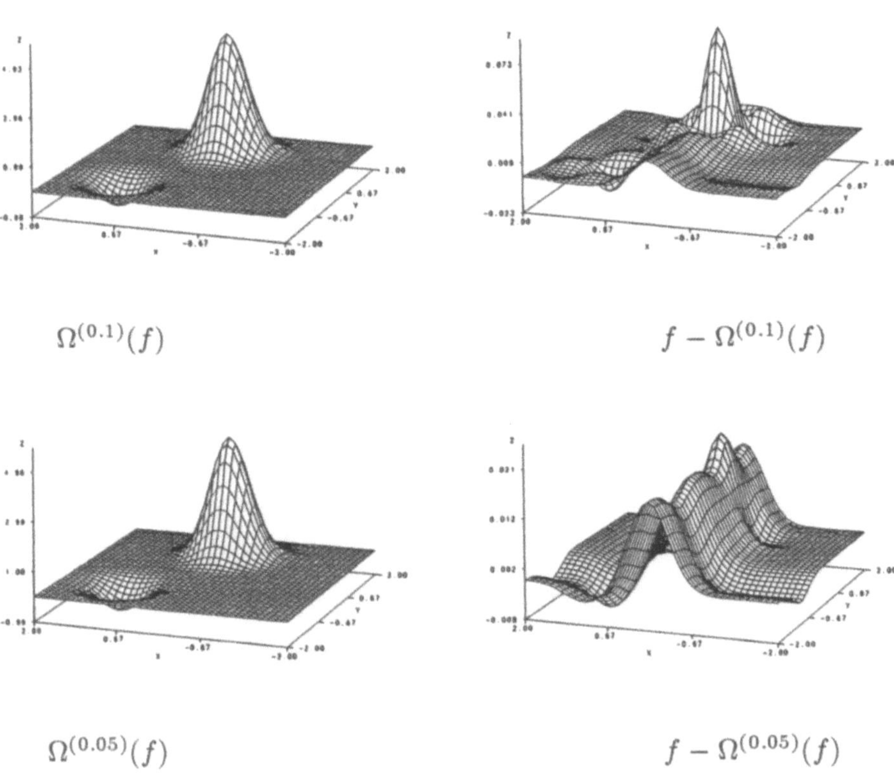

$\Omega^{(0.1)}(f)$ $\qquad\qquad$ $f - \Omega^{(0.1)}(f)$

$\Omega^{(0.05)}(f)$ $\qquad\qquad$ $f - \Omega^{(0.05)}(f)$

Figure 3.

finite number of hidden sigma-pi units. Although f is partially rather steep, the results obtained by our approximation operators are really satisfactory as Figure 3 (for $h = 0.1$ (with $(-20, -20) \leq (j_1, j_2) < (20, 20)$) and $h = 0.05$ (with $(-40, -40) \leq (j_1, j_2) < (40, 40)$)) shows.

Note added in proof: In the case that for fixed dimension $n \in \mathbb{N}$ the sigmoidal function σ satisfies $\sigma(\xi) = 0$ for $\xi \leq -2^{-n}$ and $\sigma(\xi) = 1$ for $\xi \geq 2^{-n}$ (for example, think of scaled integrated B-splines) the operators defined in (3.5) interpolate $f \in BVC_0(\mathbb{R}^n)$ on the cardinal h-grid, i.e., $\Omega^{(h)}(f)(jh) = f(jh), j \in \mathbb{Z}^n, h > 0$. Especially, in the case that $f \in BVC_0(\mathbb{R}^n)$ vanishes outside a compact set the operators imply explicit realizations of perfectly trained three-layer feedforward sigma-pi neural networks. Moreover, σ may be allowed to be even discontinuous in order to include perceptron-type networks, too. A proof and a more detailed discussion of these interpolation results will be given elsewhere.

Acknowledgement: It is a special pleasure for me to thank Charles Chui (Texas A&M University) very much for not only sending me the preprints [9,10], but also for encouraging me to try to apply my theoretical results in the interesting field of neural networks. Moreover, much thanks to Will Light (University of Leicester) for drawing reference [5] to my attention.

References

1. S. Bochner, *Vorlesungen über Fouriersche Integrale*, Akademische Verlagsgesellschaft, Leipzig, 1932.
2. J. Boman, Saturation problems and distribution theory, in *Topics in Approximation Theory*, Lecture Notes in Mathematics Vol. **187**, H. S. Shapiro (ed.), Springer-Verlag, Berlin, Appendix I, 1971, 249–266.
3. D. Braess and A. Pinkus, Interpolation by ridge functions, J. Approx. Th., to appear.
4. R. Brause, *Neuronale Netze*, B.G. Teubner Verlag, Stuttgart, 1991.
5. P.C. Bressloff and D.J. Weir, Neural Networks, GEC Journal of Research **8** (1991), 151–169.
6. M.D. Buhmann, Multivariate cardinal interpolation with radial basis functions, Constr. Approx. **6** (1990), 225–255.
7. P.L. Butzer and R.J. Nessel, Contributions to the theory of saturation for singular integrals in several variables I, General Theory, Indag. Math. **28**, Ser. A (1966), 515–531.
8. C.K. Chui, K. Jetter, and J.D. Ward, Cardinal interpolation with differences of tempered functions, Computers Maths. Applics., to appear.

9. C.K. Chui and X. Li, Approximation by ridge functions and neural networks with one hidden layer, CAT Report 222, Texas A&M University, College Station, 1990.

10. C.K. Chui and X. Li, Realization of neural networks with one hidden layer, CAT Report 244, Texas A&M University, College Station, 1991.

11. G. Cybenko, Approximation by superpositions of a sigmoidal function, Mathematics of Control, Signals, and Systems **2** (1989), 303–314.

12. W. Dahmen and C.A. Micchelli, Some remarks on ridge functions, J. Appr. Theory and its Appl. **3:2** (1987), 139–143.

13. N. Dyn, Interpolation of scattered data by radial basis functions, in *Topics in Multivariate Approximation*, C.K. Chui, L.L. Schumaker, and F.I. Utreras (eds.), Academic Press, New York, 1987, 47–61.

14. N. Dyn, W.A. Light, and E.W. Cheney, Interpolation by piecewise-linear radial basis functions I, J. Appr. Theory **59** (1989), 202–223.

15. K.-I. Funahashi, On the approximate realization of continuous mappings by neural networks, Neural Networks **2** (1989), 183–192.

16. C.L. Giles and T. Maxwell, Learning, invariance, and generalization in high-order neural networks, Applied Optics **26** (1987), 4972–4978.

17. C.L. Giles, R.D. Griffin, and T. Maxwell, Encoding geometric invariances in higher-order neural networks, in *Neural Information Processing Systems – Natural and Synthetic –*, D.Z. Anderson (ed.), Amer. Inst. of Physics, New York, 1988, 301–309.

18. R.L. Hardy, Multiquadric equations of topography and other irregular surfaces, J. Geophysical Research **76** (1971), 1905–1915.

19. R. Hecht-Nielsen, Kolmogorov's mapping neural network existence theorem, in *Proceedings of the International Conference on Neural Networks III*, IEEE-Press, New York, 1987, 11–13.

20. R. Hecht-Nielsen, *Neurocomputing*, Addison-Wesley, Reading, Massachusetts, 1990.

21. K. Hornik, M. Stinchcombe, and H. White, Multilayer feedforward networks are universal approximators, Neural Networks **2** (1989), 359–366.

22. E. Kamke, *Das Lebesgue-Stieltjes-Integral*, B.G. Teubner Verlagsgesellschaft, Leipzig, 1956.

23. M. Köhle, *Neurale Netze*, Springer-Verlag, Wien–New York, 1990.

24. A. Lapedes and R. Farber, How neural nets work, in *Neural Information Processing Systems – Natural and Synthetic –*, D.Z. Anderson (ed.), Amer. Inst. of Physics, New York, 1988, 442–456.

25. B. Lenze, Mehrdimensionale einseitige Approximationsoperatoren, Habilitationsschrift, FernUniversität Hagen, 1989, in: Seminarber. des Fb. Math. & Inf. der FernUniversität Hagen **35** (1989), 1–104.

26. B. Lenze, On multidimensional Lebesgue-Stieltjes convolution operators, in *Multivariate Approximation Theory IV*, C.K. Chui, W. Schempp, and K. Zeller (eds.), ISNM 90, Birkhäuser Verlag, Basel, 1989, 225–232.

27. B. Lenze, On constructive one-sided approximation of multivariate functions of bounded variation, Numer. Funct. Anal. and Optimiz. **11** (1990), 55–83.

28. B. Lenze, A hyperbolic modulus of smoothness for multivariate functions of bounded variation, J. Appr. Theory and its Appl. **7:1** (1991), 1–15.

29. B. Lenze and F. Locher, On ridge-type functions with elliptic contour lines, in *Multivariate Approximation and Interpolation* , W. Haußmann and K. Jetter (eds.), ISNM 94, Birkhäuser Verlag, Basel, 1990, 193–204.

30. V.Y. Lin and A. Pinkus, Fundamentality of ridge functions, J. Approx. Th., submitted.

31. E.J. McShane, *Integration*, Princeton University Press, Princeton, 1974, eighth edition.

32. C.A. Micchelli, Interpolation of scattered data: distance matrices and conditionally positive definite functions, Constr. Approx. **2** (1986), 11–22.

33. R.J. Nessel, Contributions to the theory of saturation for singular integrals in several variables III, Radial Kernels, Indag. Math. **29**, Ser. A (1967), 65–73.

34. M.J.D. Powell, Radial basis functions for multivariable interpolation: a review, in *Algorithms for Approximation*, J.C. Mason and M.G. Cox (eds.), Clarendon Press, Oxford, 1987, 143–167.

35. D.E. Rumelhart and J.L. McClelland, *Parallel Distributed Processing: Explorations in the Microstructure of Cognition*, Vols. I&II, MIT-Press, Cambridge, Massachusetts, 1986.

36. S. Saks, *Theory of the Integral*, Hafner Publishing Company, New York, 1937, second edition.

Burkhard Lenze
FernUniversität Hagen
Fb. Mathematik
D-5800 Hagen
Germany
ma112@dhafeu11.bitnet

Numerical Methods of Approximation Theory, Vol. 9

Dietrich Braess and Larry L. Schumaker (eds.), pp. 177–194.

International Series of Numerical Mathematics, Vol. 105

Copyright © 1992 by Birkhäuser Verlag, Basel

ISBN 3-7643-2746-4.

Spline-Wavelets of Minimal Support

T. Lyche and K. Mørken

Dedicated to the memory of Lothar Collatz

Abstract. For any finite dimensional univariate spline spaces S_0 and S_1 of the same degree, with $S_0 \subset S_1$, we determine a basis for the orthogonal complement of S_0 in S_1. This basis has minimal support and is of interest in wavelet decompositions.

§1. Introduction and Background Material

There has recently been considerable interest in various types of "wavelets", see [4,5,8,9]. Such functions arise when one considers a subspace V_0 of a Hilbert space V_1. The space of wavelets in this case is the orthogonal complement of V_0 in V_1. In most cases the spaces are assumed to be translation invariant, so that a basis is given by translates of a fixed function. Also, it is assumed that the space V_1 is obtained from V_0 by a binary scaling $V_1 = \{f(2\cdot) : f \in V_0\}$. An important example of this is provided by splines with uniform knots, e.g., let V_0 have knots at the integers $\mathbb{Z}$; the "refined" space V_1 then has knots at the half integers $\mathbb{Z}/2$. For an account of this, see [4].

The construction of (spline) wavelets has been generalized to the case where V_0 has arbitrary simple knots, and V_1 is obtained from V_0 by inserting a fixed number of simple knots between each pair of neighbouring knots, see Buhmann and Micchelli [3]. The construction has also been extended to the case where the knots of V_0 are uniform, but have multiplicity greater than one, see Goodman and Lee [6]. For an early discussion of orthogonality in spline spaces, see [1].

In this paper we consider the general case where $V_0 = S_{k,\tau}$ is a spline space of some order k with knots τ, and $V_1 = S_{k,t}$ is a spline space with $\tau \subset t$. Our purpose is to show that the existence of a minimal support basis for the orthogonal complement of V_0 in V_1 is a simple consequence of elementary properties of splines. We show how such a basis can be constructed and give several examples. We consider primarily finite knot vectors, and do not discuss properties of wavelet expansions.

In the rest of this section we introduce our notation and review some basic results about splines that we will use later.

1.1. Knot Sequences, Matrices and Determinants

Suppose τ is a nondecreasing, finite or infinite sequence of real numbers. For each real number x, we let $m_\tau(x)$ denote the multiplicity of x in τ, i.e., the number of times the number x occurs in τ. We call τ a *knot sequence* if $m_\tau(x)$ is finite for all real numbers x.

If τ and t are knot sequences, the notation $\tau \subset t$ means that τ is a subsequence of t. Formally, this means that $m_\tau(x) \leq m_t(x)$ for all real numbers x in τ. If τ, s, and t are nondecreasing knot sequences with $s \subset t$, then we define the sequence $\tau \cup s$ by the characterization

$$m_{\tau \cup s}(x) = m_\tau(x) + m_s(x),$$

for all real numbers x.

We define the *right* and *left* multiplicity of each τ_i in a knot sequence τ by

$$\rho_\tau(i) = \max\{j \in \mathbb{N} : \tau_{i+j-1} = \tau_i\}, \quad \lambda_\tau(i) = \max\{j \in \mathbb{N} : \tau_{i-j+1} = \tau_i\}.$$

We say that two sufficiently well behaved functions f and g *agree on* τ if

$$D^{\lambda_\tau(i)} f(\tau_i) = D^{\lambda_\tau(i)} g(\tau_i), \qquad \text{for all } \tau_i \in \tau.$$

Here D^n denotes differentiation n times. For piecewise functions with possible jump discontinuities, the symbol D^n will denote one sided differentiation from the *right*. A function is said to *vanish on* τ if it agrees with the zero function there.

If $\mathbf{x} = (x_1, \ldots, x_m)$ consists of m real numbers with $a \leq x_1 \leq \cdots \leq x_m \leq b$ and $f_1, \ldots, f_n$ are n sufficiently smooth functions on $[a, b]$, we let

$$M \left(\begin{array}{c} x_1, \ldots, x_m \\ f_1, \ldots, f_n \end{array} \right)$$

denote the $m \times n$ collocation matrix with element $D^{\lambda \mathbf{x}(i)} f_j(x_i)$ in row i and column j. If $m = n$, we let

$$D \left(\begin{array}{c} x_1, \ldots, x_n \\ f_1, \ldots, f_n \end{array} \right) \tag{1.1}$$

denote the determinant of this matrix.

1.2. Some Basic Spline Theory

The central theme of this work is splines and B-splines, and in the following we review some of their properties. The B-spline $B_{i,k} = B_{i,k,\tau}$ of order k (degree $k - 1$) on τ is defined by

$$B_{i,k}(x) = \begin{cases} (\tau_{i+k} - \tau_i)[\tau_i, \tau_{i+1}, \ldots, \tau_{i+k}](\cdot - x)_+^{k-1}, & \text{if } \tau_i < \tau_{i+k}; \\ 0, & \text{otherwise}; \end{cases} \tag{1.2}$$

where

$$(y - x)_+^{k-1} = \begin{cases} (y - x)^{k-1}, & \text{if } y > x; \\ 0, & \text{otherwise}; \end{cases}$$

is a truncated power, and the square brackets denote for each x the divided difference of the function $f(y) = (y - x)_+^{k-1}$. These B-splines are piecewise polynomials of order k with joins between the pieces at the knots (τ_j); they are normalized to sum to one; they are nonnegative and the B-spline $B_{i,k}$ is positive in (τ_i, τ_{i+k}); and if $\tau_i < \tau_{i+k}$ for all i then the B-splines $\{B_{i,k}\}_i$ are linearly independent.

Given the knot sequence τ, we consider the space X of piecewise polynomials which are polynomials of order k on each (nonempty) interval (τ_i, τ_{i+1}) and joined together with continuity at least $C^{k-1-m_\tau(x)}$ at a knot $x = \tau_i$ of multiplicity $m_\tau(x)$. Furthermore, we stipulate that for $x < \inf_i \tau_i$ and for $x > \sup_i \tau_i$, the functions in X are identically zero. The Curry-Schoenberg Theorem states that the space of B-splines $\{B_{i,k}\}_i$ provide a basis for X, i.e.,

$$X = S_{k,\tau} = \operatorname*{span}_i \{B_{i,k,\tau}\}.$$

From the above definition of B-splines, their close relation with divided differences is very explicit. Another example of this is provided by the Peano integral representation of divided differences; for each $f \in C^{k-1}[\tau_i, \tau_{i+k}]$ with $f^{(k)}$ piecewise continuous, we have

$$(\tau_{i+k} - \tau_i)[\tau_i, \ldots, \tau_{i+k}]f = \frac{1}{(k-1)!} \int_{\tau_i}^{\tau_{i+k}} B_{i,k}(x) D^k f(x)\, dx. \qquad (1.3)$$

We will also need the following well known result which provides information about the sign of the determinant of the B-spline collocation matrix.

Lemma 1. *Let n and k be positive integers and suppose that the knot vector $\tau = (\tau_i)_{i=1}^{n+k}$ satisfies $\tau_i < \tau_{i+k}$ for all i, and that $x = (x_i)_{i=1}^{p}$ is a nondecreasing sequence of real numbers with $p \le n$ and $x_i < x_{i+k}$ for all i. For any integers $1 \le \sigma_1 < \sigma_2 < \cdots < \sigma_p \le n$, the determinant*

$$D\begin{pmatrix} x_1, \ldots, x_p \\ B_{\sigma_1,k}, \ldots, B_{\sigma_p,k} \end{pmatrix}$$

is nonnegative, and it is positive if and only if

$$x_i \in (\tau_{\sigma_i}, \tau_{\sigma_i+k}) \cup \{x : D^{\lambda_x(i)} B_{\sigma_i,k}(x) \ne 0\}.$$

For a proof of the quoted results on splines we refer to [10], Chapter 4. We will also need a general result about Gram matrices. Since we usually will be working on an interval $[a, b]$, we will use the inner product

$$(f, g) = \int_a^b fg$$

for L^2-functions f and g. Given two families of functions $\{f_i\}_{i=1}^{n}$ and $\{g_j\}_{j=1}^{n}$ on (a, b), their Gram matrix is defined to be

$$G\begin{pmatrix} f_1, \ldots, f_n \\ g_1, \ldots, g_n \end{pmatrix},$$

with $\int_a^b f_i g_j$ as its entry in row i and column j. We then have the basic composition formula

$$\det G\begin{pmatrix} f_1, \ldots, f_n \\ g_1, \ldots, g_n \end{pmatrix} = \int_\Omega D\begin{pmatrix} y_1, \ldots, y_n \\ f_1, \ldots, f_n \end{pmatrix} D\begin{pmatrix} y_1, \ldots, y_n \\ g_1, \ldots, g_n \end{pmatrix} dy_n \cdots dy_1, \qquad (1.4)$$

where $\Omega = \{(y_1, \ldots, y_n) : a \le y_1 < y_2 < \cdots < y_n \le b\}$, see [7], pages 16–17. In particular, this identity shows that the Gram matrix of B-splines is totally positive.

The final result that we quote here shows that a function which is orthogonal to a spline space must have a certain number of sign changes. This result is a modification of Corollary XIII.1 in [2] in that the interval (a, b) does not need to contain all the knots.

Lemma 2. *Let $\tau = (\tau_j)_{j=1}^{n+k}$ be a knot vector with $\tau_j < \tau_{j+k}$ for $j = 1, \ldots,$ n, and let a and b be two real numbers with $a < \tau_{k+1}$ and $\tau_n < b$. Let ψ be a piecewise continuous function with support in (a, b), and not identically zero on any open subinterval of (a, b). If ψ is orthogonal to $S_{k,\tau}$, i.e.,*

$$\int_a^b f(x)\psi(x) = 0, \qquad \text{for all } f \in S_{k,\tau},$$

there exists a strictly increasing sequence $\boldsymbol{\xi} = (\xi_j)_{j=1}^n$ in (a, b) such that ψ changes sign at each ξ_j, i.e.,

$$\psi(\xi_j - \epsilon)\psi(\xi_j + \epsilon) < 0, \qquad \text{for } j = 1, \ldots, n$$

and for all $\epsilon > 0$ sufficiently small.

Proof: Consider functions of the form

$$\theta(x) = p(x) + \int_a^b (x - v)_+^{k-1}\psi(v)dv/(k-1)!, \tag{1.5}$$

where p is any polynomial of order at most k. Note that for any order k polynomial p we have $D^k\theta = \psi$. Fix p as the polynomial of order k such that θ vanishes on $\tau_1, \ldots, \tau_k$. Taking divided differences with respect to $\tau_i, \ldots,$ τ_{i+k} for $i = 1, \ldots, n$, we see by (1.3) that θ vanishes on all of τ, i.e., it has (at least) $n + k$ zeros of which at most k coincide. By repeated application of Rolle's theorem we find that $D^{k-1}\theta$ must have $n + 1$ distinct zeros $(\eta_i)_{i=1}^{n+1}$ in the interval $[\tau_1, \tau_{n+k}]$.

But we know more than this; all the η_i must be in the interval $[a, b]$. To show that all the η_i are in $[a, \tau_{n+k}]$, note that $\theta(x)$ is equal to the polynomial $p(x)$ for $x < a$. If $p \equiv 0$, we see that θ has a zero of multiplicity k at a which can be used in the applications of Rolle's theorem instead of the k zeros at $\tau_1,$ $\ldots, \tau_k$. Thus all the zeros of $D^{k-1}\theta$ must be in $[a, \tau_{n+k}]$. If the exact degree of p is $\sigma > 0$, we have

$$D^\sigma\theta = c + \int_a^b (x - v)_+^{k-\sigma-1}\psi(v)dv/(k - \sigma - 1)!$$

for some $c \neq 0$. But then we see that $D^\sigma\theta$ has $n + k - \sigma$ zeros in $(a, \tau_{n+k}]$. Continued application of Rolle's theorem shows that $D^{k-1}\theta$ has $n + 1$ sign changes in (a, τ_{n+k}).

Similarly, we must have that $\theta(x)$ reduces to a polynomial $p_2(x)$ for $x > b$, since $D^k\theta(x) \equiv 0$ for $x > b$. The same argument as above then shows that all the $n + 1$ zeros of $D^{k-1}\theta$ must be in $[\tau_1, b)$. Together, we therefore have that $D^{k-1}\theta$ has (at least) $n + 1$ distinct zeros $(\eta_i)_{i=1}^{n+1}$ in $[a, b]$. From this it follows that $D^k\theta = \psi$ has n strong sign changes in (a, b). ∎

§2. Spline Wavelets

In this section we show how to construct minimally supported wavelets. We are primarily concerned with finite knot vectors, but the construction also applies in many cases with infinite knot vectors. Initially it is convenient to assume that the knot vectors are infinite.

2.1. Construction of a Minimally Supported Wavelet

Let τ and t be infinite knot vectors with τ a subsequence of t and $t_i < t_{i+k}$ for all i. Let s be the sequence such that $\tau \cup s = t$. We think of t as obtained from τ by inserting the knots s in τ, so we call s the *new* knots. Our aim is to determine a local basis for the orthogonal complement W of $S_{k,\tau}$ in $S_{k,t}$, the space of "wavelets". We start by picking an arbitrary new knot s and then show how to construct a wavelet in W of minimal support centered around s. For this purpose it will be convenient to assume that s is also infinite. We will then show later how the construction can be carried out when all sequences are finite.

Let s be a new knot. We seek integers ℓ and r with $r - \ell$ as small as possible, and a nonzero function $\psi \in W$ such that

$$\psi(x) = 0, \quad \text{for } x \notin [t_\ell, t_r], \tag{2.1}$$

and s is somehow centered in $[t_\ell, t_r]$. Since $W \subset S_{k,t}$, we require $\psi \in V_1$, where V_1 is given by

$$V_1 = \mathrm{span}\{B_{\ell,k,t}, B_{\ell+1,k,t}, \ldots, B_{r-k,k,t}\}.$$

(By the minimal support property of splines, we must have $r \geq \ell + k$, so that V_1 is nonempty.) Note that the requirement $\psi \in V_1$ is stronger than what is in general required to satisfy (2.1). The following notation will be convenient.

Definition 3. *Let f be a spline in $S_{k,t}$. We say that the support of f is $[\ell : r]$ if f is a linear combination of $B_\ell, \ldots, B_{r-k}$.*

Let $\tau_{\mu+1}, \tau_{\mu+2}, \ldots, \tau_{\mu+\nu}$ be the only knots from τ in the open interval (t_ℓ, t_r). In order to have $\psi \perp S_{k,\tau}$, it is sufficient to require that $\psi \perp V_0$, where V_0 is the space of splines in $S_{k,\tau}$ that intersect V_1, i.e.,

$$V_0 = \mathrm{span}\{B_{\mu+1-k,k,\tau}, B_{\mu+2-k,k,\tau}, \ldots, B_{\mu+\nu,k,\tau}\}.$$

To be able to enforce the orthogonality conditions and obtain a nonzero ψ, we must have $\dim V_1 > \dim V_0$, which is equivalent to $r - \ell + 1 > \nu + 2k$.

Therefore, if ψ is to have minimal support, then we show below that r must be selected as the smallest integer, and ℓ as the greatest integer such that

$$r - \ell = \nu + 2k. \tag{2.2}$$

To simplify the notation in what follows, let us set $q = \dim V_1 = r - \ell - k + 1$ and assume that (2.2) holds, so that $\dim V_0 = q - 1$. Let us also define $\{\phi_j\}$ and $\{\gamma_i\}$ by

$$\phi_j = B_{\mu-k+j,k,\tau}, \qquad \text{for } j = 1, 2, \ldots, q - 1,$$
$$\gamma_i = B_{\ell+i-1,k,t}, \qquad \text{for } i = 1, 2, \ldots, q.$$

We then have

$$V_0 = \operatorname{span}\{\phi_1, \phi_2, \ldots, \phi_{q-1}\}$$

and

$$V_1 = \operatorname{span}\{\gamma_1, \gamma_2, \ldots, \gamma_q\}.$$

There is a simple formula for a function ψ in the orthogonal complement W of V_0 in V_1 with support $[\ell : r]$:

$$\psi(x) = \det \begin{pmatrix} (\phi_1, \gamma_1) & \cdots & (\phi_1, \gamma_{q-1}) & (\phi_1, \gamma_q) \\ (\phi_2, \gamma_1) & \cdots & (\phi_2, \gamma_{q-1}) & (\phi_2, \gamma_q) \\ \vdots & \ddots & \vdots & \vdots \\ (\phi_{q-1}, \gamma_1) & \cdots & (\phi_{q-1}, \gamma_{q-1}) & (\phi_{q-1}, \gamma_q) \\ \gamma_1(x) & \cdots & \gamma_{q-1}(x) & \gamma_q(x) \end{pmatrix}. \tag{2.3}$$

In order to get minimal support, we must require more than (2.2), or equivalently that $\dim V_1 = \dim V_0 + 1$. The exact condition is given in the following definition.

Definition 4. *A wavelet with support $[\ell : r]$ is said to have* minimal support *if the following holds: If there is a nonzero wavelet with support $[u : v]$ with $\ell \leq u < v \leq r$, then $u = \ell$ and $v = r$. A wavelet of minimal support is called a* B-wavelet.

To see what minimal support entails, we need to study the implications of (2.2) more carefully. Suppose that $\rho_t(\ell) = \rho$ is the right multiplicity of t_ℓ and that $\lambda_t(r) = \lambda$ is the left multiplicity of t_r. The identity (2.2) can be interpreted as saying that in (t_ℓ, t_r), there should be $r - \ell - \nu = 2k$ new knots, not counting s, but counting ρ knots at t_ℓ and λ knots at t_r as new knots. For symmetry reasons, it is then natural to have k new knots between t_ℓ and s and k new knots between s and t_r. Minimal support means that we should find the largest ℓ and smallest r that satisfy these conditions. This is summarized in the next proposition.

Proposition 5. *Let s_i be a new knot, let ℓ be the greatest integer that satisfies*

$$\#\{j : \ s_j \in (t_\ell, s_i] \text{ and } j < i\} + \rho_t(t_\ell) = k,$$

and let r be the smallest integer that satisfies

$$\#\{j : \ s_j \in [s_i, t_r) \text{ and } j > i\} + \lambda_t(t_r) = k.$$

Then there is a B-wavelet with support $[\ell : r]$. Here $\#A$ denotes the cardinality of A.

Proof: That $[\ell : r]$ is minimal as the support of a wavelet follows from the foregoing discussion. That there in fact is a nonzero wavelet with this support follows from the proof of Proposition 7. ■

We emphasize that the assumption $\dim V_1 = \dim V_0 + 1$ is not sufficient to ensure minimal support of a wavelet. We must have ℓ and r such that $r - \ell$ is minimal with this property. For an example, suppose that

$$t_\ell < t_{\ell+1} = \tau_{\mu+1} = \cdots = \tau_{\mu+u} < \tau_{\mu+u+1}, \tag{2.4}$$

for some positive integer u. If we set $\ell' = \ell + u$, we obtain a new space V_1' of splines in $S_{k,t}$ that are nonzero on $[\ell' : r]$ and we clearly have $\dim V_1' = \dim V_1 - u$. On the other hand, the number of B-splines in $S_{k,\tau}$ with support intersecting $(t_{\ell'}, t_r)$ is also reduced by u, so that the corresponding space V_0' satisfies $\dim V_0' = \dim V_0 - u$. We therefore have $\dim V_1' = \dim V_0' + 1$. Using the same construction as before, it is therefore possible to construct a ψ' in W that is nonzero only on $[\ell' : r]$. In other words, in the situation described by (2.4), we would not have minimal support of the wavelet ψ. The following lemma formalizes and extends this example.

Lemma 6. *Let a be a knot in τ with multiplicity u, and let $[\ell : r]$ be the support of a B-wavelet with $t_\ell < a < t_r$. Then ℓ satisfies*

$$\#\{i : \ s_i \in (t_\ell, a)\} + \rho_t(t_\ell) \geq u + 1, \tag{2.5}$$

and r satisfies

$$\#\{i : \ s_i \in (a, t_r)\} + \lambda_t(t_r) \geq u + 1. \tag{2.6}$$

Proof: Suppose contrary to the lemma that there is a B-wavelet ψ with support $[\ell : r]$ and that $t_\ell < a$, for which

$$\#\{i : \ s_i \in (t_\ell, a)\} + \rho_t(t_\ell) = v \leq u.$$

Note that if we write (2.2) in the form $r - \ell - \nu = 2k$, it says that if we do not count the τ-knots in (t_ℓ, t_r) then the sequence $(t_i)_{i=\ell}^r$ should contain exactly $2k + 1$ knots. In our case we have $v \le u \le k$ knots to the left of a, at most k knots at a, and therefore at least one knot to the right of a so that $t_r > a$.

To complete the proof, we define w by

$$w = \#\{i : \ \tau_i \in (t_\ell, a)\},$$

and set $\ell' = \ell + u + w$. The space of splines V_1' with support $[\ell' : r]$ has dimension $\dim V_1' = \dim V_1 - u - w$, while the number of B-splines in $S_{k,\tau}$ with support intersecting (t_ℓ', t_r) is $\dim V_0 - u - w$, so that $\dim V_0' = \dim V_0 - u - w$. This means that we have $\dim V_1' = \dim V_0' + 1$ so that we can form a B-wavelet on (possibly a subinterval of) $[\ell' : r]$. The function ψ therefore cannot be a B-wavelet. ∎

With this more precise knowledge of minimal support, we can derive some properties of the B-wavelet defined by Proposition 5.

Proposition 7. *Suppose that ψ is a B-wavelet. Then the coefficients (c_i) in the expansion*

$$\psi = \sum_{i=1}^{q} c_i \gamma_i$$

oscillate strictly in sign and ψ is nonzero in (t_ℓ, t_r), apart from $q - 1$ points where it changes sign.

Proof: In this proof we use the notation $\mathrm{supp}_0\, f$ to denote the interior of the support of f so that $\mathrm{supp}_0\, B_{i,k} = (\tau_i, \tau_{i+k})$. If we set

$$d_i = \det G \begin{pmatrix} \gamma_1, \ldots, \gamma_{q-1} \\ \phi_1, \ldots, \phi_{i-1}, \phi_{i+1}, \ldots, \phi_q \end{pmatrix}, \tag{2.7}$$

we see from (2.3) that $c_i = (-1)^{q-i} d_i$. To prove that the c_i oscillate in sign, it is therefore sufficient to prove that $d_i > 0$ for $i = 1, 2, \ldots, q$.

From (1.4) we have

$$d_i = \int_\Omega D \begin{pmatrix} y_1, \ldots, y_{q-1} \\ \phi_1 \ldots, \phi_{q-1} \end{pmatrix} D \begin{pmatrix} y_1, \ldots, y_{q-1} \\ \gamma_1, \ldots, \gamma_{i-1}, \gamma_{i+1}, \ldots, \gamma_q \end{pmatrix}, \tag{2.8}$$

where $\Omega = \{(y_1, \ldots, y_{q-1}) \in \mathbb{R}^{q-1} : \ y_1 < y_2 \cdots < y_{q-1}\}$, so we clearly have $d_i \ge 0$. Since B-splines are continuous within their support, we have strict positivity if the integrand is positive at one point in the interior of the domain of integration Ω. But by Lemma 1, this happens if the two diagonals of the matrices in (2.8) are positive for some point in Ω, i.e., if and only if

$$\operatorname{supp}_0 \phi_j \cap \operatorname{supp}_0 \gamma_j \neq \emptyset, \qquad \text{for } j = 1, 2, \ldots, i-1,$$
$$\operatorname{supp}_0 \phi_j \cap \operatorname{supp}_0 \gamma_{j+1} \neq \emptyset, \qquad \text{for } j = i, i+1, \ldots, q-1. \tag{2.9}$$

For the purpose of this proof, let us relabel the knots so that $\mu = k$ and $\ell = 1$, which also gives $r - k = q = k + \nu + 1$. Then $\operatorname{supp}_0 \phi_j = (\tau_j, \tau_{j+k})$ and $\operatorname{supp}_0 \gamma_j = (t_j, t_{j+k})$, and the interior τ-knots in $(t_\ell, t_r) = (t_1, t_{q+k})$ are $\tau_{k+1}, \ldots, \tau_{q-1}$. By construction we know that $\operatorname{supp}_0 \phi_1 \cap \operatorname{supp}_0 \gamma_1 \neq \emptyset$ since $\tau_k \leq t_1 < \tau_{k+1}$. Likewise, we have $\operatorname{supp}_0 \phi_{q-1} \cap \operatorname{supp}_0 \gamma_q \neq \emptyset$ since $\tau_{k+\nu} < t_r \leq \tau_{k+\nu+1}$ or $\tau_{q-1} < t_{q+k} \leq \tau_q$.

Suppose now that $\operatorname{supp}_0 \phi_\sigma \cap \operatorname{supp}_0 \gamma_\sigma = \emptyset$ for some $\sigma > 1$. Since (2.9) holds for $j = 1$ and τ is a subsequence of t, we must then have $t_{p+k} \leq \tau_p$ for $p \geq \sigma$ and in particular $t_{q+k-1} \leq \tau_{q-1}$. Together with $\tau_{q-1} < t_{q+k}$, this means that

$$t_{q+k-1} \leq \tau_{q-1} < t_{q+k} = t_r.$$

But by Lemma 6 this contradicts the minimality of the support of ψ. A similar contradiction is obtained if $\operatorname{supp}_0 \phi_j \cap \operatorname{supp}_0 \gamma_{j+1} = \emptyset$ for some j. This shows that $d_i > 0$ for all i and therefore that the sequence (c_i) oscillates strictly in sign.

The fact that ψ has $q - 1$ strong sign changes in (t_ℓ, t_r) follows from Lemma 2 (since the B-spline coefficients of ψ oscillate strictly in sign it cannot vanish identically on any subinterval of (t_ℓ, t_r)). ∎

With the assumption that s contains infinitely many knots, the above construction can always be carried out. But if we relax this assumption, it is easy to find examples where it does not work. Suppose, for example, that $k = 2$ and $\tau = (0, 2, 3)$ and $t = (0, 1, 2, 3)$. Then there is only one new knot $s = 1$. Since there is only one knot to the left of s it is not possible to satisfy Proposition 5. We consider this problem in the next section.

2.2. A Basis of B-wavelets

The above discussion has shown how to construct a wavelet associated with a new knot. To make the construction work, we assumed that all three sequences involved were infinite. This is not the only assumption that makes the construction feasible. Suppose instead that all three sequences are finite,

$$\tau = (\tau_i)_{i=1}^{n+k}, \quad s = (s_i)_{i=1}^{m}, \quad t = (t_i)_{i=1}^{n+m+k}, \quad t = \tau \cup s, \tag{2.10}$$

and that

$$\tau_1 = \tau_k, \quad \tau_{n+1} = \tau_{n+k}, \qquad \text{and } s_i \in (\tau_k, \tau_{n+1}) \text{ for } i = 1, 2, \ldots, m. \quad (2.11)$$

Then the construction in Proposition 5 will always work since we can always find a knot of multiplicity k to the left and right of any new knot. In this way we can construct m wavelets $\{\psi_i\}_{i=1}^m$, one for each new knot s_i. The next result shows that these functions form a basis for W, the orthogonal complement of $S_{k,\tau}$ in $S_{k,t}$.

Proposition 8. *Suppose that the three sequences τ, s and t satisfy (2.10) and (2.11). For each new knot s_i, Proposition 5 gives a B-wavelet ψ_i with support $[\ell_i : r_i]$ such that $\ell_{i-1} < \ell_i$ and $r_{i-1} < r_i$, for $i = 2, \ldots, m$. The functions $\{\psi_i\}_{i=1}^m$ form a basis for the orthogonal complement of $S_{k,\tau}$ in $S_{k,t}$.*

Proof: The inequalities $\ell_{i-1} < \ell_i$ and $r_{i-1} < r_i$ are immediate from the construction. To show that the ψ_i are linearly independent, suppose that

$$f = \sum d_i \psi_i \equiv 0;$$

we must show that $d_i = 0$ for all i. Let $x = t_{\ell_1}$ and $\rho = \rho_t(\ell_1)$. Since $\ell_i > \ell_1$ for $i > 1$, we have $D^{k-1-\rho}\psi_1(x) \neq 0$ and $D^{k-1-\rho}\psi_i(x) = 0$ for $i > 1$. By taking the $(k-1-\rho)$-th derivative of f and evaluating at x, we therefore obtain $d_1 = 0$. Continuing like this with $t_{\ell_2}, \ldots, t_{\ell_m}$, we find that all the d_i must be zero. We thus have m linearly independent functions in W. Since $\dim W = m$, we therefore have a basis. ∎

2.3. B-wavelets for General, Finite Knot Vectors

The natural question now is how to construct a basis of minimally supported wavelets when $\tau = (\tau_j)_{j=1}^{n+k}$ and $t = (t_i)_{i=1}^{n+m+k}$ are arbitrary, finite knot vectors, with $\tau \subset t$. The above construction breaks down at the ends because there we cannot always find k new knots to the left and right of any new knot. This is similar to the situation in evaluation of spline functions. The standard algorithms assume that there are k nonzero B-splines at the point where the spline is to be evaluated. If this is not the case, one typically adds extra knots at the ends but sets the extra coefficients to zero. Here we will follow a similar procedure.

Let τ and t be general, finite knot vectors with $\tau \subset t$, and let $s = (s_j)_{j=1}^m$ be the new knots. Extend t to t' and τ to τ' by inserting k knots at a and b, where a and b are numbers satisfying $a < t_1$ and $b > t_{n+m+k}$. Note that we have $\boldsymbol{\tau}' \cup \boldsymbol{s} = \boldsymbol{t}'$. We will show that a wavelet basis for the orthogonal complement W of $S_{k,\tau}$ in $S_{k,t}$ can be found quite easily from a wavelet basis for the orthogonal complement W' of $S_{k,\tau'}$ in $S_{k,t'}$. Note that once we know the support of a minimally supported wavelet, the wavelet itself is given by (2.3). The complete construction is given in the following algorithm and proposition.

Algorithm 9. *Let* $\boldsymbol{\tau} = (\tau_j)_{j=1}^{n+k}$ *and* $\boldsymbol{t} = (t_i)_{i=1}^{n+m+k}$ *be two finite knot vectors with* $\boldsymbol{\tau} \subseteq \boldsymbol{t}$. *The following procedure determines* m *intervals* $[\ell_i : r_i]$ *on which a basis of B-wavelets can be constructed:*

1) *Form* $\boldsymbol{\tau}'$ *and* $\boldsymbol{t}'$ *by extending* $\boldsymbol{\tau}$ *and* $\boldsymbol{t}$ *with* k *knots at* a *and* b, *where* a *and* b *are real numbers satisfying* $a < t_1$ *and* $b > t_{n+m+k}$, *i.e.,*

$$\boldsymbol{\tau}' = (\tau_j)_{j=1-k}^{n+2k} \qquad \text{and} \qquad \boldsymbol{t}' = (t_i)_{i=1-k}^{n+m+2k}$$

where $\tau_j = t_j = a$ *for* $j < 1$ *and* $\tau_j = t_{m+j} = b$ *for* $j > n + k$.

2) *Determine a B-wavelet basis* $(\psi_i')_{i=1}^{m}$ *for the orthogonal complement of* $S_{k,\tau'}$ *in* $S_{k,t'}$ *as in Proposition 8, and let* (ℓ_i', r_i') *be the support of* ψ_i'.

3) *The intervals* $[\ell_i : r_i]$ *for* $i = 1, 2, \ldots, m$ *are then given by*

$$
\begin{aligned}
\ell_i &= \begin{cases} \ell_i', & \text{if } t_{\ell_i'} \geq \tau_k; \\ \ell_i' + j, & \text{if } t_{\ell_i'} \in [\tau_{k-j}, \tau_{k-j+1}) \text{ for some } j \in [1, k]; \end{cases} \\
r_i &= \begin{cases} r_i', & \text{if } t_{r_i'} \leq \tau_{n+1}; \\ r_i' - j & \text{if } t_{r_i'} \in (\tau_{n+j}, \tau_{n+j+1}] \text{ for some } j \in [1, k]. \end{cases}
\end{aligned}
\tag{2.12}
$$

Proposition 10. *The sequences* (ℓ_i) *and* (r_i) *defined in Algorithm 9 are strictly increasing, and corresponding to each interval* $[\ell_i : r_i]$ *there is a B-wavelet* ψ_i. *Each* ψ_i *has* $r_i - \ell_i - k$ *strict sign changes in* (t_{ℓ_i}, t_{r_i}), *but is otherwise nonzero in this interval, and* $\{\psi_i\}_{i=1}^{m}$ *provides a minimal support basis for the orthogonal complement of* $S_{k,\tau}$ *in* $S_{k,t}$.

Proof: The conclusions all follow as in Proposition 5, Proposition 7 and Proposition 8 if we can show that $[\ell_i : r_i]$ is a minimal interval for obtaining a nonzero function in the orthogonal complement W, and that the sequences (ℓ_i) and (r_i) are strictly increasing. Let us first show that $[\ell_i : r_i]$ is minimal for each i.

If both $\ell_i = \ell_i'$ and $r_i = r_i'$, we are in the same situation that was treated earlier, so there is nothing to prove. Suppose therefore that $\ell_i = \ell_i' + j$ and $r_i = r_i'$ for some i and some $j \in [1, k]$, i.e., that

$$t_{\ell_i'} \in [\tau_{k-j}, \tau_{k-j+1}).
\tag{2.13}$$

Let V_1' be the set of splines in $S_{k,t'}$ with support $[\ell_i' : r_i']$, and let V_0' be the set of splines in $S_{k,\tau'}$ whose supports intersect $I' = (t_{\ell_i'}, t_{r_i'})$. We then know that $\dim V_1' = \dim V_0' + 1$ and that there are no nonzero splines in W' with support strictly inside I'. Similarly, let V_1 be the subspace of $S_{k,t}$ with support $[\ell_i : r_i]$, and let V_0 be the subspace of of splines in $S_{k,\tau}$ whose supports intersect $I = (t_{\ell_i}, t_{r_i})$. We must show that $\dim V_1 = \dim V_0 + 1$ and that $r_i - \ell_i$ is minimal with this property.

When the k extra knots at a and b are removed, we lose the first k and the last k B-splines in $S_{k,\tau'}$ and end up with $S_{k,\tau}$. Of these lost B-splines in $S_{k,\tau'}$, only $B_{1-j}', \ldots, B_0'$ are nonzero in $(\tau_{k-j}, \tau_{k-j+1})$ since we have assumed that $r_i' = r_i$. But now we claim that $t_{\ell_i} < \tau_k$, from which it follows that $\dim V_0 = \dim V_0' - j$ (the support of the first B-spline on τ intersects both I and I'). To establish the inequality $t_{\ell_i} < \tau_k$, we start with $t_{\ell_i'} < \tau_{k-j+1}$, which follows from (2.13). From Lemma 6 we know that $t_{\ell_i'+1} < \tau_{k-j+1}$, and hence that $t_{\ell_i'+j} < \tau_k$, which is the required inequality.

To show that $r_i - \ell_i$ is minimal, we first note that if r_i could be smaller, then r_i' could have been smaller, which contradicts the minimality of $r_i' - \ell_i'$. At the other end, if ℓ_i is increased, we decrease $\dim V_1$, and $\dim V_0$ must therefore decrease correspondingly for this to work. But this would require $t_{\ell_i} \geq \tau_{k+1}$, in which case we should have had $\ell_i' = \ell_i$ in the first place.

This completes the proof that $[\ell_i : r_i]$ is minimal when $\ell_i = \ell_i' + j$ and $r_i = r_i'$. The case $\ell_i = \ell_i'$ and $r_i = r_i' - j$ is similar. In the remaining case $\ell_i = \ell_i' + j_1$ and $r_i = r_i' - j_2$ for suitable j_1 and j_2. This can also be treated similarly by removing the k knots at the two ends of τ' in two steps.

It remains to show that the two sequences $(\ell_i)_{i=1}^m$ and $(r_i)_{i=1}^m$ are strictly increasing, and it clearly suffices to consider the first of them. We already know that $(\ell_i')_{i=1}^m$ is increasing, so that we only need to consider the beginning of the sequence where $\ell_i \neq \ell_i'$. Suppose first that ℓ_i' is the greatest number such that $t_{\ell_i'} \in [\tau_{k-1}, \tau_k)$, or $\ell_i = \ell_i' + 1$. From Lemma 6 we see that we must have $t_{\ell_i'+1} < \tau_k$, so that there is "space" to increase by one all of the ℓ_i' belonging to this interval.

Consider next that ℓ_i' is the greatest number such that $t_{\ell_i'} \in [\tau_{k-2}, \tau_{k-1})$, or $\ell_i = \ell_i' + 2$. If $\tau_{k-1} = \tau_k$, then Lemma 6 shows that there must be two free "spaces" in t, i.e., that $t_{\ell_i'+2} < \tau_{k-1}$ which ensures that (ℓ_i) will be increasing from this interval onwards. If $\tau_{k-1} < \tau_k$, we know that the ℓ_i' in the interval $[\tau_{k-1}, \tau_k)$ have already been moved one to the right, leaving one gap at the left of the interval, and Lemma 6 once again assures us that there is one more space immediately preceding τ_{k-1}. Together this gives us the necessary two spaces. This argument can clearly be continued until we have considered all of the intervals. ∎

2.4. An Alternative Representation of B-wavelets

The proof of Lemma 2 suggests that by applying k-fold integration to a function which is orthogonal to a spline space $S_{k,\tau}$, one obtains a function which vanishes at the knots τ. In this way we can find an alternative representation for a B-wavelet ψ.

Lemma 11. *Let τ and t be two knot vectors with $\tau \subset t$ and $t_i < t_{i+k}$ for all i. Let $[\ell : r]$ be the support of a B-wavelet ψ such that $t_\ell \geq \tau_k$ and $t_r \leq \tau_{n+1}$. Then $\psi(x) = D^k \theta(x)$, where*

$$\theta(x) = \gamma D \left(\begin{matrix} x, \tau_{\mu+1}, \ldots, \tau_{\mu+\nu} \\ B_{\ell,2k,t}, B_{\ell+1,2k,t}, \ldots, B_{\ell+\nu,2k,t} \end{matrix} \right). \qquad (2.14)$$

Here γ is a suitable constant, while $\tau_{\mu+1}, \ldots, \tau_{\mu+\nu}$ are the knots from τ in (t_ℓ, t_r) and $\ell + \nu = r - 2k$.

Proof: From the proof of Lemma 2, we know that the function

$$\theta(x) = \int_{t_\ell}^{t_r} (x - w)_+^{k-1} \psi(w) dw / (k - 1)!$$

vanishes on τ and $D^k \theta = \psi$. This means that θ must be a piecewise polynomial of order $2k$, and from continuity considerations we see that we must have $\theta \in S_{2k,t}$. From this it also follows that the support of θ must be $[\ell : r]$, i.e., it must be a linear combination of $\{B_{i,2k,t}\}_{i=\ell}^{r-2k}$, which is a total of $r - \ell - 2k + 1 = \nu + 1$ B-splines. Since the number of τ's in (t_ℓ, t_r) is ν, it is clear that θ must be given by (2.14) provided that the determinant is not identically zero. To show that this is not the case, it is sufficient to show that one of the coefficients in the expansion of the determinant is nonzero, e.g.,

$$D \left(\begin{matrix} \tau_{\mu+1}, \ldots, \tau_{\mu+\nu} \\ B_{\ell+1,2k,t}, \ldots, B_{\ell+\nu,2k,t} \end{matrix} \right) \neq 0.$$

From Lemma 1 we see that this holds if $\tau_{\mu+j} \in \operatorname{supp} B_{\ell+j,2k,t}$ for $j = 1, 2, \ldots, \nu$. It is therefore sufficient to show that $t_{\ell+1} < \tau_{\mu+1}$ and $\tau_{\mu+\nu} < t_{\ell+\nu+2k}$. The last inequality is trivial since $r = \ell + \nu + 2k$. For the first one, note that $t_\ell < \tau_{\mu+1}$, so that the only case that can cause problems is $t_\ell < t_{\ell+1} = \tau_{\mu+1}$. But this is impossible by Lemma 6. ∎

A representation of ψ similar to that given in Lemma 11 exists in general. However, since $D^k \theta = \psi$, it requires taking k antiderivatives of a general spline space and must be formulated in terms of natural splines.

§3. Examples

In this section we give some examples and plots of wavelets for different knot configurations. To compute the wavelets, we have used the representation (2.14) (more general than shown here, using natural splines).

Example 1 (Cardinal splines). Let us first consider the cardinal case and make sure that our construction coincides with that of [4], Chapter 6. In this case $\boldsymbol{\tau}$ and $\boldsymbol{t}$ are given by $\tau_i = i$ and $t_i = i/2$ for $i \in \mathbb{Z}$. Because of the uniform knot vectors, there is essentially only one B-wavelet. For the ψ corresponding to the new knot $s = k - 1/2$, the ℓ and r in Proposition 5 become $\ell = 0$ and $r = 2k - 1$. This is in agreement with the result in [4, Theorem 6.4].

For $k = 3$, the θ in Lemma 11 is easily found to be

$$\theta(x) = \gamma \det \begin{pmatrix} \phi_0(x) & \phi_1(x) & \phi_2(x) & \phi_3(x) & \phi_4(x) \\ 26 & 1 & 0 & 0 & 0 \\ 26 & 66 & 26 & 1 & 0 \\ 0 & 1 & 26 & 66 & 26 \\ 0 & 0 & 0 & 1 & 26 \end{pmatrix}, \qquad (3.1)$$

where $\phi_i(x) = 120 B_{i,6,t}$, and γ is a scaling constant. Plots of θ and $\psi = D^3\theta$ for a suitable γ are shown in Figures 1 and 2 (cf. Figure 6.2.2 in [4]).

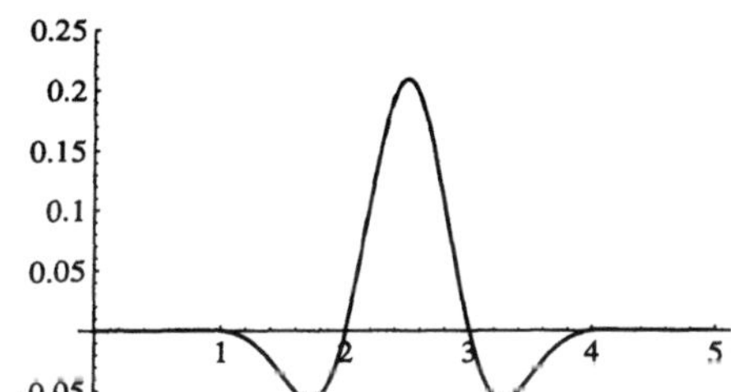

Figure 1. The function θ in (2.14).

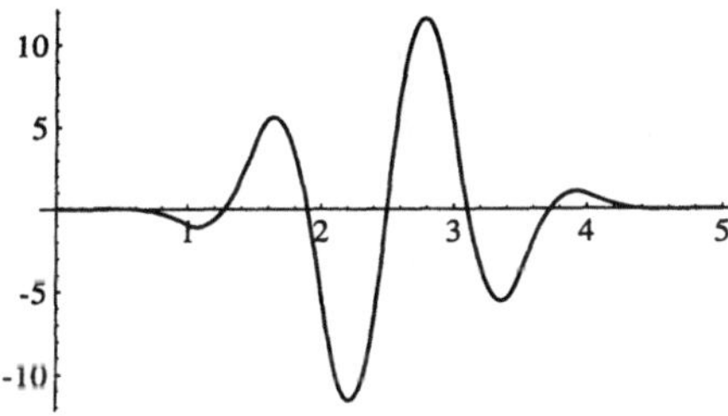

Figure 2. The quadratic B-wavelet for cardinal splines.

Example 2 (Bernstein case). Suppose next that $\boldsymbol{\tau}$ contains k knots at -1 and k knots at 1 and that there are k new knots at 0 so that $\boldsymbol{t}$ contains k knots at -1, 0, and 1. Following the recipe above, we find k B-wavelets $(\psi_i)_{i=1}^{k}$ and the support of ψ_i is (t_i, t_{i+2k}), i.e., the same as $B_{i,2k,t}$. But this is no surprise since (2.14) now simplifies to $\theta = B_{i,2k,t}$ so that $\psi = \gamma D^k B_{i,2k,t}$. The three B-wavelets in the quadratic case are shown in Figure 3. Note that they are all discontinuous.

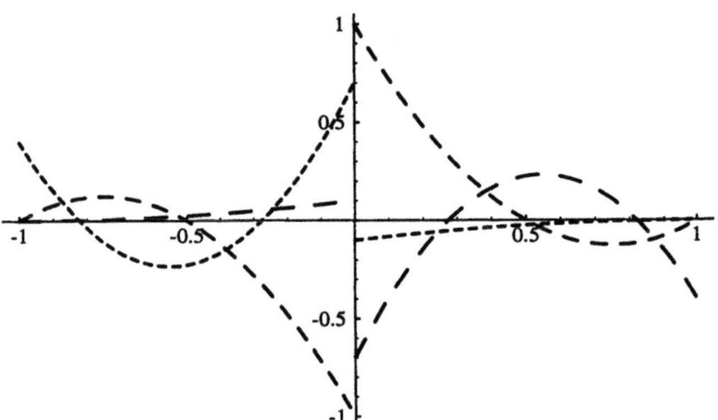

Figure 3. The three B-wavelets in the quadratic Bernstein case.

Example 3. Consider the knot vectors

$$\boldsymbol{\tau} = (0, 0, 0, 3, 3, 4, 4, 4)$$
$$\boldsymbol{t} = (0, 0, 0, 1, 1, 3, 3, 3, 3.1, 4, 4, 4).$$

There are 4 new knots at 1, 1, 3, and 3.1. From Proposition 5 we find the support of the 4 corresponding B-wavelets to be $[1 : 7]$, $[2 : 8]$, $[3 : 11]$, and $[6 : 12]$. Plots of the second and third wavelets ψ_2 and ψ_3, together with the corresponding θ_2 and θ_3 in Lemma 11, are shown in Figures 4–7.

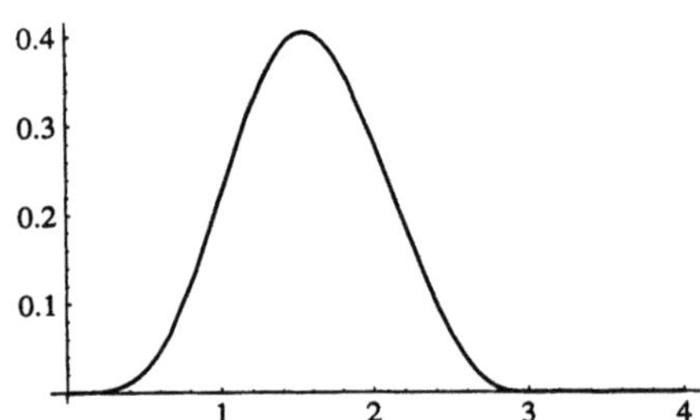

Figure 4. The function θ_2 of example 3.

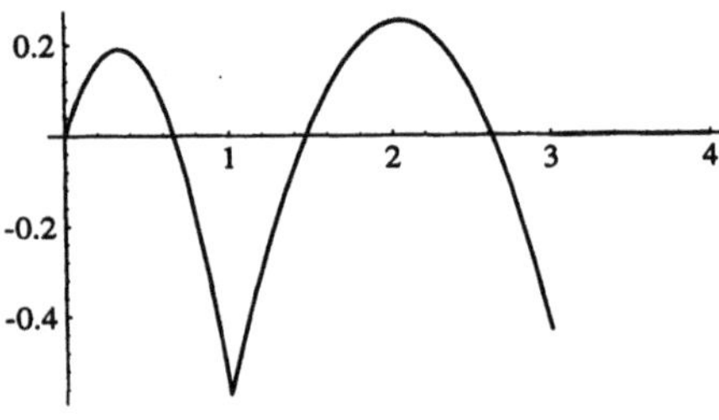

Figure 5. The B-wavelet ψ_2 of example 3.

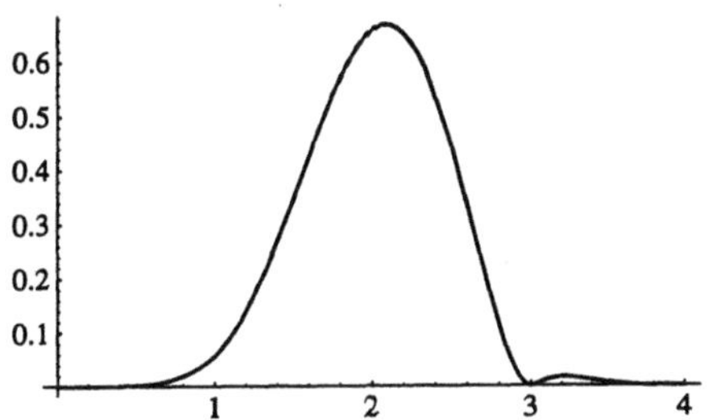
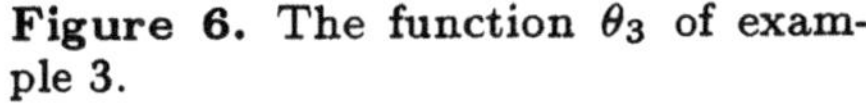

Figure 6. The function θ_3 of example 3.

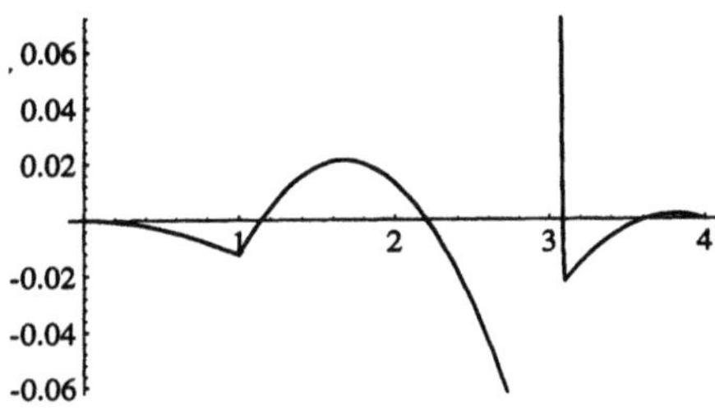

Figure 7. The B-wavelet ψ_3 of example 3.

References

1. Ahlberg, J. H., and E. N. Nilson, Orthogonality properties of spline functions, J. Math. Anal. Appl. **11**, (1965), 321–337.

2. de Boor, C., *A Practical Guide to Splines*, Springer-Verlag, New York, 1978.

3. Buhmann, M., and C. A. Micchelli, Spline pre-wavelets for non-uniform knots, Numer. Math. , to appear.

4. Chui, C., *An Introduction to Wavelets*, Academic Press, Boston, 1992.

5. Daubechies, I., Orthonormal bases of compactly supported wavelets, Comm. Pure Appl. Math. **41** (1988), 909–996.

6. Goodman, T. N. T, and S. L. Lee, Wavelets of multiplicity r, Report AA/921, Dept. of Math. and Comp. Science, University of Dundee, Scotland, February 1992.

7. Karlin, S., *Total Positivity, Vol. I*, Stanford University Press, Stanford, California, 1968.

8. Mallat, S., Multiresolution approximations and wavelet orthonormal bases of $L^2(\mathbb{R})$, Trans. Amer. Math. Soc. **315** (1989), 69–87.

9. Meyer, Y., *Ondelettes et Opérateurs I*, Hermann, Paris, 1990.

10. Schumaker, L. L., *Spline functions: Basic Theory*, Wiley, New York, 1981.

Acknowledgements. The first author was supported in part by the Royal Norwegian Council for Scientific and Industrial Research - NTNF and Center for Industrial Research - SI.

T. Lyche and K. Mørken
Institutt for Informatikk, University of Oslo
P.O.Box 1080, Blindern
0316 Oslo 3, Norway
tom @ ifi.uio.no, knutm @ ifi.uio.no

Numerical Methods of Approximation Theory, Vol. 9
Dietrich Braess and Larry L. Schumaker (eds.), pp. 195–206.
International Series of Numerical Mathematics, Vol. 105

195

Necessary Conditions for Local Best Chebyshev Approximations by Splines with Free Knots

Bernd Mulansky

Dedicated to the memory of Lothar Collatz

Abstract. This paper is concerned with Chebyshev approximation of real continuous functions from the class $S_{n,k}$ of polynomial splines of degree n with at most k free knots. Introducing the notion of an extended tangent cone, which also contains discontinuous splines, we derive an improved necessary condition for local best approximations from $S_{n,k}$. This condition is equivalently formulated as an alternant criterion. It shows that the corresponding error function must have an alternant of a certain length with a prescribed sign on a subinterval.

§1. Introduction

Although the nonlinear approximation problem of best Chebyshev approximation by polynomial spline functions with free knots has been thoroughly investigated in a number of papers (see *e.g.*, [7,8] and references therein), several questions still remain open. The aim of this paper is to improve the known necessary conditions for local best approximations. The considerations are based on the notion of an extended tangent cone, which also contains discontinuous splines as tangent directions. The derived necessary characterization may be equivalently formulated as an alternant condition.

Let E be a normed linear space, $V \subset E$ be a subset of E, and suppose we are given $f \in E$. An element $v_0 \in V$ is a *best approximation to f from V* if

$$\|f - v_0\| \le \|f - v\| \tag{1}$$

is satisfied for all $v \in V$. The element v_0 is called a *local best approximation to f from V* if the inequality (1) holds for all v from a neighbourhood $U \cap V$ of v_0 in V.

Here we consider best Chebyshev approximation, *i.e.*, we deal with the space $E = C[0,1]$ of real continuous functions defined on the interval $[0,1]$, which is endowed with the maximum norm

$$\|g\| := \max\{|g(t)| : \ t \in [0,1]\}. \tag{2}$$

For convenience, we define the approximation set $S_{n,k}$ of spline functions by an explicit representation. Therefore, the truncated power functions are introduced by

$$(x)_+^j := x^j \cdot (x)_+^0, \quad j > 0,$$

where

$$(x)_+^0 := \begin{cases} 0 & \text{if } x < 0 \\ 1 & \text{otherwise.} \end{cases}$$

Let integers $n \ge 1$ and $k \ge 1$ be given. The set $S_{n,k}$ of *polynomial splines of degree n with at most k free knots* is defined by

$$S_{n,k} := \left\{ s : [0,1] \to \mathbb{R} : \ s(t) = \sum_{i=0}^{\lambda} \sum_{j=1}^{\mu_i} \alpha_{ij} \cdot (t - \xi_i)_+^{n+1-j} : \ \lambda \in \mathbb{N}, \right.$$

$$\mu_1, \ldots, \mu_\lambda \in \{1, \ldots, n\} : \textstyle\sum_{i=1}^{\lambda} \mu_i \le k; \ \xi_1, \ldots, \xi_\lambda \in \mathbb{R} :$$

$$\left. 0 < \xi_1 < \ldots < \xi_\lambda < 1; \ \alpha_{ij} \in \mathbb{R} : j = 1, \ldots, \mu_i, i = 0, \ldots, \lambda \right\},$$

where the notations $\xi_0 := 0$, $\mu_0 := n+1$ are used for abbreviation. All the other greek letters denote free parameters of the class $S_{n,k}$. The numbers ξ_i are called the *knots* and μ_i their corresponding *multiplicities*, which are related to the smoothness of the splines at the knots. The integer λ denotes the *number of distinct knots*. The total number of knots counting multiplicities is bounded by the given integer k. The real coefficients α_{ij} may be considered as the *linear* parameters of the class $S_{n,k}$.

Since splines with multiple knots occur as limits of splines having simple knots, it is necessary to allow for multiple knots in the definition of $S_{n,k}$ in order to assure the existence of best approximations. Indeed, it was proved by Schumaker that a best approximation from $S_{n,k}$ exists to each function $f \in C[0,1]$, see [1,8]. However, the occurence of multiple knots complicates the local structure of the approximation set $S_{n,k}$.

The usual approach to nonlinear approximation problems involves a local linearization based on the notion of the tangent cone, which reflects the local differential structure of the approximation manifold. A general necessary condition states that in order for $v_0 \in V$ to be a local best approximation of a given function f, the zero function 0 must be a best approximation to $f - v_0$ from the tangent cone $T(V, v_0)$ to V at v_0 (for definitions see the next section).

It was established by Cromme [3,4] (see also [6]) that the tangent cone $T(S_{n,k}, s_0)$ to $S_{n,k}$ at a nondegenerate spline $s_0 \in S_{n,k}$ is a convex subcone, defined by sign restrictions on some coefficients, of a linear space of spline functions with fixed knots.

As already pointed out in [7], the resulting conditions are not satisfactory if knots of multiplicity $m_i \geq n - 1$ occur, which is a common situation in the most interesting cases of splines of low degree, *e.g.*, $n = 1, 2, 3$. The reason seems to be that the tangent cone is limited to contain only continuous splines. This remains true even if the space $C[0, 1]$ is embedded into the space $L_\infty[0, 1]$ of measureable, almost everywhere bounded functions on $[0, 1]$ endowed with the norm

$$\|g\| := \operatorname{ess\,sup}\{|g(t)| : t \in [0, 1]\}, \tag{3}$$

and the tangent cone is computed with respect to $L_\infty[0, 1]$. Let us note that for continuous functions g, the norm defined in (3) reduces to the maximum norm (2). Therefore, we use the same notation for both norms.

Our main idea consists of modifying the definition of the tangent cone in $L_\infty[0, 1]$. The usual tangent cone describes an *inner* property of the approximation set and does not depend on the approximation problem. Our modification refers to the considered approximation problem, since it requires the approximated function as well as the members of the approximation set to be *continuous* functions.

The extended tangent cone to $S_{n,k}$ at s_0 obtained in this way also contains discontinuous splines, but it still serves to give a necessary condition for local best approximations. Based on minor extensions of the results in [7], the improved necessary conditions may be formulated as alternant conditions.

Our characterization result combines and improves the conditions given in [7] and [9]. The idea of using discontinuous splines as tangent directions is contained implicitly in [9], although no reference is made to an extension of the tangent cone.

§2. An Extended Tangent Cone

Let V be a non-empty set in $E = C[0,1]$, and let v_0 be an element of V.

Definition 1. *An element $h \in E$ is a tangent direction to V at v_0 if there is a $\lambda_0 > 0$ and a mapping $(0, \lambda_0) \to V : \lambda \mapsto v_\lambda$ such that*

$$\|\frac{1}{\lambda}(v_\lambda - v_0) - h\| \to 0 \text{ as } \lambda \to 0. \tag{4}$$

The set of all tangent directions is called the tangent cone $T(V, v_0)$ to V at v_0.

Obviously, $T(V, v_0) \subset E = C[0,1]$. This also remains true if the space $C[0,1]$ is embedded into the space $L_\infty[0,1]$, and the tangent cone to V at v_0 is defined with respect to $E = L_\infty[0,1]$. In order to allow for discontinuous tangent directions, we have to modify the definition of the tangent cone.

Definition 2. *An element $h \in E = L_\infty[0,1]$ is an extended tangent direction to V at v_0, if there is a $\lambda_0 > 0$ and a mapping $(0, \lambda_0) \to V : \lambda \mapsto v_\lambda$ such that $\|v_\lambda - v_0\| \to 0$ as $\lambda \to 0$ and*

$$\|\frac{1}{\lambda}(v_\lambda - v_0) - g\| \to \|h - g\| \text{ as } \lambda \to 0 \tag{5}$$

for all $g \in C[0,1]$. The set of all extended tangent directions is called the extended tangent cone $T_e(V, v_0)$ to V at v_0.

It is easily established that all extended tangent directions form a cone, i.e., if $h \in T_e(V, v_0)$, then also $\alpha h \in T_e(V, v_0)$ for all $\alpha \geq 0$. Furthermore, the extended tangent cone has the interesting property

$$T_e(V, v_0) \cap C[0,1] = T(V, v_0). \tag{6}$$

The following theorem shows that the extended tangent cone can also be used to give a necessary condition for local best approximations.

Theorem 1. *Let $V \subset C[0,1]$, $v_0 \in V$ and $f \in C[0,1]$ be given. If v_0 is a local best approximation to f from V (in $C[0,1]$), then 0 is a best approximation to $f - v_0$ from $T_e(V, v_0)$ (in $L_\infty[0,1]$).*

Proof: We consider an arbitrary function $h \in T_e(V, v_0)$. By definition there exists a mapping $(0, \lambda_0) \to V : \lambda \mapsto v_\lambda$ such that

$$\|h_\lambda - g\| \to \|h - g\| \text{ as } \lambda \to 0 \tag{7}$$

for all $g \in C[0,1]$, where $h_\lambda := \frac{1}{\lambda}(v_\lambda - v_0)$. For $0 < \lambda < 1$ we get

$$\|f - v_\lambda\| = \|(1 - \lambda)(f - v_0) + \lambda(f - v_0 - h_\lambda\|$$
$$\leq (1 - \lambda)\|f - v_0\| + \lambda\|f - v_0 - h_\lambda\|,$$

and therefore

$$\|f - v_0 - h_\lambda\| \geq \frac{1}{\lambda}\Big(\|f - v_\lambda\| - (1 - \lambda)\|f - v_0\|\Big). \tag{8}$$

Since v_0 is a local best approximation to f in V and $v_\lambda \to v_0$ as $\lambda \to 0$, for sufficiently small $\lambda > 0$ it holds

$$\|f - v_\lambda\| \geq \|f - v_0\|,$$

and from (8) we obtain

$$\|f - v_0 - h_\lambda\| \geq \|f - v_0\|. \tag{9}$$

Taking the limit as $\lambda \to 0$ in (9) and applying (7) to $g = f - v_0 \in C[0,1]$, we get

$$\|f - v_0 - h\| \geq \|f - v_0\|. \quad \blacksquare$$

The following lemma may be useful for the computation of the extended tangent cone in special applications. It allows us to separate the discontinuity of an element h (e.g., as a step function) in the course of proving that $h \in T_e(V, v_0)$.

Lemma 1. *Let mappings $(0, \lambda_0) \to C[0,1] : \lambda \mapsto h_{1,\lambda}$ and $(0, \lambda_0) \to C[0,1] : \lambda \mapsto h_{2,\lambda}$ be given having the properties*

$$\|h_{1,\lambda} - g\| \to \|h_1 - g\| \text{ as } \lambda \to 0$$

for all $g \in C[0,1]$ and

$$\|h_{2,\lambda} - h_2\| \to 0 \text{ as } \lambda \to 0,$$

where $h_2 \in C[0,1]$, $h_1 \in L_\infty[0,1]$. Then

$$\|h_{1,\lambda} + h_{2,\lambda} - g\| \to \|h_1 + h_2 - g\| \text{ as } \lambda \to 0$$

for all $g \in C[0,1]$.

Proof: The proof follows immediately from

$$\|h_{1,\lambda}+h_2-g\|-\|h_{2,\lambda}-h_2\| \le \|h_{1,\lambda}+h_{2,\lambda}-g\| \le \|h_{1,\lambda}+h_2-g\|+\|h_{2,\lambda}-h_2\|$$

by taking the limit as $\lambda \to 0$. $\blacksquare$

Discussion:

1. The notion of the extended tangent cone was already introduced in [6] and applied to approximation by linear splines (of degree 1).

2. Although the definition of the extended tangent cone seems to depend strongly on the special situation considered here, in view of other applications, it may be worthwhile to point out the main idea more abstractly. The usual tangent cone defines an inner property of the approximation set independently of the approximation problem. By utilizing special properties of the approximation (minimization) problem, *i.e.*, by restricting the class of considered objective functionals, it may be possible to formulate improved necessary optimality conditions with respect to an extended tangent cone.

§3. Improved Necessary Conditions

We are now ready to derive the improved necessary conditions for local best approximations from $S_{n,k}$. Let $s_0 \in S_{n,k}$ be given by the representation

$$s_0(t) = \sum_{i=0}^{\ell} \sum_{j=1}^{m_i} a_{ij} \cdot (t - x_i)_+^{n+1-j}. \tag{10}$$

We assume that s_0 is a nondegenerate spline in $S_{n,k}$ having k active knots, *i.e.*, $s_0 \in S_{n,k} \setminus S_{n,k-1}$. That means $a_{im_i} \ne 0$ for $i = 1,\dots,\ell$ and $\sum_{i=1}^{\ell} \mu_i = k$. To s_0 we assign a convex cone S_0 of spline functions with fixed knots by the definition

$$S_0 := \left\{ s \in L_\infty[0,1] : \ s(t) = \sum_{i=0}^{\ell} \sum_{j=1}^{\tilde{m}_i} \alpha_{ij} \cdot (t - x_i)_+^{n+1-j}; \right.$$
$$\left. \alpha_{ij} \in \mathbb{R} : \alpha_{i\tilde{m}_i} \cdot a_{im_i} \ge 0 \text{ for } \tilde{m}_i = m_i + 2 \right\}, \tag{11}$$

where

$$\tilde{m}_i := \begin{cases} 2, & \text{if } m_i = 1 \\ \min\{m_i + 2, n + 1\}, & \text{otherwise} \end{cases}, \ i = 1,\dots,\ell, \tag{12}$$

and $\tilde{m}_0 = n + 1$.

Theorem 2. *If s_0 is a local best approximation to a given function f from $S_{n,k}$, then 0 is a best approximation to $f - s_0$ from S_0.*

Proof: We prove the theorem by showing that $S_0 \subset T_e(S_{n,k}, s_0)$. It is sufficient to consider the case $\ell = 1$ and to omit the polynomial part corresponding to $i = 0$, i.e.,

$$s_0(t) = \sum_{j=1}^{m} a_j \cdot (t - x)_+^{n+1-j}, \tag{13}$$

see [5]. For $\xi_1, \ldots, \xi_j \in \mathbb{R}$, let

$$B_j(\xi_1, \ldots, \xi_j; t) := [\xi_1, \ldots, \xi_j](t - x)_+^{n+1-j}, \quad t \in [0, 1], \tag{14}$$

denote the $(j - 1)$st order divided difference of the function $(t - x)_+^n$ with respect to x for fixed t, see [10]. The spline function s_0 given by (13) may be written in the form

$$s_0(t) = \sum_{j=1}^{m} b_j B_j(x, \ldots, x; t), \tag{15}$$

where

$$b_j = (-1)^{j-1} a_j \binom{n}{j-1}^{-1} \quad \text{for } j = 1, \ldots, m. \tag{16}$$

From the assumption on s_0, it follows that $b_m \neq 0$. Analogously, let $h \in S_0$ be given by

$$h(t) = \sum_{j=1}^{\tilde{m}} d_j B_j(x, \ldots, x; t). \tag{17}$$

We have to distinguish two cases.

Case 1: Suppose that $\tilde{m} = m + 1$ (i.e., $m = 1$ or $m = n$). For sufficiently small $\lambda > 0$, we define s_λ by splitting the m-fold knot at x of s_0 into a $(m - 1)$-fold knot at x and a simple knot near x,

$$s_\lambda(t) := \sum_{j=1}^{m-1} b_j^\lambda B_j(x, \ldots, x; t) + b_m^\lambda B_m(x, \ldots, x, x + \lambda y_\lambda; t). \tag{18}$$

By an application of the recurrence relations for divided differences we obtain

$$\begin{aligned}
h_\lambda(t) :=& \frac{1}{\lambda}(s_\lambda(t) - s_0(t)) \\
=& \sum_{j=1}^{m} \frac{1}{\lambda}(b_j^\lambda - b_j) B_j(x, \ldots, x; t) \\
& + b_m^\lambda y_\lambda B_{m+1}(x, \ldots, x, x + \lambda y_\lambda; t).
\end{aligned} \tag{19}$$

Choosing $y_\lambda, b_1^\lambda, \ldots, b_m^\lambda$ according to

$$y_\lambda := \frac{d_{m+1}}{b_m + \lambda d_m},$$

$$b_j^\lambda := b_j + \lambda d_j, \quad j = 1, \ldots, m, \tag{20}$$

equation (18) reduces to

$$h_\lambda(t) = \sum_{j=1}^m d_j B_j(x, \ldots, x; t) + d_{m+1} B_{m+1}(x, \ldots, x, x + \lambda y_\lambda; t). \tag{21}$$

Since y_λ converges and $b_j^\lambda \to b_j$ as $\lambda \to 0$, it follows from the properties of divided differences (see [10]) that $\|s_\lambda - s_0\| \to 0$ as $\lambda \to 0$. Furthermore, for $m \le n - 1$ we immediately get $\|h_\lambda - h\| \to 0$ as $\lambda \to 0$, where h_λ as well as h are continuous functions. In the case $m = n$, we refer to Lemma 1 by putting

$$\begin{aligned}
h_{1,\lambda}(t) &= d_{n+1} B_{n+1}(x, \ldots, x, x + \lambda y_\lambda; t), \\
h_1(t) &= d_{n+1} B_{n+1}(x, \ldots, x; t) \qquad \text{and} \\
h_2(t) &= h_{2,\lambda}(t) = \sum_{j=1}^n d_j B_j(x, \ldots, x; t).
\end{aligned} \tag{22}$$

The function h_1 is a step function,

$$h_1(t) = \begin{cases} 0, & \text{if } t < x \\ d_{n+1}, & \text{if } t \ge x. \end{cases} \tag{23}$$

Suppose $y_\lambda > 0, d_{n+1} > 0$; the other cases may be handled similarly. Then we get $h_{1,\lambda}(t) = h_1(t)$ for $t \notin [x, x + \lambda y_\lambda]$ and $\lim_{t \to x-0} h_1(t) = 0 \le h_{1,\lambda}(t) \le h(x) = d_{n+1}$ on $[x, x + \lambda y_\lambda]$. It is easily proved that

$$\|h_{1,\lambda} - g\| \to \|h_1 - g\| \text{ as } \lambda \to 0 \tag{24}$$

for all $g \in C[0, 1]$.

Case 2: Suppose that $\tilde{m} = m + 2$ (*i.e.*, $2 \le m \le n - 1$). It follows from the definition of S_0 that $d_{m+2} \cdot b_m \ge 0$. For sufficiently small $\lambda > 0$, we define s_λ by splitting the m-fold knot at x of s_0 into a $(m - 2)$-fold knot at x and two simple knots near x,

$$\begin{aligned}
s_\lambda(t) := &\sum_{j=1}^{m-2} b_j^\lambda B_j(x, \ldots, x; t) \\
&+ b_{m-1}^\lambda B_{m-1}(x, \ldots, x, x + \lambda y_\lambda - \sqrt{\lambda z_\lambda}; t) \\
&+ (b_m^\lambda + b_{m-1}^\lambda \sqrt{\lambda z_\lambda}) \\
&\quad \times B_m(x, \ldots, x, x + \lambda y_\lambda - \sqrt{\lambda z_\lambda}, x + \lambda y_\lambda + \sqrt{\lambda z_\lambda}; t),
\end{aligned} \tag{25}$$

see [2]. By repeated applications of the recurrence relations for divided differences, we obtain

$$
\begin{aligned}
h_\lambda(t) :=& \frac{1}{\lambda}(s_\lambda(t) - s_0(t)) \\
=& \sum_{j=1}^{m} \frac{1}{\lambda}(b_j^\lambda - b_j)B_j(x,\ldots,x;t) \\
& + b_{m-1}^\lambda y_\lambda B_m(x,\ldots,x,x+\lambda y_\lambda - \sqrt{\lambda z_\lambda};t) \\
& + b_m^\lambda y_\lambda B_{m+1}(x,\ldots,x,x+\lambda y_\lambda - \sqrt{\lambda z_\lambda};t) \\
& + (b_{m-1}^\lambda y_\lambda \sqrt{\lambda z_\lambda} + b_{m-1}^\lambda z_\lambda + b_m^\lambda y_\lambda) \\
& \quad \times B_{m+1}(x,\ldots,x,x+\lambda y_\lambda - \sqrt{\lambda z_\lambda}, x + \lambda y_\lambda + \sqrt{\lambda z_\lambda};t) \\
& + (b_m^\lambda y_\lambda \sqrt{\lambda z_\lambda} + b_m^\lambda z_\lambda) \\
& \quad \times B_{m+2}(x,\ldots,x,x+\lambda y_\lambda - \sqrt{\lambda z_\lambda}, x + \lambda y_\lambda + \sqrt{\lambda z_\lambda};t).
\end{aligned}
\tag{26}
$$

Now let us define $b_1^\lambda,\ldots,b_{m-1}^\lambda, y_\lambda, b_m^\lambda, z_\lambda$ successively as follows:

$$
\begin{aligned}
b_j^\lambda &:= b_j + \lambda d_j, \quad j = 1,\ldots,m-1, \\
y_\lambda &:= \frac{d_{m+1} - b_{m-1}\frac{d_{m+2}}{b_m}}{2b_m}, \\
b_m^\lambda &:= b_m + \lambda(d_m - b_{m-1}^\lambda y_\lambda), \\
z_\lambda &:= \frac{d_{m+2}}{b_m^\lambda}.
\end{aligned}
\tag{27}
$$

Since $b_m \neq 0$ and $d_{m+2} \cdot b_m \geq 0$, for sufficiently small $\lambda > 0$ the parameters y_λ and z_λ are well defined, and $z_\lambda \geq 0$ as is necessary for the definition of s_λ. Again, since z_λ, y_λ converge and $b_j^\lambda \to b_j$ as $\lambda \to 0$, it follows that $\|s_\lambda - s_0\| \to 0$ as $\lambda \to 0$. Due to (27), it can be established that $\|h_\lambda - h\| \to 0$ as $\lambda \to 0$, provided that $m \leq n - 2$. In the case $m = n - 1$, we again refer to Lemma 2 by setting (notice $d_{n+1} = b_{n-1}^\lambda z_\lambda$ independently of λ)

$$
\begin{aligned}
h_{1,\lambda}(t) &:= d_{n+1}B_{n+1}(x,\ldots,x,x+\lambda y_\lambda - \sqrt{\lambda z_\lambda}, x + \lambda y_\lambda + \sqrt{\lambda z_\lambda};t), \\
h_1(t) &:= d_{n+1}B_{n+1}(x,\ldots,x;t),
\end{aligned}
\tag{28}
$$

and

$$
\begin{aligned}
h_{2,\lambda} &:= h_\lambda - h_{1,\lambda}, \\
h_2 &:= h - h_1.
\end{aligned}
\tag{29}
$$

For sufficiently small $\lambda > 0$ the inequalities

$$x + \lambda y_\lambda - \sqrt{\lambda z_\lambda} \le x \le x + \lambda y_\lambda + \sqrt{\lambda z_\lambda}$$

hold, and assuming $d_{n+1} > 0$ we obtain $h_{1,\lambda}(t) = h_1(t)$ for $t \notin [x + \lambda y_\lambda - \sqrt{\lambda z_\lambda}, x + \lambda y_\lambda + \sqrt{\lambda z_\lambda}]$ and $\lim_{t \to x - 0} h_1(t) = 0 \le h_{1,\lambda}(t) \le h(x) = d_{n+1}$ on $[x + \lambda y_\lambda - \sqrt{\lambda z_\lambda}, x + \lambda y_\lambda + \sqrt{\lambda z_\lambda}]$. Therefore, it is seen that again

$$\|h_{1,\lambda} - g\| \to \|h_1 - g\| \text{ as } \lambda \to 0 \tag{30}$$

for all $g \in C[0,1]$, and the proof is completed by the observation that $\|h_{2,\lambda} - h_2\| \to 0$ as $\lambda \to 0$. ∎

Discussion:

1. We want to give a short account on earlier results on necessary conditions for local best approximations by *polynomial* splines with free knots. A first version of Theorem 2 was proved by Braess in 1974 with (12) replaced by $\tilde{m}_i = \min\{m_i + 1, n\}$, see [1]. Under the assumption $m_i \le n - 2$, the necessary condition given in Theorem 2 was presented by Cromme [3,4] and equivalently formulated as an alternant condition in [7]. It was already noted in [6] that the assumption $m_i \le n - 2$ may be avoided by replacing (12) by $\tilde{m}_i := \begin{cases} \min\{m_i + 1, n\}, & \text{if } m_i = 1 \\ \min\{m_i + 2, n\}, & \text{if } m_i \ge 2 \end{cases}$. In all of these conditions, the assigned set S_0 consists of continuous splines only. Later on in [9] the necessary conditions were proved with $\tilde{m}_i = m_i + 1$. The statement of Theorem 2 combines and improves these earlier results.

2. We restricted our considerations to polynomial splines. It seems to be possible to improve the known necessary conditions also for some classes of generalized splines, see [1,9].

§4. Formulation as Alternant Conditions

The purpose of this section is to present an equivalent formulation of the necessary conditions as alternant conditions. Therefore, we have to consider alternants with prescribed signs on subintervals. A function $g \in C[0,1]$ is said to have an *alternant of length M* on $[a,b] \subset [0,1]$ with the left-hand sign $\sigma \in \{-1, 1\}$, if there exist $t_1 < \cdots < t_M \in [a,b]$ such that

$$g(t_j) = \sigma(-1)^{j-1}\|g\| \qquad \text{for} \qquad j = 1, \ldots, M,$$

where $\|g\|$ denotes the norm of g on $[0,1]$.

Let $s_0 \in S_{n,k}$ be given by (10), and suppose the cone S_0 is defined by (11), (12). To each interval $[x_p, x_q] \subset [0,1]$, which contains no knot with $m_i = n$ in its interior (*i.e.*, $s_0 \in C^1[x_p, x_q]$), we assign a number M_{pq} and a sign set Z_{pq} as follows.

For the interval $[x_p, x_q]$ let $x_{i_1} < \cdots < x_{i_r}$ be only the knots in (x_p, x_q) with $\tilde{m}_i = m_i + 2$ (*i.e.*, $2 \le m_i \le n-1$). Introducing the notations

$$\sigma_\nu := \operatorname{sgn} a_{i_\nu m_{i_\nu}} \quad \text{for } \nu = 1, \ldots, r \tag{31}$$

and

$$z_\nu := \begin{cases} 0, & \text{if } (-1)^{m_{i_\nu}+1}\sigma_{\nu+1}\sigma_\nu > 0 \\ 1, & \text{otherwise} \end{cases} \quad \text{for } \nu = 1, \ldots, r-1, \tag{32}$$

we define

$$M_{pq} := n + \sum_{i=p+1}^{q-1} (m_i + 1) + \sum_{\nu=1}^{r-1} z_\nu + 1 \tag{33}$$

and

$$Z_{pq} := \begin{cases} \{-1, 1\}, & \text{if } r = 0 \\ \{\sigma_1(-1)^{n+1}\}, & \text{otherwise.} \end{cases} \tag{34}$$

Theorem 3. *Given $f \in C[0,1]$, the zero function is a best approximation to $f - s_0$ in S_0 if and only if there is an interval $[x_p, x_q]$, where $s_0 \in C^1[x_p, x_q]$ and $f - s_0$ has an alternant of length $M_{pq} + 1$ with a left-hand sign from Z_{pq}.*

Proof: The cone S_0 has the structure considered in [7], but now knots of multiplicity $\tilde{m}_i = n + 1$ may also occur. A careful analysis of the proofs given in [7] shows that the sign rule for spline functions as well as the alternant criterion can be extended to the case of knots of multiplicity $n + 1$. Following the definitions given in [7], a number M_{pq} and a sign set Z_{pq} is assigned to each subinterval $[x_p, x_q] \subset [0,1]$.

Now let us consider an interval $[x_p, x_q]$ which contains a knot x_r of multiplicity $\tilde{m}_r = m_r + 1 = n + 1$ (*i.e.*, there is no sign restriction at x_r) in its interior. Suppose that $f - s_0$ has an alternant of length $M_{pq} + 1$ with a left-hand sign from Z_{pq} on the subinterval $[x_p, x_q]$. By case distinction it can be proved that $f - s_0$ must also have an alternant of length $M_{pr} + 1$ with a left-hand sign from Z_{pr} on the subinterval $[x_p, x_r]$ or an alternant of length $M_{rq} + 1$ with a left-hand sign from Z_{rq} on the subinterval $[x_r, x_q]$. To give more details for at least one case, let us assume $Z_{rq} = \{-1, 1\}$. In this case we get $Z_{pq} = Z_{pr}$ and $M_{pq} = M_{pr} + n + 1 + (M_{rq} - n - 1) = M_{pr} + M_{rq}$, and the statement follows immediately.

The assertion stated above allows us to restrict the formulation of the alternant condition to subintervals $[x_p, x_q]$, where $s_0 \in C^1[x_p, x_q]$. As explained in [7], for such subintervals the assigned numbers M_{pq} and sign sets Z_{pq} are given by (31)–(34). ∎

References

1. Braess, D., *Nonlinear Approximation Theory*, Springer-Verlag, Berlin, 1986.
2. Brink-Spalink, K., Tschebyscheff-Approximation durch γ-Polynome mit teilweise fixierten Frequenzen, J. Approx. Th. **15** (1975), 60–77.
3. Cromme, L. J., Regular C^1-parametrizations for exponential sums and splines, J. Approx. Th. **35** (1982), 30–44.
4. Cromme, L. J., A unified approach to differential characterizations of local best approximations by exponential sums and splines, J. Approx. Th. **36** (1982), 294–303.
5. Mulansky, B., L_p-Approximation durch Splines mit freien Knoten, Math. Nachr. **131** (1987), 73–81.
6. Mulansky, B., Charakterisierungsaussagen für nichtlineare Approximationsaufgaben, Dissertation, Technische Universität Dresden, 1987.
7. Mulansky, B., Chebyshev approximation by spline functions with free knots, IMA J. Numer. Anal. **12** (1992), 95–105.
8. Nürnberger, G., *Approximation by Spline Functions*, Springer-Verlag, Berlin, 1989.
9. Nürnberger, G., L. L. Schumaker, M. Sommer, and H. Strauss, Uniform approximation by generalized splines with free knots, J. Approx. Th. **59** (1989), 150–169.
10. Schumaker, L. L., *Spline Functions: Basic Theory*, Wiley, New York, 1981.

Bernd Mulansky
Institut für Numerische Mathematik
Technische Universität
Mommsenstr. 13
D-8027 Dresden, Germany
mulansky@urzdfn.mathematik.tu-dresden.dbp.de

Numerical Methods of Approximation Theory, Vol. 9
Dietrich Braess and Larry L. Schumaker (eds.), pp. 207–220.
International Series of Numerical Mathematics, Vol. 105
Copyright © 1992 by Birkhäuser Verlag, Basel
ISBN 3-7643-2746-4.

C^1 Interpolation on Higher-Dimensional Analogues of the 4-Direction Mesh

A. Neff and J. Peters

Dedicated to the memory of Lothar Collatz

Abstract. The well-known square mesh with both diagonals drawn in is generalized to the m-dimensional mesh Δ_m generated by $\{(v_1,\ldots,v_m) : v_i \in \{-1,0,1\}, i = 1,\ldots,m\}$. Sharp necessary and sufficient conditions on data at the vertices of Δ_m are given that allow interpolation of the data by m-variate, C^1 piecewise polynomials of degree $m+1$. For degree $m+2$ and higher, values and normals at the vertices can be stably interpolated and a unit-norm C^2 Lagrange function for each vertex is exhibited.

§1. Introduction

As the simplest symmetric triangulation of the square mesh, the mesh generated by the vectors $(1,0)$, $(0,1)$, $(1,1)$, $(1,-1)$ has repeatedly received attention in the literature, where it is called *four-direction mesh*, *criss-cross partition* or *type-2 triangulation* ([1,5,3]). In particular, the dimensions and spanning sets of polynomial spline spaces of various degree and smoothness on the mesh are known [13,4,14] and spline functions of minimal support have been identified [15,1,12]. The study of C^1 piecewise cubics on the mesh has been fruitful, yielding a characterization of the approximation order [6] and proving that interpolation with C^1 cubics is in general unstable [9].

This paper analyses C^1 interpolation on triangulations Δ_m in $\mathbb{R}^m$ that generalize the 4-direction mesh Δ_2 to m dimensions. In view of interpolation at the vertices, the 4-direction mesh can be thought of as arising from a uniform square lattice by inserting an additional vertex at the center of each square. The m-dimensional analogue is to start with the canonical decomposition of $\mathbb{R}^m$ into cubes and to insert additional vertices at the centers of all cubes of dimension greater than or equal to 2. The corresponding 'triangulation' (simplicial decomposition) Δ_m of $\mathbb{R}^m$ defined by (S) below has several desirable properties that make it a 'natural' generalization of the 4-direction mesh:

(1) Shift-invariance: all cubes are subdivided in the same fashion.

(2) Symmetry: the reflection across any facet plane of a simplex in Δ_m yields again Δ_m. The triangulation contains the directions $\{(v_1, \ldots, v_m) : v_i \in \{-1, 0, 1\}, i = 1, \ldots, m\}$ and, for each cube of sidelength greater than or equal to 2, a center vertex.

(3) Isomorphy of the simplices.

(4) Recursive structure: the intersection of Δ_m with any plane spanned by $m - 1$ coordinate axes is isomorphic to Δ_{m-1}.

(5) Aspect ratio $\frac{1}{m}$: every unit simplex of Δ_m contains a cube of sidelength $\frac{1}{m}$.

The periodic and recursive structure of Δ_m allows a complete characterization of data that can be boundedly interpolated by $S^1_{m+1}(\Delta_m)$, the C^1 piecewise polynomials of degree $m + 1$ in m variables. In particular, if the data are sampled from a function of bounded variation [8, Eq (2.3)], bounded interpolation of value and normals is always possible. Locally supported (and therefore bounded) interpolation is always possible if the polynomial pieces are of degree $m + 2$ or higher. This second result is established in Section 4 by exhibiting a Lagrange function of compact support. The linear growth of the degree of this function as the dimension grows stands in contrast to the exponential growth of the interpolants suggested in [2,7] for general triangulations.

The main work of this paper goes into establishing sharp necessary and sufficient bounds for interpolation from $S^1_{m+1}(\Delta_m)$. The reduction of the degree from $m + 2$ is of practical importance and also leads to the study of the function space that best exhibits the properties of Δ_m. In [10] it was already shown that C^1 interpolation to *periodic data* at the vertices of Δ_m with polynomial pieces of degree $m+1$ is possible if and only if the alternating sum of the values at the vertices of any 1-cube in Δ_m is zero:

$$\sum_{\alpha \in \{0,1\}^m} (-)^\alpha f(\alpha) = 0, \qquad \text{where } (-)^\alpha := (-1)^{\alpha_1} \cdots (-1)^{\alpha_m}. \qquad (A)$$

In this paper, a shorter proof of this fact follows from a general theorem that connects Stokes' Theorem with the multivariate identity

$$\sum_{i=0}^{m} (-1)^{m-i} \binom{m}{i} f(v_i) = m! \int_{S_m} \frac{\partial^m f}{\partial x_1 \ldots \partial x_m},$$

where v_i are the vertices of a simplex S_m in Δ_m. The necessary condition for periodic data implies that the alternating sum of the values prescribed at the vertices of any finite subtriangulation of Δ_m with diameter r must vanish faster than r^m if a bounded interpolant is to exist. Using Markov estimates, we refine this necessary bound for interpolation of *arbitrary data* at the vertices of Δ_m to

$$\sum_{|\alpha| \leq r} (-)^\alpha f(\alpha) \leq K_m \|f\|_\infty r^{m-1}. \qquad (A^r)$$

Here K_m is a constant that depends only on m. The bound is shown to be sharp by constructing, for any data satisfying (A^r), a bounded interpolant as a linear combination of functions with local support on exactly two adjacent mesh points.

Notation and the definition of the triangulation. We use standard multi-index notation: for $\alpha := (\alpha_1, \ldots, \alpha_m) \in \mathbb{Z}^m$, and $x := (x_1, \ldots, x_m) \in \mathbb{R}^m$,

$$|\alpha| := |\alpha_1| + \ldots + |\alpha_m|, \qquad \alpha! := \alpha_1! \cdots \alpha_m! \qquad x^\alpha := x_1^{\alpha_1} \ldots x_m^{\alpha_m}.$$

For $\alpha, \beta \in \mathbb{Z}^m$, a *lattice box* is defined as

$$[\alpha, \beta] := \{\gamma \in \mathbb{Z}^m : \alpha_i \leq \gamma_i \leq \beta_i \text{ for } 1 \leq i \leq m\}.$$

The lattice box $[-\alpha, \alpha]$ with $\alpha_i = r$ for $1 \leq i \leq m$ is denoted by $B(r)$. Depending on the context, we use 0 and 1 also for a vector of zeros and a vector of ones respectively, and denote the kth unit vector by e_k, where $e_0 := 0$.

The *triangulation* (simplicial mesh) Δ_m is defined by the following construction. Divide the m-dimensional 1-cube, $\square_0$, with vertices $(\{0,1\}^m)$ into $m!$ distinct simplices such that the vertices v_l of each simplex satisfy:

$$v_0 = 0, \quad v_{l+1} = v_l + e_{\kappa_l}, \quad v_m = 1, \tag{S}$$

where $\{\kappa_1,\ldots,\kappa_m\} = \{1,\ldots,m\}$. Compute $\square_{i+1}$ as the union of $\square_i$ and its reflection across the hyperplane $x_{i+1} = 0$ for $i = 0,\ldots,m-1$. Then $\square_m$ fills the m-dimensional 2-cube defined by the vertices $(\{-1,1\}^m)$ with $2^m m!$ simplices. Shift the 2-cube over the grid $2\mathbb{Z}^m$ to cover $\mathbb{R}^m$.

The subdivision of $\square_0$ is the one obtained from slicing the cube by all $\binom{m}{2}$ hyperplanes $\{x : x_i = x_j\}$ with normals $e_j - e_i$, $j > i$ that pass through 0 and 1. Given $k+1$ points, $v_0,\ldots,v_k$, in $\mathbb{R}^m$, the oriented k-dimensional simplex which they span is denoted by $[v_0,\ldots,v_k]$. Of particular interest in this paper is the simplex

$$S_m := [v_0, v_1, v_2, \ldots, v_m] := [0, e_1, e_1 + e_2, \ldots, \sum_{i=1}^{m} e_i],$$

with the $m+1$ faces $F_i := [v_0, \ldots, v_{i-1}, v_{i+1}, \ldots, v_m]$ and corresponding outward normal vectors

$$n_0 := e_1, \quad n_i := e_{i+1} - e_i, \quad \text{for } i = 1,\ldots,m-1, \quad n_m := -e_m.$$

To prove Property (5), we show that S_m, and therefore any simplex in Δ_m, contains an m-dimensional cube of sidelength $\frac{1}{m}$. Explicitly, this cube has coordinates

$$w(\gamma) := \frac{1}{m} \begin{pmatrix} m-1 \\ m-2 \\ \vdots \\ 0 \end{pmatrix} + \frac{1}{m} \begin{pmatrix} \gamma_{m-1} \\ \gamma_{m-2} \\ \vdots \\ \gamma_0 \end{pmatrix} \qquad \text{where } \gamma_i \in \{0,1\}.$$

One checks that $w(\gamma)$ has barycentric coordinates $\lambda_i := \frac{1}{m}(1 + \gamma_i - \gamma_{i+1})$ with respect to v_i for $i = 0,\ldots,m$ if one defines $\gamma_{-1} = 0$ and $\gamma_m = 1$ and observes that $0 \leq \lambda_i \leq 1$ and $\sum_{i=0}^{m} \lambda_i = 1$.

§2. Periodic Interpolation

If the data at the vertices of Δ_m are periodic, i.e., symmetric and shift-invariant with period 2, and if a bounded interpolant exists, then there exists a periodic interpolant by the argument in [9,10]. If the interpolant is additionally C^1, then the normal derivatives across the facets of each simplex must vanish. This observation motivates the following surprising identity between alternately summed function values and the integral of a derivative of a function over S_m.

Lemma 2.1. *Let f be a C^m function defined on S_m such that for each $i = 0, \ldots, m$,*

$$\left. \frac{\partial}{\partial n_i} f \right|_{F_i} \equiv 0.$$

Then

$$\sum_{i=0}^{m} (-1)^{m-i} \binom{m}{i} f(v_i) = m! \int_{S_m} \frac{\partial^m f}{\partial x_1 \ldots \partial x_m}. \tag{2.2}$$

Proof: The case $m = 1$ is the fundamental theorem of calculus. For $m \geq 2$, consider the vector field

$$W := (g_1, \ldots, g_m), \qquad g_i := \frac{\partial^{m-1} f}{\partial x_1 \ldots \partial x_{i-1} \partial x_{i+1} \ldots \partial x_m},$$

which is chosen so that

$$\nabla \cdot W = m \frac{\partial^m f}{\partial x_1 \ldots \partial x_m},$$

and, for $i = 1, \ldots, m - 1$,

$$\begin{aligned}
W \cdot n_i &= \frac{\partial^{m-2}}{\partial x_1 \ldots \partial x_{i-1} \partial x_{i+2} \ldots \partial x_m} \left(\frac{\partial}{\partial x_i} - \frac{\partial}{\partial x_{i+1}} \right) f \\
&= \frac{\partial^{m-2}}{\partial x_1 \ldots \partial x_{i-1} \partial x_{i+2} \ldots \partial x_m} \left(\frac{\partial f}{\partial n_i} \right),
\end{aligned}$$

implying that $W \cdot n_i \big|_{F_i} \equiv 0$. By Stokes' Theorem,

$$m \int_{S_m} \frac{\partial^m f}{\partial x_1 \ldots \partial x_m} = \int_{S_m} \nabla \cdot W = \int_{\partial S_m} W \cdot n.$$

The integral on the right can be split into the sum of integrals over the $m+1$ faces of S_m:

$$\int_{\partial S_m} W \cdot n = \sum_{i=0}^{m} \int_{F_i} W \cdot n_i = \int_{F_0} g_1 - \int_{F_m} g_m$$

so that

$$m! \int_{S_m} \frac{\partial^m f}{\partial x_1 \ldots \partial x_m} = (m-1)! \int_{F_0} g_1 - (m-1)! \int_{F_m} g_m.$$

By induction,

$$(m-1)! \int_{F_0} g_1 = \sum_{i=0}^{m-1} (-1)^{m-1-i} \binom{m-1}{i} f(v_{i+1}) \tag{2.3}$$

and

$$(m-1)! \int_{F_m} g_m = \sum_{i=0}^{m-1} (-1)^{m-1-i} \binom{m-1}{i} f(v_i). \tag{2.4}$$

Since the difference of the two terms on the right in (2.3) and (2.4) is exactly the term on the left in (2.2), the proof is complete. ∎

Corollary 2.5. *Suppose that f and S_m are as in Lemma 2.1, with the added assumption that f is a polynomial of degree $m+1$. Then*

$$\sum_{i=0}^{m} (-1)^{m-i} \binom{m}{i} f(v_i) = 0.$$

Proof: The function

$$h := \frac{\partial^m f}{\partial x_1 \ldots \partial x_m}$$

is linear, because f is of degree $m+1$. Since

$$h\Big|_{F_0} = h\Big|_{F_m} = 0,$$

$h \equiv 0$. ∎

    A consequence of Lemma 2.1 is that interpolation of periodic data at the vertices of Δ_m with C^1 piecewise polynomials of degree $m+1$ is stable only if (A) holds [10, Thm 3.11].

§3. Necessary Constraints for Bounded Interpolation from $S_{m+1}^1(\Delta_m)$

Using the averaging techniques in [9], and applying Lemma 2.1, it is straightforward to derive the following necessary condition for the existence of a bounded interpolant to arbitrary data.

Theorem 3.1. *Let $\{z_\alpha\}_{\alpha \in \mathbb{Z}^m}$ be a collection of real values (the data). A necessary condition for the existence of a bounded piecewise polynomial C^1 function f on Δ_m of degree $m + 1$ with $f(\alpha) = z_\alpha$ is that*

$$\lim_{r \uparrow \infty} \left[r^{-m} \sum_{\alpha \in B(r)} (-)^\alpha z_\alpha \right] = 0. \tag{3.2}$$

In this section, we want to improve the quantitative aspect of the estimate by showing how quickly the left hand side of (3.2) must tend to zero in terms of $\|f\|_\infty$. This is of use for interpolation on finite grids, and complements the sufficient condition for the existence of an interpolant in Section 4. Two lemmas precede the main theorem.

Lemma 3.3. *Let p be a polynomial in m variables of degree d and $\alpha \in \mathbb{N}^m$, $|\alpha| = k \leq d$. Then there exists a constant $M_{m,d,k} > 0$ depending only on m, d and k such that*

$$\left\| \frac{\partial^\alpha p}{\partial x^\alpha} \right\|_{L^\infty(S_m)} \leq M_{m,d,k} \|p\|_{L^\infty(S_m)}.$$

Proof: We use Markov's inequality for a univariate polynomial q,

$$\|q'\|_{L^\infty([0,1])} \leq 2d^2 \|q\|_{L^\infty([0,1])},$$

and the fact that among all polynomials of degree d bounded by 1 on the interval $[0, 1]$, the Chebyshev polynomial (of degree d) grows fastest outside the interval (cf. [11]):

$$|q(x)| \leq |c_d(x)| \, \|q\|_{L^\infty([0,1])} \text{ for } x \notin [0, 1].$$

Iterating Markov's inequality one variable at a time and applying the Chebyshev estimate outside the cube of sidelength $\frac{1}{m}$ that is contained in S_m (cf. 1.0), we estimate

$$\left\| \frac{\partial^\alpha p}{\partial x^\alpha} \right\|_{L^\infty(S_m)} \leq \left\| \frac{\partial^\alpha p}{\partial x^\alpha} \right\|_{L^\infty(\square_0)}$$

$$\leq 2^k \left(\frac{d!}{(d-k)!} \right)^2 \|p\|_{L^\infty(\square_0)}$$

$$\leq M_{m,d,k} \|p\|_{L^\infty(\frac{1}{m}\square_0)},$$

$$\leq M_{m,d,k} \|p\|_{L^\infty(S_m)},$$

where $M_{m,d,k} := 2^k \left(\frac{d!}{(d-k)!} \right)^2 \left(\max_{x \in [0,m]} |c_d(x)| \right)^m.$ ∎

The second lemma is similar to Lemma 2.1. We define $Q := \alpha + \square_0$ to be the translate of the unit m-cube by the integer lattice vector α, denote by S_m^j, $j = 1, \ldots, m!$ the $m!$ simplices contained in $\Delta_m \cap Q$, and denote by $V(X)$ the set of vertices of the lattice box, cube or simplex X.

Lemma 3.4. *Let f be a C^1 piecewise polynomial (of arbitrary degree) on $\Delta_m \cap Q$. Then*

$$\sum_{\beta \in V(Q)} (-)^{\beta - \alpha} f(\beta) = \sum_{j=1}^{m!} \int_{S_m^j} \frac{\partial^m f}{\partial x_1 \ldots \partial x_m}. \tag{3.5}$$

Proof: As in Lemma 2.1, the integral is decomposed into a sum of integrals over the faces of the simplices S_m^j. For arbitrary data the summands $W \cdot n_i$ for $i = 0, \ldots, m$ are in general not zero on the faces; however, the overall sum includes for each summand $W \cdot n_i$ corresponding to an interior face of Q also one summand $W \cdot (-n_i)$. Since $W \cdot n_i$ is a C^0 function (all directions of differentiation except for one are parallel to the face), all terms except for the values at the vertices cancel when the recursion of Lemma 2.1 is applied.   ∎

For the main theorem of this section it is convenient to define for a lattice block $B := [\alpha, \beta]$ and a lattice point γ contained in B

$$d_B(\gamma) := \min \left\{ \dim(F) : F \text{ is a face of } B \text{ and } \gamma \in F \right\}.$$

For example, if γ is a vertex of B, then $d_B(\gamma) = 0$, and if γ is in the interior of B, then $d_B(\gamma) = m$.

Theorem 3.6. *Let B be a lattice block, and let f be a C^1 function on B that is a piecewise polynomial of degree $m + 1$ on $\Delta_m \cap B$. Then*

$$\sum_{\alpha \in B} (-)^\alpha 2^{d_B(\alpha)} f(\alpha) \leq \#(\partial B)(m - 1)! M \|f\|_\infty,$$

where $\#(\partial B)$ is the number of lattice points on the boundary of B and $M := M_{m,m+1,m}$.

Proof: Define α_Q by $Q = \alpha_Q + [0,1]^m$. Then by Lemma 3.4, and the fact that $\frac{\partial^m f}{\partial x_1 \ldots \partial x_m}$ is a linear function over each simplex,

$$\sum_{\alpha \in B} (-)^\alpha 2^{d_B(\alpha)} f(\alpha) = \sum_{Q \subset B} (-)^{\alpha_Q} \sum_{\beta \in V(Q)} (-)^{\beta - \alpha_Q} f(\beta)$$

$$= \sum_{Q \subset B} (-)^{\alpha_Q} \sum_{S_m^j \in Q} \int_{S_m^j} \frac{\partial^m f}{\partial x_1 \ldots \partial x_m}$$

$$= \sum_{Q \subset B} (-)^{\alpha_Q} \sum_{S_m^j \in Q} \frac{\mathrm{vol}(S_m^j)}{m+1} \sum_{v \in V(S_m^j)} \frac{\partial^m f}{\partial x_1 \ldots \partial x_m}(v).$$

In the last equation, we used the fact that $\int_{S_m} l(x)dx = \frac{\mathrm{vol}(S_m)}{m+1} \sum_{v \in V(S_m)} l(v)$ for a linear function l in m variables. Next we observe that $\frac{\partial^m f}{\partial x_1 \ldots \partial x_m}$ is continuous across the faces orthogonal to one of the coordinate axes (since f is C^1 and all but one direction of differentiation are parallel to the face). As in Lemma 3.4, most of the terms cancel due to the alternating sign. Viewing the boundary of B as consisting of those $m-1$-dimensional faces of simplices in B, where the matching terms are missing, we find that there is exactly one contribution from each simplex that has a face of dimension $m-1$ on the boundary of B. There are $(m-1)!$ such simplices for each box on the boundary. Using Lemma 3.3, the sum on the right is therefore bounded above by

$$\#(\partial B)(m-1)! M \|f\|_\infty. \quad \blacksquare$$

If $B := B(r)$, then Theorem 3.6 gives the bound

$$\sum_{\alpha \in B(r)} (-)^\alpha f(\alpha) \leq M \sup_{\alpha \in B(r)} |z_\alpha| (m-1)! \left[(2r+1)^m - (2r-1)^m \right]$$

$$= M \|f\|_\infty 2^m m! r^{m-1} + \text{ lower order terms in } r.$$

This shows that the bracketed term in (3.2) goes to zero at least as fast as k_m/r for some constant k_m depending only on the dimension.

§4. Sufficient Conditions for Stable Interpolation on Δ_m

This section gives an explicit scheme for computing a bounded interpolant to arbitrary data on Δ_m. We show that interpolation with polynomials of degree $m+2$ is always possible and stable, and that interpolation with polynomials of degree $m+1$ is stable if the alternating sum of the data at the vertices of a lattice box in Δ_m is bounded by a constant times r^{m-1}, where r is the diameter of the lattice box.

The key to both constructions is the polynomial piece χ_{01}. To define χ_{01}, we denote by

$$b(\alpha), \quad \alpha := (\alpha_1, \ldots, \alpha_m) \in \{-d, \ldots, d\}^m,$$

the Bernstein-Bézier (BB) coefficient associated with the coordinates $d^{-1}\alpha \in d^{-1}\mathbb{Z}^m$ (which correspond to the vertices of $(2d)^m$ cubes of edge-length d^{-1} that fill $\square_m$). Choosing one coefficient per coordinate implies that the function χ_{01} defined by

$$b(\alpha) := \begin{cases} 1 & \text{if } \alpha_j \in \{0, \ldots, m+1-j\}, j = 1, \ldots, m \text{ and } \alpha_1 \geq 0 \\ 0 & \text{else} \end{cases},$$

is C^0 continuous and compactly supported in $\square_0$.

To check C^1 continuity of a piecewise polynomial f with coefficients $b(\alpha)$ and supported on $I \subset \square_m$, denote any face in a cutting hyperplane with normal $e_j - e_i$ by F_{ij}. We may assume $j > i$. Due to the linearity of the C^1 conditions, it is sufficient to prove continuity of the derivative in a single direction, e_i or e_j not in F_{ij} since any other transversal direction can be written as a linear combination of directions in the hyperplane and that particular direction. Now let S^1 and S^2 be two simplices that share an $m-1$-dimensional face. There are two cases.

If $i = 0$, then S^2 is the reflection of S^1 across the hyperplane $x_j = 0$. If $v_m := \sum_{i=1}^m e_i$, then v_m, $v_m - e_j$ and $v_m - 2e_j$ lie on a straight line in the union of S^1 and S^2. Therefore $D_j f$ is continuous across F_{j0} if and only if

$$b(\alpha) - b(\alpha - e_j) = b(\alpha + e_j) - b(\alpha) \quad \text{for } \alpha_j = 0, \alpha \in I \tag{4.1}$$

($d^{-1}\alpha$ is on F_{j0}).

If $i > 0$, then, by Definition (S) there must exist an index ℓ such that for the index sequences κ^1 and κ^2 of S^1 and S^2,

$$i = \kappa_\ell^1 \neq \kappa_\ell^2 = j, \quad \kappa_{\ell+1}^1 = \kappa_\ell^2, \kappa_{\ell+1}^2 = \kappa_\ell^1, \text{ and } \kappa_l^2 = \kappa_l^1, \text{ for } \ell \neq l \neq \ell+1.$$

Therefore $D_j f$ is continuous across F_{ji} if and only if

$$b(\alpha + e_i) - b(\alpha) = b(\alpha + e_j + e_i) - b(\alpha + e_j) \quad \text{for } \alpha_j = \alpha_i, \alpha \in I. \tag{4.2}$$

By symmetry and shift-invariance of Δ_m, it is sufficient to establish continuity across the hyperplanes that bound the simplex S_m. The maximal number of continuity constraints have to be checked at v_0, since here the maximal number of hyperplanes meet. Continuity of functions supported at $v_i, i = 1, \ldots, m$, is established analogously using a subset of the cutting hyperplanes, and hence considering continuity constraints. In the proofs below, it is therefore sufficient to show validity of the construction for v_0.

Theorem 4.3. *Arbitrary values at the vertices of Δ_m can be boundedly interpolated by a C^2 piecewise polynomial of degree $d \geq m + 2$. Arbitrary values and normals at the vertices of Δ_m can be boundedly interpolated by a C^1 piecewise polynomial of degree $d \geq m + 2$.*

Proof: Define ϕ_{01} as the reflection of χ_{01} across all hyperplanes $x_i = 0$, $i = 1, \ldots, m$, analogous to the construction of $\square_m$ from $\square_0$. That is, ϕ_{01} is defined by

$$b(\alpha) := \begin{cases} 1 & \text{if } \alpha \in I := \{\alpha : |\alpha_j| \in \{0, \ldots, m + 1 - j\}, j = 1, \ldots, m\} \\ 0 & \text{otherwise.} \end{cases}$$

We need only check Equations 4.1 and 4.2 for $\alpha \in I$ since otherwise all coefficients are zero. For $\alpha \in I$ with $\alpha_i = \alpha_j$, the decreasing width of I with the index implies $b(\alpha + e_i) = b(\alpha) = 1$ and $b(\alpha + e_j + e_i) = b(\alpha + e_j)$. The latter expression is either 1 or 0 depending on whether $\alpha + e_j \in I$. Due to the symmetry of I, we have

$$\begin{array}{llll}
b(\alpha) - b(\alpha - e_j) & b(\alpha + e_j) - b(\alpha) & \text{for} & \\
1 - 1 & 1 - 1 & & \alpha_j = 0 \\[4pt]
b(\alpha + e_i) - b(\alpha) & b(\alpha + e_j + e_i) - b(\alpha + e_j) & & \\
1 - 1 & 1 - 1 & \alpha_j < m + 1 - j & \alpha_i = \alpha_j \\
1 - 1 & 0 - 0 & \alpha_j = m + 1 - j & \alpha_i = \alpha_j
\end{array}$$

This shows that ϕ_{01} is a C^1 Lagrange function of compact support at v_0. Averaging ϕ_{01} with its reflections across all hyperplanes, one obtains a C^1 Lagrange function that is symmetric with respect to all hyperplanes and is therefore in C^2. Alternatively, one can form a linear combination of ϕ_{01} and its reflections to match an arbitrary gradient at v_0. Since at most

$$K(m) := \sum_{i=1}^{m} \binom{m}{i} i = \text{number of edges of } \square_0$$

functions ϕ_{kl} overlap on any point in $(m+2)^{-1}\mathbb{Z}^m$, the norm of the Lagrange interpolation operator is at most $K(m)$. $\blacksquare$

We now construct a second function from χ_{01} to show sharpness of the $O(r^{m-1})$ bound on $\sum_{\alpha \in B(r)} (-)^\alpha f(\alpha)$ established in Theorem 3.6.

Lemma 4.4. *For any edge-adjacent pair of vertices v_k and v_l of Δ_m there exists a C^1 piecewise polynomial function τ_{kl} of degree $d := m + 1$ and of* *compact support that has value 1 at v_k and $(-1)^{\|v_l - v_k\| - 1}$ at v_l and value 0 at all other vertices.*

Proof: By symmetry of Δ_m, it is sufficient to argue for the vertices of the simplex S_m. Furthermore, it is sufficient to construct the function f for vertices v_k and v_{k+1} since one can alternately add and subtract the compactly supported functions. The maximal number of continuity constraints have to be checked if $k = 0$, since here the maximal number of hyperplanes meet. Define ψ_{01} as the reflection of χ_{01} across all hyperplanes $x_i = 0$, $i = 2, \ldots, m$ analogous but not equal to the construction of ϕ_{01}. Define ψ_{10} as the reflection of ψ_{01} across the plane $x_1 = (v_0 + v_1)/2$. Then $\tau_{01} := \psi_{01} + \psi_{10}$ is defined by

$$b(\alpha) := \begin{cases} 0 & \text{if } \alpha \notin I \\ 1 & \text{if } \alpha_1 = 0 \text{ or } \alpha_1 = d \text{ and } \alpha \in I \\ 2 & \text{if } 0 < \alpha_1 < d \text{ and } \alpha \in I \end{cases},$$

where I is the lattice box

$$I := \{\alpha \in \mathbb{Z}^m : 0 \le \alpha_1 \le d, |\alpha_j| \in \{0, \ldots, d - j\} \text{ for } j = 2, \ldots, m, \}.$$

We check 4.1 for $\alpha \in I$ and $\alpha_j = 0$,

$$\begin{array}{ccccc}
b(\alpha) - b(\alpha - e_j) & b(\alpha + e_j) - b(\alpha) & & \text{for} & \\
1 - 0 & 2 - 1 & & & j = 1 \\
1 - 1 & 1 - 1 & \alpha_1 = 0 & & j > 1 \\
2 - 2 & 2 - 2 & \alpha_1 > 0 & & j > 1
\end{array}$$

and 4.2 for $\alpha \in I$ and $\alpha_i = \alpha_j$, $i < j$,

$$\begin{array}{cccc}
b(\alpha + e_i) - b(\alpha) & b(\alpha + e_j + e_i) - b(\alpha + e_j) & & \\
2 - 1 & 2 - 1 & \alpha_1 = 0 & i = 1 \\
1 - 1 & 1 - 1 & \alpha_1 = 0 & i > 1 \\
2 - 2 & 2 - 2 & \alpha_1 > 0 & 0 \le \alpha_j < d - j \\
2 - 2 & 0 - 0 & \alpha_1 > 0 & \alpha_j = d - j.
\end{array}$$

This establishes first order continuity. Due to symmetry, τ_{01} is smoother than C^1 in most directions. ∎

Given data $f(v)$ at the vertices v of Δ_m, the task of interpolating the data by a function g that is a piecewise polynomial of degree $m + 1$ on Δ_m is equivalent to interpolating the data 0 at v_i and $f(v_j) + (-1)^{\|v_j - v_i\|_1} f(v_i)$ at v_j by $g + f(v_i)\tau_{i,j}$. In analogy to a Lagrange (or Cardinal) function, which can be used to "remove data" by including an appropriate multiple in the interpolant, a function τ_{kl} can thus be used to "move data". This suggests calling a function supported at exactly two mesh points a *transfer function.* Thus, for any lattice box $[a, b]$ of Δ_m of dimension m with $\#([a, b])$ points, interpolation of the data in $[a, b]$ is equivalent to interpolating $1/\#([a, b])$ times

$$\sum_{\alpha \in [a,b]} (-)^\alpha f(\alpha) \tag{A*}$$

at every vertex.

Theorem 4.5. *If the alternating sum* (A*) *is bounded by* $const\ r^{m-1}$*, where* $0 < r := \min\{|a_1 - b_1|, \ldots, |a_m - b_m|\}$*, then the data* $f(\alpha)$ *can be boundedly interpolated by a linear combination of maps* τ_{ij}*.*

Proof: There are at least r^{m-1} vertices one unit outside the boundary of $[a, b]$. Distributing the alternating sum (A*) uniformly on these vertices yields a constant bound on the value at each vertex independent of r. The boundedness of the interpolant follows from a careful transfer of the values from the lattice box and $\|\tau_{01}\|_\infty = 2$. ∎

Corollary 4.6. *If the alternating sum* (A*) *is zero, then the data* $f(\alpha)$ *can be boundedly interpolated by a linear combination of maps* τ_{ij} *such that the interpolant has support only in* $[a, b]$*.*

Acknowledgements. Part of the research was done while the second author enjoyed a postdoctoral stay at the IBM research center in Yorktown Heights.

References

1. de Boor, C., and K. Höllig, Minimal support for bivariate splines, Approx. Theory Appl. **3** (1987), 11–23.
2. Chui, C. K., M.-J. Lai, Multivariate vertex splines and finite elements, J. of Approx. Theory **60** (19xx), 245–343.
3. Chui, C. K., and R.-H. Wang, Bivariate B-splines on triangulated rectangles, *Approximation Theory IV*, C. K. Chui, L. L. Schumaker, and J. Ward (eds.), Academic Press, New York, 1983, 413–418.
4. Chui, C. K., and R.-H. Wang, Multivariate spline spaces, J. Math. Anal. & Appl. **94** (1983), 197–221.
5. Dahmen, W., and C. A. Micchelli, Recent progress in multivariate splines, *Approximation Theory IV*, C. K. Chui, L. L. Schumaker, and J. Ward (eds.), Academic Press, New York, 1983, 27–121.

6. Lai, M.-J., Approximation order from bivariate C^1 cubics on a four directional mesh is full, Dept. Math., U. of Utah, preprint.

7. Le Méhauté, A. A finite element approach to surface reconstruction, in *Computation of Curves and Surfaces*, W. Dahmen et al. (eds.), Kluwer Academic Publishers, Dordrecht, 237–274.

8. Lenze, B., On constructive one-sided approximation of multivariate functions of bounded variation, Numer. Funct. Anal. and Opt. **11** (1990), 55–83.

9. Peters, J. and M. Sitharam, Stability of Interpolation from C^1 Cubics at the Vertices of an Underlying Triangulation, SIAM J. Num. Anal. **20** (1992), 528–533.

10. Peters, J. and Sitharam, M., Stability of m-variate C^1 interpolation, RC 16732, IBM, TJ Watson Res. Ctr., April, 1991.

11. Rivlin, T. J., *Chebyshev Polynomials: from Approximation Theory to Algebra and Number Theory*, New York: J. Wiley, 1990.

12. Sablonière, P. De l'existence de splines a support borné sur une triangulation équilaterale du plan, Publ. ANO-30, UER d'IEEA–Informatique Univ. de Lille I, 1981.

13. Schumaker, L. L., On the dimension of spaces of piecewise polynomials in two variables, in *Multivariate Approximation Theory*, W. Schempp, and K. Zeller (eds.), Birkhäuser, Basel, 1979, 396–412.

14. Schumaker, L. L., Recent progress on multivariate splines, in *Mathematics of Finite Elements VII*, J. Whiteman (ed.), Academic Press, London, 1991, 535–562.

15. Zwart, Philip B., Multi-variate splines with non-degenerate partitions, SIAM J. Numer. Anal. **10** (1973), 665–673.

Andy Neff
IBM – TJ Watson Research Center
PO Box 218
Yorktown Heights NY 10598, USA
caneff@watson.ibm.com

Jörg Peters
Department of the Mathematical Sciences
Rensselaer Polytechnic Institute
Troy, NY 12180, USA
jorg@cs.rpi.edu

Numerical Methods of Approximation Theory, Vol. 9
Dietrich Braess and Larry L. Schumaker (eds.), pp. 221–244.
International Series of Numerical Mathematics, Vol. 105
Copyright © 1992 by Birkhäuser Verlag, Basel
ISBN 3-7643-2746-4.

Tabulation of Thin Plate Splines on a Very Fine Two-Dimensional Grid

M. J. D. Powell

Dedicated to the memory of Lothar Collatz

Abstract. A thin plate spline approximation has the form

$$s(x) = \sum_{j=1}^{n} \lambda_j \, \|x - x_j\|_2^2 \, \log \|x - x_j\|_2 + p(x), \quad x \in \mathbb{R}^2,$$

where $\{\lambda_j \in \mathbb{R} : j = 1, 2, \ldots, n\}$ and $\{x_j \in \mathbb{R}^2 : j = 1, 2, \ldots, n\}$ are parameters and where p is a linear polynomial. There exist several applications that require s to be tabulated at all the lattice points of a very fine square grid. For example, 10^8 grid points and $n = 500$ can occur, and then the direct evaluation of s at every grid point would be impracticable. Fortunately each thin plate spline term is smooth away from its centre x_j, so it is possible to apply a scheme that subtabulates by finite differences provided that special attention is given to those terms whose centres are close to the current x. Thus the total work is bounded by a small constant multiple of the number of grid points plus a constant multiple of $n \epsilon^{-1/3} |\log h|$, where ϵ is a given tolerance on the calculated values of $s(x)$ and where h is the mesh size of the fine grid. We will find that the exponent $-1/3$ is due to the order of the differences that are employed. An algorithm for this calculation is described and discussed and some numerical results are presented. The errors of the subtabulation procedures are studied in an appendix.

§1. Introduction

When a time-dependent system is under observation, one may wish to compare pictures of the system that are taken at regular intervals. Examples include monitoring the possible growth of a tumour in a hospital patient and observing the silting-up of a shipping channel. These two applications are mentioned because in both cases two kinds of differences may occur between one picture and the next. Of course one kind of change is due to the physical property that is being explored. On the other hand, some ambient properties of the picture may alter too, such as the amount of air in the patient's lungs and the state of the tide in the shipping lane. Other differences in the ambient conditions can include the position of the patient and of the vessel that is collecting data, perhaps by means of side-scan sonar measurements.

Therefore, techniques that allow for the changes in the ambient conditions are needed. They depend on known relations between one picture and the next. The bone structure of the patient, for example, can provide suitable information, as can the positions of buoys, wrecks and rocks in the shipping channel. We address the case where such information can be identified precisely in the sequence of pictures that is being compared. Specifically, we let $\{(s_j, t_j) : j = 1, 2, \ldots, n\}$ be the coordinates of points in one picture that are known to correspond to the points $\{(x_j, y_j) : j = 1, 2, \ldots, n\}$ in another picture. Then we employ a transformation from the (x, y) picture to the (s, t) picture that satisfies the interpolation equations

$$s(x_j, y_j) = s_j \quad \text{and} \quad t(x_j, y_j) = t_j, \quad j = 1, 2, \ldots, n. \tag{1}$$

Two questions have arisen. One is the choice of the mapping functions s and t, each being from $\mathbb{R}^2$ to $\mathbb{R}$, and the other is the purpose of these functions. Of course s and t must have the ability to interpolate data in general position. Therefore, as in Barrodale, Berkley and Skea (1992), we assume that they are "thin plate splines", which means that they have the form

$$s(x, y) = \sum_{j=1}^{n} \lambda_j \left[(x - x_j)^2 + (y - y_j)^2\right] \log[(x - x_j)^2 + (y - y_j)^2]^{1/2}$$
$$+ \, ax + by + c,$$
$$t(x, y) = \sum_{j=1}^{n} \mu_j \left[(x - x_j)^2 + (y - y_j)^2\right] \log[(x - x_j)^2 + (y - y_j)^2]^{1/2}$$
$$+ \, dx + ey + f, \tag{2}$$

where $\{\lambda_j : j = 1, 2, \ldots, n\}$, a, b, c, $\{\mu_j : j = 1, 2, \ldots, n\}$, d, e and f are parameters that are chosen to satisfy the interpolation equations (1). Usually the six remaining degrees of freedom are fixed by the conditions

$$\sum_{j=1}^{n} \lambda_j = \sum_{j=1}^{n} \lambda_j x_j = \sum_{j=1}^{n} \lambda_j y_j = \sum_{j=1}^{n} \mu_j = \sum_{j=1}^{n} \mu_j x_j = \sum_{j=1}^{n} \mu_j y_j = 0, \quad (3)$$

because then the interpolants are solutions to variational problems that minimize second derivative norms of s and t (Duchon, 1977). Further, the variational principle ensures that the parameters are well-defined, provided that the points $\{(x_j, y_j) : j = 1, 2, \ldots, n\}$ are distinct and not collinear. Our work does not depend on the constraints (3).

Having picked the transformation (2), we apply it to a discretization of the whole (x, y) picture. The resultant image in (s, t) space becomes suitable for comparison with the (s, t) picture, because the mapping functions provide exact agreement between the interpolation points. The construction of the image in (s, t) space may require the functions (2) to be calculated for very many values of (x, y), perhaps as many as 10^8, while $n = 200$ is a typical value of n. Our purpose is to show that, by taking advantage of the smoothness properties of thin plate splines, the amount of computation to perform this task can be much less than the total time of direct evaluation. We assume that the values $\{s(\ell h, mh) : 0 \leq \ell, m \leq M\}$ are required, where M is a large integer and $h = 1/M$, and that it is sufficient to approximate these values to about 6–9 decimal places of accuracy. We make no further reference to the second half of expression (2), because t can be treated in the same way as s.

Ian Barrodale suggested that I study this problem because he knew of my interest in radial basis function methods. The first application that he mentioned was a comparison of satellite pictures of the forests of British Columbia. Here very small values of h can occur because the resolution of the pictures is so fine that the positions of road junctions and confluences of rivers are available as interpolation points. I provided Barrodale Computing Services with a Fortran package that performs the tabulation. We are going to consider the method that is used.

We let the initial value of h be 2^k times its required final value for some integer k, which is chosen so that, if the number of points of the initial grid is between about 20×20 and 40×40, then this grid covers the (s, t) picture with enough overlap to allow for some edge effects that will be explained later. The function $s(x, y)$ is calculated from formula (2) at the mesh points of the initial grid, except for a modification that will also be explained later, that is invoked when (x, y) is within a certain distance of at least one of the interpolation points. Then an iterative procedure is employed k times, where each iteration halves the value of h by applying the method that is described in Sections 2 and 3. Here the old and the new mesh sizes are $2h$ and h, respectively. When ℓ and m are odd integers and when the point $(\ell h, mh)$ is sufficiently far from all the interpolation points, then we approximate $s(\ell h, mh)$ by a linear combination of the 16 function values of the coarser grid that are contained in the square $[(\ell-3)h, (\ell+3)h] \times [(m-3)h, (m+3)h]$, the coefficients of the linear combination being chosen so that the error of the approximation is of magnitude $\mathcal{O}(h^6)$. Another procedure provides $\mathcal{O}(h^6)$ estimates of $s(\ell h, mh)$ when $\ell+m$ is odd. These techniques are given in Section 2. They provide approximations of sufficient accuracy when the point $(\ell h, mh)$ of the finer mesh satisfies the conditions

$$\max\left[\,|\ell h - x_j|,\ |mh - y_j|\,\right] \geq \rho h, \quad j=1,2,\ldots,n, \tag{4}$$

where ρ is a constant that depends on the required precision, a typical value being $\rho = 15$. Otherwise some modifications to the calculation of $s(\ell h, mh)$ are necessary to provide sufficient accuracy. They are the subject of Section 3.

Condition (4) implies an upper bound on the total number of explicit evaluations of thin plate spline terms when the current mesh size is halved by the procedure of Sections 2 and 3. We will find that this number is of magnitude $[3\rho^2 + \mathcal{O}(\rho)]\,n$. Further, each new value of $s(\ell h, mh)$ requires a simple linear combination of at most 16 old values to be formed, whatever the number of terms in the sums of expression (2). Further, for each j, the Section 3 procedure includes some corrections at the boundary of the j-th constraint (4), the work of this task being at most a constant times ρ for every j. Thus, the computational effort of halving the mesh size is bounded above by a small multiple of the number of new grid points plus a large multiple of n. Since the number of halvings is $k \approx |\log_2 h|$, where h is now the final mesh size, it follows that the total work of our procedure is of magnitude $\mathcal{O}(h^{-2}+n|\log_2 h|)$, which usually provides very substantial gains over the direct use of formula (2) when h is tiny, because the number of operations of the direct approach is of magnitude $\mathcal{O}(h^{-2}n)$. Some numerical results illustrate this important point in Section 4, and they are followed by a brief discussion of the given algorithm.

The choice of ρ for condition (4) depends on the accuracy of the subtabulation schemes of Section 2. Indeed, a recommended value of ρ is derived from the $\mathcal{O}(h^6)$ terms of the subtabulation errors. We should ask, however, whether higher order terms can invalidate these $\mathcal{O}(h^6)$ error estimates for typical values of h. This question is addressed in an appendix. Fortunately we find that the recommended choice is suitable. Further, the analysis provides some explicit formulae for derivatives and Taylor series expansions of thin plate splines.

§2. The Subtabulation Procedures

This section addresses the problem of estimating a function f on a square grid of mesh size h, when $f(\ell h, mh)$ is available for even values of the integers ℓ and m. Further, we assume that f satisfies the biharmonic equation

$$\frac{\partial^4 f(x,y)}{\partial x^4} + 2\frac{\partial^4 f(x,y)}{\partial x^2 \partial y^2} + \frac{\partial^4 f(x,y)}{\partial y^4} = 0, \tag{5}$$

because thin plate splines have this property away from their interpolation points. It is helpful to regard f as a substitute for s that does not have any singularities, because we are going to derive local error estimates from Taylor series expansions. First we pick an approximation to $f(\ell h, mh)$ when both ℓ and m are odd integers.

This approximation is shown in Figure 1. One should relate the figure to a 7×7 grid of mesh size h, where 16 of the points belong to a grid of size $2h$ and where $(\ell h, mh)$ is the central point of the picture. Further, the numbers in the boxes are the factors that multiply the given values of f in the approximation to $f(\ell h, mh)$. Therefore the figure depicts the estimate

$$f(\ell h, mh) \approx \sum_{\pm,\pm} \left[\tfrac{39}{128} f(\ell h \pm h, mh \pm h) - \tfrac{3}{128} f(\ell h \pm 3h, mh \pm h) \right.$$
$$\left. - \tfrac{3}{128} f(\ell h \pm h, mh \pm 3h) - \tfrac{1}{128} f(\ell h \pm 3h, mh \pm 3h) \right]. \tag{6}$$

The numbers in Figure 1 were chosen in the following way to maximize the order of the error of this approximation to $f(\ell h, mh)$.

Let the factors $\tfrac{39}{128}$, $\tfrac{-3}{128}$ and $\tfrac{-1}{128}$ in formula (6) be called α, β and γ. The symmetry implies that the error is zero when f is any quadratic polynomial if and only if it is zero when $f(x,y) = 1$ and when $f(x,y) = x^2$. Therefore we require the equations

$$4\alpha + 8\beta + 4\gamma = 1 \quad \text{and} \quad \alpha + 10\beta + 9\gamma = 0 \tag{7}$$

to be satisfied. Further, by identifying $(\ell h, mh)$ with the origin in the Taylor series expansion

$$f(x,y) = \sum_{j=0}^{\infty}\sum_{k=0}^{\infty} \frac{x^j}{j!}\frac{y^k}{k!}\frac{\partial^{j+k} f(0,0)}{\partial x^j \partial y^k}, \tag{8}$$

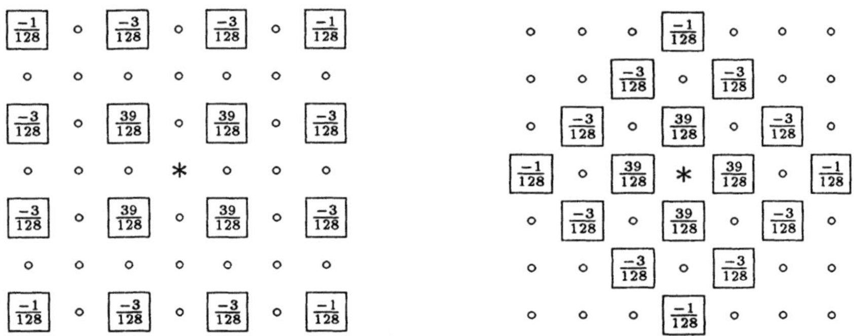

Figure 1:
The 16-point subtabulation stencil.

Figure 2:
The rotated 16-point stencil.

we deduce that the fourth order part of the error is the expression

$$4\alpha\left[\frac{h^4}{4!}\frac{\partial^4 f(0,0)}{\partial x^4}+\frac{h^2}{2!}\frac{h^2}{2!}\frac{\partial^4 f(0,0)}{\partial x^2\partial y^2}+\frac{h^4}{4!}\frac{\partial^4 f(0,0)}{\partial y^4}\right]$$

$$+4\beta\left[\frac{(3h)^4}{4!}\frac{\partial^4 f(0,0)}{\partial x^4}+\frac{(3h)^2}{2!}\frac{h^2}{2!}\frac{\partial^4 f(0,0)}{\partial x^2\partial y^2}+\frac{h^4}{4!}\frac{\partial^4 f(0,0)}{\partial y^4}\right.$$

$$\left.+\frac{h^4}{4!}\frac{\partial^4 f(0,0)}{\partial x^4}+\frac{h^2}{2!}\frac{(3h)^2}{2!}\frac{\partial^4 f(0,0)}{\partial x^2\partial y^2}+\frac{(3h)^4}{4!}\frac{\partial^4 f(0,0)}{\partial y^4}\right]$$

$$+4\gamma\left[\frac{(3h)^4}{4!}\frac{\partial^4 f(0,0)}{\partial x^4}+\frac{(3h)^2}{2!}\frac{(3h)^2}{2!}\frac{\partial^4 f(0,0)}{\partial x^2\partial y^2}+\frac{(3h)^4}{4!}\frac{\partial^4 f(0,0)}{\partial y^4}\right]$$

$$=\tfrac{1}{6}h^4\left[(\alpha+82\beta+81\gamma)\left(\frac{\partial^4 f(0,0)}{\partial x^4}+\frac{\partial^4 f(0,0)}{\partial y^4}\right)\right.$$

$$\left.+(6\alpha+108\beta+486\gamma)\frac{\partial^4 f(0,0)}{\partial x^2\partial y^2}\right]. \tag{9}$$

Therefore, in view of equation (5), this expression is zero if we force the relation

$$\alpha+82\beta+81\gamma = 3\alpha+54\beta+243\gamma. \tag{10}$$

The conditions (7) and (10) define the coefficients that are shown in Figure 1 and formula (6).

Therefore, if the error of the approximation (6) is expanded in powers of h, the leading nonzero term is of magnitude $\mathcal{O}(h^6)$. Specifically, the analogue

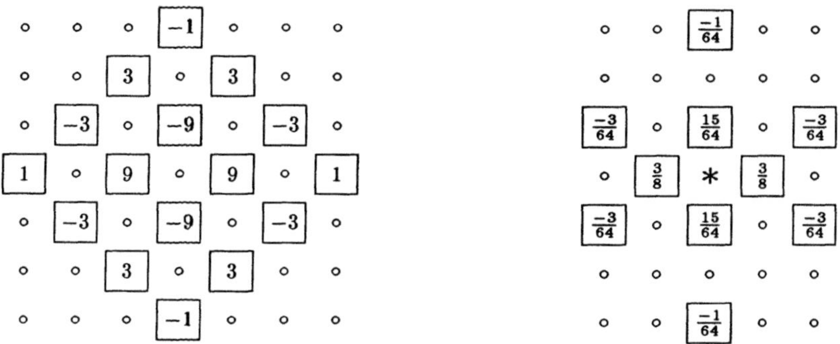

Figure 3:
An $\mathcal{O}(h^6)$ functional.

Figure 4:
A 10-point subtabulation stencil.

of expression (9) gives the term

$$\frac{1}{180}h^6\left[(\alpha+730\beta+729\gamma)\left(\frac{\partial^6 f(0,0)}{\partial x^6}+\frac{\partial^6 f(0,0)}{\partial y^6}\right)\right.$$
$$\left.+\;(15\alpha+1350\beta+10935\gamma)\left(\frac{\partial^6 f(0,0)}{\partial x^4\partial y^2}+\frac{\partial^6 f(0,0)}{\partial x^2\partial y^4}\right)\right]$$
$$=-\frac{1}{8}h^6\left(\frac{\partial^6 f(0,0)}{\partial x^6}+5\frac{\partial^6 f(0,0)}{\partial x^4\partial y^2}+5\frac{\partial^6 f(0,0)}{\partial x^2\partial y^4}+\frac{\partial^6 f(0,0)}{\partial y^6}\right). \quad (11)$$

Invoking the identity

$$\frac{\partial^6 f(0,0)}{\partial x^6}+3\frac{\partial^6 f(0,0)}{\partial x^4\partial y^2}+3\frac{\partial^6 f(0,0)}{\partial x^2\partial y^4}+\frac{\partial^6 f(0,0)}{\partial y^6}=0, \quad (12)$$

which is obtained by applying the Laplacian operator to equation (5), we write expression (11) in the form

$$\frac{1}{12}h^6\left(\frac{\partial^6 f(0,0)}{\partial x^6}+\frac{\partial^6 f(0,0)}{\partial y^6}\right). \quad (13)$$

It follows that, if f is any smooth function that satisfies the biharmonic equation, then the approximation (6) has the error

$$\frac{1}{12}h^6\left(\frac{\partial^6 f(\ell h,mh)}{\partial x^6}+\frac{\partial^6 f(\ell h,mh)}{\partial y^6}\right)+\mathcal{O}(h^8), \quad (14)$$

which will be used in the next section to pick the constant ρ that occurs in condition (4).

We also require a procedure that provides a sufficiently accurate approximation to $f(\ell h, mh)$ when $\ell + m$ is odd. We assume now that $f(\ell h, mh)$ is available whenever $\ell + m$ is even, the values when ℓ and m are both even being data and the values when ℓ and m are both odd having been generated by formula (6). Therefore the stencil in Figure 2 is suitable, which is obtained by rotating the stencil of Figure 1 through the angle $\pi/4$ and scaling it by the factor $2^{-1/2}$. There are advantages in using a narrower stencil, however. Therefore we derive the stencil in Figure 4 by adding $1/128$ times the stencil in Figure 3 to the one in Figure 2, which gives the algebraic formula

$$f(\ell h, mh) \approx \sum_{\pm} \left[\tfrac{3}{8} f(\ell h \pm h, mh) + \tfrac{15}{64} f(\ell h, mh \pm h) - \tfrac{1}{64} f(\ell h, mh \pm 3h) \right]$$

$$- \tfrac{3}{64} \sum_{\pm, \pm} f(\ell h \pm 2h, mh \pm h). \tag{15}$$

The error of this approximation is also of magnitude $\mathcal{O}(h^6)$, because it can be verified that Figure 3 depicts a linear combination of function values that vanishes for all quintic polynomials, which is easy to see if Figure 3 is rotated through the angle $\pi/4$. Further, by rotating Figure 4 through $\pi/2$, we find the $\mathcal{O}(h^6)$ approximation

$$f(\ell h, mh) \approx \sum_{\pm} \left[\tfrac{3}{8} f(\ell h, mh \pm h) + \tfrac{15}{64} f(\ell h \pm h, mh) - \tfrac{1}{64} f(\ell h \pm 3h, mh) \right]$$

$$- \tfrac{3}{64} \sum_{\pm, \pm} f(\ell h \pm h, mh \pm 2h). \tag{16}$$

The Fortran software applies formula (6) when ℓ and m are both odd, formula (15) when ℓ is even and m is odd, and formula (16) when ℓ is odd and m is even. Thus most of the function values on the right hand sides of expressions (15) and (16) are data from previous calculations, rather than estimates that have been generated by equation (6) for the current h. This strategy is particularly convenient at the grid boundaries. Indeed, Figure 5 depicts the top left hand corner of a large square grid, where "$\oplus$" denotes the points $(\ell h, mh)$ at which both ℓ and m are even. Therefore the asterisks indicate the points $(\ell h, mh)$ for which all the function values on the right hand side of formula (6) are available. Having calculated these function values, formulae (15) and (16) are applicable and provide estimates of $f(\ell h, mh)$ at the points that are indicated by "$\Leftrightarrow$" and "$\Updownarrow$" respectively.

It is important to note that, if the subtabulation procedure is applied recursively to halve the grid size many times, then the given formulae can

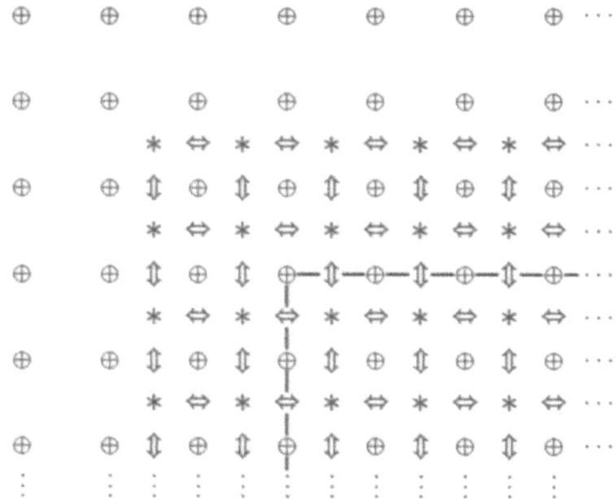

Figure 5:
Subtabulation at a corner of the grid.

provide estimates of all the required function values to the right of and below the solid lines of Figure 5. Indeed, the figure shows that the finer grid inherits from the coarser grid three complete lines of function values at mesh points that are outside the solid lines, and of course this property can be passed on to all subsequent calculations that halve the mesh size. Therefore, because we are supposing that the final grid should cover the square $[0, 1] \times [0, 1]$, we let the coverage of each intermediate grid of mesh size $2h$ be exactly the square $[-6h, M_h h] \times [-6h, M_h h]$, where M_h is the least even integer that satisfies the inequality $M_h h \geq 1 + 6h$. This construction provides the overlap that is mentioned in Section 1.

§3. Removal of the Singularities

We modify the obvious choice $f = s$ in the subtabulation formulae (6), (15) and (16) when the integers ℓ and m fail to satisfy all the conditions (4), where the constant ρ will be specified in this section. In order to describe the procedure, we let $\mathcal{N}_j(h)$ be the neighbourhood

$$\mathcal{N}_j(h) = \{(x, y) : \max[\, |x - x_j|, \, |y - y_j| \,] < \rho h\} \subset \mathbb{R}^2 \tag{17}$$

of (x_j, y_j), we let $\mathcal{J}_h(x, y)$ be the set

$$\mathcal{J}_h(x, y) = \{j : (x, y) \notin \mathcal{N}_j(h)\} \cap \{1, 2, \ldots, n\}, \tag{18}$$

and we let s_h be the function that takes the values

$$s_h(x,y) = \sum_{j \in \mathcal{J}_h(x,y)} \lambda_j [(x-x_j)^2+(y-y_j)^2] \log[(x-x_j)^2+(y-y_j)^2]^{1/2}+ax+by+c. \tag{19}$$

In other words, $s_h(x,y)$ is the same as the first line of expression (2), except that we have deleted the contributions to $s(x,y)$ from any interpolation points whose ∞-norm distance to (x,y) is less than ρh. Thus s_h usually has discontinuities at the boundaries of the neighbourhoods $\{\mathcal{N}_j(h) : j = 1, 2, \ldots, n\}$. When the Fortran implementation of the procedure of Section 2 applies the subtabulation formulae that reduce the mesh size from $2h$ to h, it works with values of s_h at the grid points instead of with values of s. Therefore it is possible to let ρ be so large that the errors that arise from each singularity of the thin plate spline are tolerable. Then the following procedure includes techniques that allow for the discontinuities in s_h and for the dependence of s_h on h.

Given the values of $s_h(\ell h, mh)$ for all even values of ℓ and m, where the grid of mesh size $2h$ covers the square $[-6h, M_h h] \times [-6h, M_h h]$ that is mentioned at the end of Section 2, the software applies the method of that section to generate function values on the grid of mesh size h that covers the square $[-3h, M_{h/2}\frac{1}{2}h] \times [-3h, M_{h/2}\frac{1}{2}h]$. Then the new function values are corrected where necessary so that they all become values of s_h to within the $\mathcal{O}(h^6)$ accuracy of the subtabulation formulae. Indeed, for every j, we have to revise the function value at $(\ell h, mh)$ if ℓ and m are not both even and if the ∞-norm distance from $(\ell h, mh)$ to the boundary of $\mathcal{N}_j(h)$ is less than $3h$. This calculation is done explicitly and is often the most expensive part of the entire computation. Thus adequate estimates of s_h are generated on the finer grid. Then they are overwritten by values of $s_{h/2}$ by adding to expression (19) the contributions from the integers j that are in the set $\mathcal{J}_{h/2}(x,y) \setminus \mathcal{J}_h(x,y)$. Here, instead of taking the view that (x,y) ranges over the points of the finer grid, one should treat the values of j in sequence, adding in all the differences between $s_{h/2}$ and s_h for each j before turning to a new value of j. Indeed, the total work of the latter approach is $\mathcal{O}(n\rho^2)$ operations for every h, but an $\mathcal{O}(h^{-2})$ overhead would occur if one considered each point of the finer grid individually. The number of thin plate spline terms that are evaluated during these operations for each h is of magnitude $[3\rho^2 + \mathcal{O}(\rho)]n$, as mentioned in Section 1. Further, when h reaches its final value, the numbers $\{s_{h/2}(\ell h, mh) : 0 \le \ell, m \le M\}$ have to be replaced by values of s. Therefore the calculation is completed by a cycle through the neighbourhoods $\{\mathcal{N}_j(\frac{1}{2}h) : j = 1, 2, \ldots, n\}$ that adds the term

$$\lambda_j [(\ell h-x_j)^2+(mh-y_j)^2] \log[(\ell h-x_j)^2+(mh-y_j)^2]^{1/2} \tag{20}$$

to $s_{h/2}(\ell h, mh)$ for every grid point $(\ell h, mh)$ that is in $\mathcal{N}_j(\frac{1}{2}h)$, so there are about another $n\rho^2$ evaluations of thin plate spline terms.

We now turn to the choice of ρ, letting h be the generic mesh size of a cycle through the calculations of Section 2. When f is the function s_h in the subtabulation formula (6), it follows from expressions (14) and (19) that the estimate of $s_h(\ell h, mh)$ has the truncation error

$$\tfrac{1}{12} h^6 \sum_{j \in \mathcal{J}_h(\ell h, mh)} \lambda_j \left(\frac{\partial^6 \phi(\ell h - x_j, mh - y_j)}{\partial x^6} + \frac{\partial^6 \phi(\ell h - x_j, mh - y_j)}{\partial y^6} \right) + \mathcal{O}(h^8), \tag{21}$$

where ϕ is the function

$$\phi(x, y) = (x^2 + y^2) \log(x^2 + y^2)^{1/2}, \quad (x, y) \in \mathbb{R}^2, \tag{22}$$

and where we are ignoring the consequences of discontinuities in s_h, because we have noted already that they are treated explicitly. Therefore ρ depends on sixth derivatives of ϕ. We calculate the values

$$\begin{aligned}
\partial \phi(x, y) / \partial x &= x \log(x^2 + y^2) + x \\
\partial^2 \phi(x, y) / \partial x^2 &= \log(x^2 + y^2) + 1 + 2x^2/(x^2 + y^2) \\
\partial^3 \phi(x, y) / \partial x^3 &= (2x^3 + 6xy^2)/(x^2 + y^2)^2 \\
\partial^4 \phi(x, y) / \partial x^4 &= (-2x^4 - 12x^2 y^2 + 6y^4)/(x^2 + y^2)^3 \\
\partial^5 \phi(x, y) / \partial x^5 &= (4x^5 + 40x^3 y^2 - 60xy^4)/(x^2 + y^2)^4 \\
\partial^6 \phi(x, y) / \partial x^6 &= (-12x^6 - 180x^4 y^2 + 540x^2 y^4 - 60y^6)/(x^2 + y^2)^5
\end{aligned} \tag{23}$$

and then symmetry provides the formula

$$\frac{\partial^6 \phi(x, y)}{\partial x^6} + \frac{\partial^6 \phi(x, y)}{\partial y^6} = (-72x^4 + 432x^2 y^2 - 72y^4)/(x^2 + y^2)^4. \tag{24}$$

Further, by combining the elementary identities

$$x^4 - 6x^2 y^2 + y^4 = (x^2 + y^2)^2 - 8x^2 y^2 = 2(x^2 - y^2)^2 - (x^2 + y^2)^2 \tag{25}$$

with equation (24), we obtain the bound

$$\left| \frac{\partial^6 \phi(x, y)}{\partial x^6} + \frac{\partial^6 \phi(x, y)}{\partial y^6} \right| \leq 72/(x^2 + y^2)^2. \tag{26}$$

Expressions (21) and (26) show that, for each $j \in \mathcal{J}_h(\ell h, mh)$, the modulus of the contribution from the (x_j, y_j) term of s_h to the truncation error of the subtabulation formula (6) when $f = s_h$ is at most the product

$$6\left\{ h/[(\ell h - x_j)^2 + (mh - y_j)^2]^{1/2} \right\}^6 \left| \lambda_j[(\ell h - x_j)^2 + (mh - y_j)^2] \right|, \quad (27)$$

where we have ignored the terms of magnitude $\mathcal{O}(h^8)$, but they are studied in the appendix. Thus formula (6) provides the relative accuracy

$$6\left\{ h/[(\ell h - x_j)^2 + (mh - y_j)^2]^{1/2} \right\}^6 / \left| \log [(\ell h - x_j)^2 + (mh - y_j)^2]^{1/2} \right| \tag{28}$$

in the j-th term of expression (19). We assume that the logarithm can also be ignored, this question being discussed in the next paragraph. Therefore, because we wish to work to a relative accuracy of ϵ, we require the definition of $\mathcal{J}_h(\ell h, mh)$ to provide the bound

$$[(\ell h - x_j)^2 + (mh - y_j)^2]^{1/2} \geq (6/\epsilon)^{1/6} h, \quad j \in \mathcal{J}_h(\ell h, mh). \tag{29}$$

It follows from condition (4) that it is suitable to set ρ to the least integer that satisfies the inequality $\rho \geq (6/\epsilon)^{1/6}$. For example, the values $\rho = 14$ and $\rho = 43$ are chosen in the cases $\epsilon = 10^{-6}$ and $\epsilon = 10^{-9}$ respectively. Therefore each application of the procedure of the second paragraph of this section requires $\mathcal{O}(n\epsilon^{-1/3})$ operations, which causes the $\mathcal{O}(n\epsilon^{-1/3}|\log h|)$ component of the total work that is mentioned in the abstract.

Ignoring the logarithm of expression (28) is valid for certain scalings of the x and y variables. We have in mind that, from a practical point of view, it should not matter if the measurements in the pictures that are being compared are expressed in centimetres or in miles, for instance. At least one of these choices would admit our assumption, provided that a change in units does not damage the accuracy of formula (6). Now, if the x and y variables are scaled by a constant, then, instead of changing the log terms of expression (19), it is equivalent to add a certain quadratic polynomial to s_h. Further, we have chosen subtabulation formulae that are exact when they are applied to quadratic polynomials. It follows that the errors of the given approximations are independent of such changes of scale, so there is some reasonable justification for the use of inequality (29) when a relative accuracy of ϵ is required.

We complete this section by showing that the sixth order terms of the other formulae of Section 2 do not demand a larger value of ρ. We seek the sixth order Taylor series terms of the stencils of Figures 2 and 3 that are analogous to the sixth order term (13) of Figure 1. Therefore we continue to let $(\ell h, mh)$ be the origin in the expansion (8). Hence, writing the stencil of Figure 2 in the form

$$f(\ell h, mh) \approx \sum_{\pm} \left[\tfrac{39}{128} f(\ell h \pm h, mh) + \tfrac{39}{128} f(\ell h, mh \pm h) - \tfrac{1}{128} f(\ell h \pm 3h, mh) \right.$$
$$\left. - \tfrac{1}{128} f(\ell h, mh \pm 3h) \right] - \tfrac{3}{128} \sum_{\pm,\pm} \left[f(\ell h \pm 2h, mh \pm h) + f(\ell h \pm h, mh \pm 2h) \right],$$

$$(30)$$

we see that it has the sixth order terms

$$\frac{h^6}{6!} \left(\frac{\partial^6 f(0,0)}{\partial x^6} + \frac{\partial^6 f(0,0)}{\partial y^6} \right) \left[2 \times \tfrac{39}{128} - 2 \times \tfrac{729}{128} - 4 \times \tfrac{3}{128} \times (64+1) \right]$$
$$- \frac{h^4}{4!} \frac{h^2}{2!} \left(\frac{\partial^6 f(0,0)}{\partial x^4 \partial y^2} + \frac{\partial^6 f(0,0)}{\partial x^2 \partial y^4} \right) \left[4 \times \tfrac{3}{128} \times (16+4) \right]$$
$$= -\tfrac{3}{128} h^6 \left(\frac{\partial^6 f(0,0)}{\partial x^6} + \frac{\partial^6 f(0,0)}{\partial y^6} \right) - \tfrac{5}{128} h^6 \left(\frac{\partial^6 f(0,0)}{\partial x^4 \partial y^2} + \frac{\partial^6 f(0,0)}{\partial x^2 \partial y^4} \right)$$
$$= -\tfrac{1}{96} h^6 \left(\frac{\partial^6 f(0,0)}{\partial x^6} + \frac{\partial^6 f(0,0)}{\partial y^6} \right), \tag{31}$$

where the last line depends on the identity (12). The magnitude of this term is one eighth of expression (13), because the Euclidean diameter of the stencil of Figure 2 is the diameter of the stencil of Figure 1 divided by $2^{1/2}$. Moreover, because the stencil of Figure 3 denotes the functional

$$\sum_{\pm} \left[9f(\ell h \pm h, mh) - 9f(\ell h, mh \pm h) + f(\ell h \pm 3h, mh) - f(\ell h, mh \pm 3h) \right]$$
$$+ \sum_{\pm,\pm} \left[3f(\ell h \pm h, mh \pm 2h) - 3f(\ell h \pm 2h, mh \pm h) \right], \tag{32}$$

it has the sixth order terms

$$\frac{h^6}{6!} \left(\frac{\partial^6 f(0,0)}{\partial x^6} - \frac{\partial^6 f(0,0)}{\partial y^6} \right) \left[2 \times 9 + 2 \times 729 + 4 \times 3 - 4 \times 3 \times 64 \right]$$
$$+ \frac{h^4}{4!} \frac{h^2}{2!} \left(\frac{\partial^6 f(0,0)}{\partial x^4 \partial y^2} - \frac{\partial^6 f(0,0)}{\partial x^2 \partial y^4} \right) \left[4 \times 3 \times 4 - 4 \times 3 \times 16 \right]$$
$$= h^6 \left(\frac{\partial^6 f(0,0)}{\partial x^6} - 3 \frac{\partial^6 f(0,0)}{\partial x^4 \partial y^2} + 3 \frac{\partial^6 f(0,0)}{\partial x^2 \partial y^4} - \frac{\partial^6 f(0,0)}{\partial y^6} \right)$$
$$= 4h^6 \left(\frac{\partial^6 f(0,0)}{\partial x^6} - \frac{\partial^6 f(0,0)}{\partial y^6} \right), \tag{33}$$

where the last line depends on the elementary observation that equation (5) implies the identity

$$\frac{\partial^6 f(0,0)}{\partial x^6} + \frac{\partial^6 f(0,0)}{\partial x^4 \partial y^2} - \frac{\partial^6 f(0,0)}{\partial x^2 \partial y^4} - \frac{\partial^6 f(0,0)}{\partial y^6} = 0. \tag{34}$$

Remembering that the approximation (15) is formed by adding $1/128$ times the stencil of Figure 3 to the stencil of Figure 2, it follows from expressions (31) and (33) that the leading term of the error of formula (15) is the quantity

$$-\frac{1}{96}h^6 \left(\frac{\partial^6 f(0,0)}{\partial x^6} + \frac{\partial^6 f(0,0)}{\partial y^6} \right) + \frac{1}{32}h^6 \left(\frac{\partial^6 f(0,0)}{\partial x^6} - \frac{\partial^6 f(0,0)}{\partial y^6} \right). \tag{35}$$

We also recall that we picked ρ so that, if the bound $x^2 + y^2 \geq \rho^2$ holds and if $f(0,0)$ is replaced by $\phi(x,y)$ in expression (13), then inequality (26) makes the modulus of the sixth order term (13) acceptably small. In other words, due to the choice of ρ, the error term (13) is small enough because it satisfies the condition

$$\left| \frac{1}{12}h^6 \left(\frac{\partial^6 \phi(x,y)}{\partial x^6} + \frac{\partial^6 \phi(x,y)}{\partial y^6} \right) \right| \leq 6h^6/(x^2 + y^2)^2. \tag{36}$$

Therefore the choice of ρ also ensures that expression (35) is suitably small if the definition (22) implies the inequality

$$\left| -\frac{1}{96}h^6 \left(\frac{\partial^6 \phi(x,y)}{\partial x^6} + \frac{\partial^6 \phi(x,y)}{\partial y^6} \right) + \frac{1}{32}h^6 \left(\frac{\partial^6 \phi(x,y)}{\partial x^6} - \frac{\partial^6 \phi(x,y)}{\partial y^6} \right) \right|$$
$$\leq 6h^6/(x^2 + y^2)^2. \tag{37}$$

Now the last of the equations (23) and symmetry give the identity

$$\frac{\partial^6 \phi(x,y)}{\partial x^6} - \frac{\partial^6 \phi(x,y)}{\partial y^6} = (48x^6 - 720x^4 y^2 + 720x^2 y^4 - 48y^6)/(x^2 + y^2)^5, \tag{38}$$

so, because the numerator of the right hand side can be expressed in the forms

$$48(x^2 + y^2)^3 - 96y^2(3x^2 - y^2)^2 \quad \text{and} \quad 96x^2(3y^2 - x^2)^2 - 48(x^2 + y^2)^3, \tag{39}$$

we have the bound

$$\left| \frac{\partial^6 \phi(x,y)}{\partial x^6} - \frac{\partial^6 \phi(x,y)}{\partial y^6} \right| \leq 48/(x^2 + y^2)^2. \tag{40}$$

Conditions (36) and (40) show that the required inequality (37) is satisfied even if we reduce the constant on the right hand side from 6 to 9/4.

We have not yet, however, allowed for the important point that the function values $f(\ell h \pm h, mh)$ of the estimate (15) include sixth order errors, because the procedure of Section 2 calculates them by applying formula (6). Therefore, in view of the derivation of expression (13), we should add the term

$$2 \times \tfrac{3}{8} \times \tfrac{1}{12} h^6 \left(\frac{\partial^6 f(0,0)}{\partial x^6} + \frac{\partial^6 f(0,0)}{\partial y^6} \right) \tag{41}$$

to the quantity (35). Thus, after replacing $f(\ell h \pm h, mh)$ by their approximations, the estimate (15) has the leading error term

$$\tfrac{5}{96} h^6 \left(\frac{\partial^6 f(0,0)}{\partial x^6} + \frac{\partial^6 f(0,0)}{\partial y^6} \right) + \tfrac{1}{32} h^6 \left(\frac{\partial^6 f(0,0)}{\partial x^6} - \frac{\partial^6 f(0,0)}{\partial y^6} \right), \tag{42}$$

which is acceptably small, because conditions (36) and (40) give the bound

$$\left| \tfrac{5}{96} h^6 \left(\frac{\partial^6 \phi(x,y)}{\partial x^6} + \frac{\partial^6 \phi(x,y)}{\partial y^6} \right) + \tfrac{1}{32} h^6 \left(\frac{\partial^6 \phi(x,y)}{\partial x^6} - \frac{\partial^6 \phi(x,y)}{\partial y^6} \right) \right|$$
$$\leq \tfrac{21}{4} h^6 / (x^2 + y^2)^2. \tag{43}$$

Similarly, the choice of ρ also ensures that the accuracy of formula (16) is adequate.

We have not investigated analytically whether a sequence of applications of the method of Section 2 can cause an unacceptably large accumulation of errors. In particular, the stability properties of the stencils when they are used recursively have not been considered, but similar questions are studied in the analysis of algorithms for "subdivision" (see Dyn, 1992, for example). The coefficients of our subtabulation formulae seem to be harmless, and no difficulties have occurred in numerical computations. Further, if the function value $f(\ell h, mh)$ is given to the procedure of Section 2, and if it includes a relatively large error due to the singularity of the thin plate spline at the interpolation point (x_j, y_j), then $(\ell h, mh)$ must be close to the boundary of the neighbourhood $\mathcal{N}_j(\kappa h)$, for some integer κ that is a power of 2 satisfying $\kappa \geq 2$. Therefore, assuming $\rho \geq 10$ for instance, we have the advantage that, for the current and future values of h, the function value $f(\ell h, mh)$ will feature only in subtabulation formulae that have relatively small new errors arising from the singularity at (x_j, y_j).

Grid size	$n=25$	$n=50$	$n=100$	$n=200$	$n=400$
100×100	0.85	1.64	3.20	6.30	12.44
200×200	1.36	2.41	4.43	8.66	16.74
400×400	2.54	3.77	6.23	11.66	21.08
800×800	6.59	7.94	10.78	16.49	27.96

Table 1: Some timings in seconds.

§4. Numerical Results and Discussion

The Fortran software has been used for many calculations. It was developed by the author in 1990, and since then has been made to run faster at Barrodale Computing Services. Some typical examples of execution times of the original version are shown in Table 1. The given figures are computation times in seconds on a Sparc workstation for a range of grid sizes and values of n with $\epsilon = 10^{-6}$, but they include some inconsistencies of at least 1% due to their dependence on the scheduling of a Unix operating system. Nevertheless, the table distinguishes the two main components of the total work, which are the use of the subtabulation formulae and the corrections that allow for the discontinuities in s_h, these components being a small multiple of h^{-2} and a large multiple of $n|\log h|$ respectively, where h is the final mesh size and n is the number of interpolation points. Indeed, because the entries in the first two rows of the table tend to be proportional to n, we deduce that the work of the correction procedure requires about $n/100$ seconds whenever the mesh size is halved. Further, the $n=25$ column provides upper bounds on the total times that are taken by the subtabulation formulae. On the other hand, if all the values of s on the final 800×800 grid are calculated separately from the definition (2), then 395.56 seconds are required when $n=25$, the time when $n=400$ being about 16 times longer. Thus it is clear that the gains that are provided by the given algorithm are very substantial.

Much finer meshes can be treated efficiently, even when there is not enough computer storage to hold the final grid of function values. The reason is that one can divide the final grid into pieces that can be accommodated in storage and one can apply the algorithm to each piece separately. Of course the overlap between pieces that is suggested by Figure 5 is needed, and it will happen that some of the interpolation points $\{(x_j, y_j) : j = 1, 2, \ldots, n\}$ will lie outside the region that is covered by the grid of a typical subcalculation. These features do not introduce any difficulties.

There are some subtabulation formulae that provide $\mathcal{O}(h^6)$ accuracy and that require less work than the stencils of Figures 1 and 4. In particular, one could apply the estimate

$$f(\ell h, mh) \approx \tfrac{1}{256} \sum_{\pm} [150 f(\ell h \pm h, mh) - 25 f(\ell h \pm 3h, mh) + 3 f(\ell h \pm 5h, mh)] \quad (44)$$

along grid lines that are parallel to the x-axis, and of course there is an analogous formula in the y-direction. Perhaps the only objection to this method is that it does not have the two-dimensional structure that is inherent in the main calculation. Further, the idea of using linear formulae of the type (44) near the edges of the regions $\{\mathcal{N}_j(h) : j = 1, 2, \ldots, n\}$ could avoid many of the time-consuming discontinuity corrections of the current algorithm. Corrections would be needed near the corners of the square neighbourhoods, however, and also where the boundaries of two different neighbourhoods intersect at right angles.

Alternatively, one can avoid the discontinuities altogether by working with values of the original function s throughout the calculation. In this case one would have to correct the values that are given by the subtabulation formulae within the neighbourhoods $\{\mathcal{N}_j(h) : j = 1, 2, \ldots, n\}$ after each cycle through the operations of Section 2. Thus the work of the modifications would be proportional to $n\rho^2$, instead of the present much larger multiple of $n\rho$ plus a smaller multiple of $n\rho^2$. In view of the entries in Table 1, it is probable that, for typical values of ρ, it would be more efficient to prefer $f = s$ instead of $f = s_h$.

In general, radial basis function methods are expensive in comparison with approximation techniques that use piecewise polynomials, because the work of calculating a value $s(x)$ of a radial basis function interpolant is proportional to the number of interpolation points. Therefore it is important that we have shown that substantial savings can be made when $s(x)$ is required for many different values of x on a regular grid. Further, techniques have been developed for the case when the points x are in general position, such as the fast multipole method of Greengard and Rokhlin (1987), which treats clusters of interpolation points that are sufficiently remote from the current x as single Laurent series. The application of this method to thin plate splines is described and analysed by Beatson and Newsam (1992). Such developments are increasing greatly the usefulness of radial basis function methods for the solution of a wide range of multivariate approximation problems.

Appendix. The Truncation Errors of the Subtabulation Formulae

When the given algorithm applies the subtabulation formula (6), the resultant truncation error is expression (21). Further, in view of inequality (36) and the definition (18), we have the bound

$$
\left| \tfrac{1}{12} h^6 \lambda_j \left(\frac{\partial^6 \phi(\ell h - x_j, mh - y_j)}{\partial x^6} + \frac{\partial^6 \phi(\ell h - x_j, mh - y_j)}{\partial y^6} \right) \right|
$$
$$
\leq 6 h^6 |\lambda_j| / [\, (\ell h - x_j)^2 + (mh - y_j)^2 \,]^2 \tag{45}
$$
$$
\leq 6 \rho^{-6} |\lambda_j| [\, (\ell h - x_j)^2 + (mh - y_j)^2 \,], \quad j \in \mathcal{J}_h(\ell h, mh).
$$

Usually, therefore, we expect our choice of ρ, namely the least integer that satisfies $\rho \geq (6/\epsilon)^{1/6}$, to provide the required relative accuracy of ϵ. The main purpose of this appendix is to investigate whether the higher order terms of expression (21) can cause serious damage to this expectation.

We derive the total contribution to the truncation error (21) from a single value of j. We assume without loss of generality that $\lambda_j = 1$ and that (x_j, y_j) is the origin. Therefore we consider the difference between the two sides of formula (6) when we have $f \equiv \phi$ and $\max[\, |\ell|, |m| \,] \geq \rho$. Since the choice of ρ is guided by inequality (36), we wish to show that the modulus of this total contribution is not much larger than $6 h^6 / [\, (\ell h)^2 + (mh)^2 \,]^2 = 6 h^2 / (\ell^2 + m^2)^2$.

We are going to employ the Taylor series expansion of ϕ about the point $(x, y) = (\ell h, mh)$. Fortunately the following procedure yields all the required derivatives of ϕ. We write the third equation of expression (23) in the form

$$
\frac{\partial^3 \phi(x, y)}{\partial x^3} = \frac{x + 2iy}{(x + iy)^2} + \frac{x - 2iy}{(x - iy)^2}, \tag{46}
$$

where $i = (-1)^{1/2}$. Then the Leibniz formula gives the higher order derivatives

$$
\frac{\partial^j \phi(x, y)}{\partial x^j} = (-1)^{j+1} (j-3)! \left(\frac{x + (j-1)iy}{(x + iy)^{j-1}} + \frac{x - (j-1)iy}{(x - iy)^{j-1}} \right), \quad j \geq 3. \tag{47}
$$

Further, the k-th derivative of this expression with respect to y has the value

$$
(-1)^j (j+k-3)! \left((-i)^k \frac{(k-1)x - (j-1)iy}{(x + iy)^{j+k-1}} + i^k \frac{(k-1)x + (j-1)iy}{(x - iy)^{j+k-1}} \right). \tag{48}
$$

It is straightforward to verify that this formula is also valid for $0 \leq j \leq 2$ provided that we have $j + k \geq 3$. We state this conclusion formally.

Lemma 1. *The thin plate spline function (22) has the derivative*

$$\frac{\partial^{j+k}\phi(x,y)}{\partial x^j \partial y^k} = 2(-1)^j (j+k-3)! \, \Re\left(i^k \frac{(k-1)x + (j-1)iy}{(x-iy)^{j+k-1}} \right), \tag{49}$$

where j and k are any nonnegative integers such that $j+k \geq 3$, and where $\Re$ denotes the real part of the term in the large brackets. ∎

Because of the smoothness properties of ϕ, it can be proved that the Taylor series expansion

$$\phi(x+\xi, y+\eta) = \sum_{j=0}^{\infty} \sum_{k=0}^{\infty} \frac{\xi^j}{j!} \frac{\eta^k}{k!} \frac{\partial^{j+k}\phi(x,y)}{\partial x^j \partial y^k} \tag{50}$$

is valid if the double sum is absolutely convergent. We will deduce from the following lemma that this property is enjoyed by all the values of x, y, ξ and η that we allow in the subtabulation formulae of Section 2.

Lemma 2. *The sum (50) is absolutely convergent if its variables satisfy the inequality*

$$|\xi| + |\eta| < (x^2+y^2)^{1/2}. \tag{51}$$

Proof: Equation (49) implies the bound

$$\left| \frac{\partial^{j+k}\phi(x,y)}{\partial x^j \partial y^k} \right| \leq \frac{2(j+k-3)! \, \max[\,k-1, j-1\,]}{(x^2+y^2)^{(j+k-2)/2}} < \frac{(j+k)!}{(x^2+y^2)^{(j+k-2)/2}} \tag{52}$$

when $j+k \geq 3$. Moreover, condition (51) implies that the sum

$$\sum_{t=0}^{\infty} \left(\frac{|\xi| + |\eta|}{(x^2+y^2)^{1/2}} \right)^t = \sum_{j=0}^{\infty} \sum_{k=0}^{\infty} \frac{(j+k)!}{j! \, k!} \frac{|\xi|^j |\eta|^k}{(x^2+y^2)^{(j+k)/2}} \tag{53}$$

is finite. Therefore the relation

$$\sum_{\substack{j=0 \\ j+k \geq 3}}^{\infty} \sum_{k=0}^{\infty} \left| \frac{\xi^j}{j!} \frac{\eta^k}{k!} \frac{\partial^{j+k}\phi(x,y)}{\partial x^j \partial y^k} \right| < (x^2+y^2) \sum_{t=0}^{\infty} \left(\frac{|\xi| + |\eta|}{(x^2+y^2)^{1/2}} \right)^t \tag{54}$$

holds, which shows that the lemma is true. ∎

We assume that the parameter ρ of the algorithm satisfies $\rho \geq 10$. Hence, because we have the conditions

$$(x, y) = (\ell h, mh), \quad (x_j, y_j) = (0, 0) \quad \text{and} \quad j \in \mathcal{J}_h(\ell h, mh), \qquad (55)$$

the right hand side of inequality (51) is bounded below by $10h$. Moreover, the values of (ξ, η) that will occur in equation (50) are the displacements from the centres to the data points of the stencils of Figures 1–4, so they satisfy the condition $|\xi| + |\eta| \leq 6h$. It follows that Lemma 2 is applicable. Therefore the Taylor series expansion (50) gives the formula

$$\phi(x+\xi, y+\eta) = \sum_{t=0}^{\infty} \left(\sum_{\substack{j,k \\ j+k=t}} \frac{\xi^j}{j!} \frac{\eta^k}{k!} \frac{\partial^{j+k}\phi(x,y)}{\partial x^j \partial y^k} \right) = \sum_{t=0}^{\infty} \psi_t(\xi, \eta), \qquad (56)$$

say. Further, we find next that the sum inside the large brackets can be calculated analytically for all $t \geq 3$.

Specifically, equation (49) implies the value

$$\psi_t(\xi, \eta) = 2(t-3)!\, \Re \left(\sum_{\substack{j,k \\ j+k=t}} \frac{(-\xi)^j}{j!} \frac{(i\eta)^k}{k!} \frac{(k-1)x+(j-1)iy}{(x-iy)^{t-1}} \right)$$
$$= \frac{2(t-3)!}{t!}\, \Re \left(\frac{(-x-iy)(-\xi+i\eta)^t + t(ix\eta-iy\xi)(-\xi+i\eta)^{t-1}}{(x-iy)^{t-1}} \right), \qquad (57)$$

where the last line is derived from the binomial expansions of $(-\xi+i\eta)^t$ and $(-\xi+i\eta)^{t-1}$. In order to simplify this expression in a way that employs the rotational symmetry of the thin plate spline function ϕ, we introduce the notation

$$d = (x^2+y^2)^{1/2}, \quad \delta = (\xi^2+\eta^2)^{1/2} \quad \text{and} \quad \theta = \arg(\xi+i\eta) - \arg(x+iy), \qquad (58)$$

so we have the elementary relations

$$x\xi+y\eta = d\delta \cos\theta \quad \text{and} \quad x\eta-y\xi = d\delta \sin\theta. \qquad (59)$$

It follows that we can write expression (57) in the form

$$2(\delta^t/d^{t-2})\, \Re\left(-(-\cos\theta+i\sin\theta)^t + it\sin\theta(-\cos\theta+i\sin\theta)^{t-1}\right) / [t(t-1)(t-2)]$$
$$= (-1)^t(\delta^t/d^{t-2})\,[(t-2)\cos(t\theta) - t\cos(t\theta-2\theta)] / [t(t-1)(t-2)], \qquad (60)$$

which gives the following assertion.

Lemma 3. *Let ϕ be the thin plate spline function (22) and let the real variables x, y, ξ and η satisfy inequality (51). Then, for every integer t such that $t \geq 3$, the sum in the large brackets of the Taylor series expansion (56) has the value*

$$\psi_t(\xi, \eta) = (-1)^t \frac{[(t-2)\cos(t\theta) - t\cos(t\theta - 2\theta)](\xi^2 + \eta^2)^{t/2}}{t(t-1)(t-2)(x^2 + y^2)^{(t-2)/2}}, \quad (61)$$

where θ is specified in the definition (58). ∎

We apply this lemma to the stencil of Figure 1, so we require the Taylor series expansion about $(x, y) = (\ell h, mh)$ of every function value on the right hand side of formula (6). We let θ_0, θ_1 and θ_2 be the values of θ in the definition (58) when (ξ, η) is $(3h, h)$, (h, h) and $(h, 3h)$ respectively. Therefore we have the relations

$$\cos(\theta_1 - \theta_0) = \cos(\theta_2 - \theta_1) = 2/5^{1/2}. \quad (62)$$

In view of the symmetry of Figure 1, the $\cos(t\theta)$ term of expression (61) causes the data values with the weights $-3/128$ to be multiplied by the factor

$$\sum_{j=0}^{3} [\cos(t\theta_0 + \tfrac{1}{2}jt\pi) + \cos(t\theta_2 + \tfrac{1}{2}jt\pi)] = 4\delta_t [\cos(t\theta_0) + \cos(t\theta_2)]$$

$$= 8\delta_t \cos(t\theta_1) \cos[t(\theta_1 - \theta_0)], \quad (63)$$

where elementary properties of the cosine function imply that δ_t has the value $\delta_t = 1$ if the integer t is a multiple of 4, but otherwise $\delta_t = 0$. The corresponding factor for the other weights of the Figure 1 stencil is $4\delta_t \cos(t\theta_1)$, and of course the factors of the $\cos(t\theta - 2\theta)$ part of expression (61) are obtained by reducing t by 2. Thus we deduce the following result from Lemma 3.

Lemma 4. *Let ϕ be the thin plate spline function (22), let the integers ℓ and m satisfy $\max[|\ell|, |m|] \geq \rho \geq 10$, let θ and $\hat{\theta}$ be the angles $\frac{1}{4}\pi - \arg(\ell + im)$ and $\cos^{-1}(2/5^{1/2})$ respectively, and let t be any integer such that $t \geq 3$. Then the t-th order term of the error of the approximation (6) is zero if t is odd, it has the value*

$$4h^2 [\tfrac{39}{128} 2^{t/2} - \tfrac{3}{64} 10^{t/2} \cos(t\hat{\theta}) - \tfrac{1}{128} 18^{t/2}] \cos(t\theta) / [t(t-1)(\ell^2 + m^2)^{(t-2)/2}] \quad (64)$$

if t is divisible by 4, and it has the value

$$4h^2 [-\tfrac{39}{128} 2^{t/2} + \tfrac{3}{64} 10^{t/2} \cos(t\hat{\theta} - 2\hat{\theta}) + \tfrac{1}{128} 18^{t/2}] \cos(t\theta - 2\theta)$$

$$/ [(t-1)(t-2)(\ell^2 + m^2)^{(t-2)/2}] \quad (65)$$

if t is twice an odd integer. ∎

It follows from this lemma that, when $t = 4$, 6, 8 and 10, the t-th order truncation errors of formula (6) are the expressions

$$0, \quad \frac{6h^2\cos(4\theta)}{(\ell^2+m^2)^2}, \quad \frac{-30h^2\cos(8\theta)}{(\ell^2+m^2)^3} \quad \text{and} \quad \frac{600h^2\cos(8\theta)}{(\ell^2+m^2)^4} \tag{66}$$

respectively. Remembering the condition $\max[\,|\ell|,\,|m|\,] \geq 10$, we deduce that the third and fourth of these expressions are at most 5% and 1% of the bound $6h^2/(\ell^2+m^2)^2$ on the sixth order truncation error, given at the end of the second paragraph of the appendix and verified by the second of the expressions (66). The lemma also implies that the modulus of the total contribution to the truncation error from all values of t that satisfy $t \geq 12$ is less than the number

$$\sum_{k=6}^{\infty} h^2 \left[\tfrac{39}{32}2^k + \tfrac{3}{16}10^k + \tfrac{1}{32}18^k \right] / \left[132(\ell^2+m^2)^{k-1} \right]$$

$$= \left[6h^2/(\ell^2+m^2)^2 \right] \sum_{k=6}^{\infty} \left[\tfrac{13}{1056}2^{k-3} + \tfrac{125}{528}10^{k-3} + \tfrac{81}{352}18^{k-3} \right] / (\ell^2+m^2)^{k-3}$$

$$\leq \left[6h^2/(\ell^2+m^2)^2 \right] \left[\tfrac{13}{1056}(0.02)^3/0.98 + \tfrac{125}{528}(0.1)^3/0.9 + \tfrac{81}{352}(0.18)^3/0.82 \right]$$

$$< 0.002 \left[6h^2/(\ell^2+m^2)^2 \right]. \tag{67}$$

Therefore, when $\rho \geq 10$, the total truncation error of formula (6) does not exceed the sixth order bound of Section 3 by more than 6.2%, so the given algorithm provides good accuracy in its calculation of $s(\ell h, mh)$, when ℓ and m are both odd.

Of course there are analogues of Lemma 4 for the stencils of Figures 2 and 3. Indeed, the modifications to Lemma 4 that make it relevant to Figure 2 are as follows. We allow for the change in orientation of the stencil by altering the definition of θ to $-\arg(\ell+im)$, and we allow for the change in scale by multiplying expressions (64) and (65) by the factor $2^{-t/2}$. It follows that the higher order errors of the Figure 2 stencil are negligible. Therefore, because formulae (15) and (16) are derived by adding and subtracting 1/128 times the Figure 3 functional to and from the Figure 2 stencil, it remains to show that, when $\rho \geq 10$, the modulus of this functional is substantially less than the quantity $128 \times 6h^2/(\ell^2+m^2)^2$. We employ the remark that, in view of the symmetry and antisymmetry properties of Figure 3, the argument that yielded Lemma 4 now provides the following assertion.

Lemma 5. *Let ϕ be the thin plate spline function (22), let the integers ℓ and m satisfy $\max[\,|\ell|, |m|\,] \geq \rho \geq 10$, let θ and $\hat{\theta}$ be the angles $-\arg(\ell+im)$ and $\cos^{-1}(2/5^{1/2})$ respectively, and let t be any integer such that $t \geq 3$. Then the t-th order term of the functional of Figure 3 is zero if t is odd, it has the value*

$$4h^2[\,-9+6{\times}5^{t/2}\cos(t\hat{\theta}-2\hat{\theta})-3^t\,]\cos(t\theta-2\theta)/[\,(t-1)(t-2)(\ell^2+m^2)^{(t-2)/2}\,] \quad (68)$$

if t is divisible by 4, and it has the value

$$4h^2[\,9-6{\times}5^{t/2}\cos(t\hat{\theta})+3^t\,]\cos(t\theta)/[\,t(t-1)(\ell^2+m^2)^{(t-2)/2}\,] \quad (69)$$

if t is twice an odd integer. ∎

Therefore, when $t=4$, 6, 8 and 10, the t-th order terms of the functional that is shown in Figure 3 are the expressions

$$0, \quad \frac{192h^2\cos(6\theta)}{(\ell^2+m^2)^2}, \quad \frac{-960h^2\cos(6\theta)}{(\ell^2+m^2)^3} \quad \text{and} \quad \frac{2688h^2\cos(10\theta)}{(\ell^2+m^2)^4} \quad (70)$$

respectively. The first two expressions are expected from the analysis of Section 3, in particular inequalities (33) and (40) imply the attainable bound $192h^2/(\ell^2+m^2)^2$ on the modulus of the sixth order term. Further, we see that the moduli of the eighth and tenth order terms are substantially less than 128 times the greatest moduli of the corresponding terms of expression (66). Further, a comparison of Lemmas 4 and 5 shows that this property is also enjoyed by all larger values of t. Hence the higher order terms of the Figure 3 stencil induce errors in the subtabulation formulae (15) and (16) that are substantially less than the corresponding errors of formula (6). We conclude from the analysis of this appendix that, when typical or high accuracy is required from the subtabulation algorithm, then it is adequate to determine ρ in the given way that depends on sixth order error estimates.

Acknowledgements. I am very grateful to Ian Barrodale for suggesting to me that it should be possible to develop an algorithm for tabulating a thin plate spline on a fine square grid that is much faster than direct evaluation. I have enjoyed developing the given procedure and contributing to its successful use by Barrodale Computing Services on a wide range of applications. I also offer my thanks to Rick Beatson and Martin Buhmann for their valuable comments on the first draft of this paper.

References

1. Barrodale, I., M. Berkley, and D. Skea (1992), Warping digital images using thin plate splines, presented at the Sixth Texas International Symposium on Approximation Theory (Austin, January, 1992).
2. Beatson, R. K., and G. N. Newsam (1992), Fast evaluation of radial basis functions: I, preprint (to be published in Comp. Maths. Appls.).
3. Duchon, J. (1977), Splines minimizing rotation-invariant seminorms in Sobolev spaces, in *Constructive Theory of Functions of Several Variables, Lecture Notes in Mathematics 571*, W. Schempp and K. Zeller (eds.), Springer-Verlag, Berlin, 85–100.
4. Dyn, N. (1992), Subdivision schemes in computer-aided geometric design, in *Advances in Numerical Analysis II: Wavelets, Subdivision Algorithms and Radial Basis Functions*, W. A. Light (ed.), Oxford University Press, Oxford, 36–104.
5. Greengard. L., and V. Rokhlin (1987), A fast algorithm for particle simulations, J. Comp. Phys. **73**, 325–348.

Department of Applied Mathematics
and Theoretical Physics
University of Cambridge
Silver Street
Cambridge CB3 9EW
England
mjdp @ amtp.cam.ac.uk

Numerical Methods of Approximation Theory, Vol. 9
Dietrich Braess and Larry L. Schumaker (eds.), pp. 245–268.
International Series of Numerical Mathematics, Vol. 105

The L_2-Approximation Orders of Principal Shift-Invariant Spaces Generated by a Radial Basis Function

Amos Ron

Dedicated to the memory of Lothar Collatz

Abstract. Approximations from the L_2-closure S of the finite linear combinations of the shifts of a radial basis function are considered, and a thorough analysis of the least-squares approximation orders from such spaces is provided. The results apply to polyharmonic splines, multiquadrics, the Gaussian kernel and other functions, and include the derivation of spectral orders. For stationary refinements it is shown that the saturation class is trivial, i.e., no non-zero function in the underlying Sobolev space can be approximated to a better rate. The approach makes essential use of recent results of de Boor, DeVore and the author.

§1. Introduction

Substantial progress in understanding the L_∞- and L_2- approximation orders of principal shift-invariant spaces was recently made in [4] and [2] respectively. While [4] discusses applications to radial functions, no such discussion can be found in [2], and the present paper is meant to fill in that gap. Thus, it is devoted to the analysis of the L_2-approximation orders associated with principal spaces generated by a radial function via the ideas, methods and results of [2].

It seems best to start our discussion with an explanation of the title. First, all functions here are assumed to be either real or complex valued and are defined on the real Euclidean space $\mathbb{R}^d$, for some $d \geq 1$. *A shift* means "an integer translate" or "an integer translation", hence *a shift-invariant space S* is a space which is invariant under the shift operation, i.e., satisfies

$$f \in S \quad \Longleftrightarrow \quad f(\cdot - \alpha) \in S, \quad \alpha \in \mathbb{Z}^d. \tag{1.1}$$

We also assume that S is a closed subspace of $L_2(\mathbb{R}^d)$. For $f \in L_2(\mathbb{R}^d)$, we denote by $S(f)$ *the space generated by f*, i.e., $S(f)$ is the L_2-closure of the *finite* linear combinations of the shifts of f:

$$S(f) := \text{closure } S_0(f),$$

with

$$S_0(f) := \{ \sum_{\alpha \in \mathbb{Z}^d} a_\alpha f(\cdot - \alpha) : \text{ almost all } a_\alpha \text{ are zero}\},$$

i.e., $S(f)$ is the smallest closed shift-invariant space that contains f. Certainly, $S(f) \subset S$ for every $f \in S$. We say that S is *principal* if it is generated by a single function, i.e., if there exists $\phi \in S$ for which

$$S = S(\phi).$$

We remark that Theorem 2.16 of [2] shows that in case $\phi \in L_2(\mathbb{R}^d)$ is compactly supported, $S(\phi)$ contains all *infinite* combinations of the shifts of ϕ (calculated pointwise) which happen to be $L_2(\mathbb{R}^d)$-functions.

In order to understand the main results and applications in this paper, no previous knowledge of radial basis functions is required. However, some knowledge of the present state-of-the-art and the present concepts in the area will help in understanding the *novelty* of the approach here. By "present state-of-the-art" we mean the survey of Powell [17], with the complement of some more recent results from [7,6,15].

What do we mean here by a *radial basis function*? In [17], Powell lists six functions as being the major examples of radial basis functions. These are $|x|$, $|x|^3$, $|x|^2 \log |x|$, $e^{-|x|^2}$, and $(|x|^2 + c^2)^{\pm 1/2}$, where $|x|$ stands for the Euclidean 2-norm

$$|x| := \sqrt{x_1^2 + x_2^2 + \cdots + x_d^2}.$$

All of these functions are, indeed, radially symmetric; some of them grow at ∞, and some of them decay at ∞. For our purposes, the radial symmetry of the basis function plays less of a role, and we use the terminology "a radial basis" function more for convenience, (as a matter of fact, a few of our examples are not radially symmetric). The two basic properties of the basis function ϕ which we employ here are (1) its *smoothness* or, more precisely, the decay at ∞ of its Fourier transform $\widehat{\phi}$, and (2) its *"ellipticity"*. By the latter we mean that the (generalized) Fourier transform of $\widehat{\phi}$ is a well-defined smooth function in the complement $\mathbb{R}^d\backslash 0$ of the origin, and does not vanish there.

Before proceeding, we now introduce the notion of *approximation orders*. For $f \in L_2(\mathbb{R}^d)$, the distance $E(f, S)$ of f from the (shift-invariant) space S is defined as usual by

$$E(f, S) := \min\{\|f - g\| : \ g \in S\}, \tag{1.2}$$

where $\|\cdot\|$ is the $L_2(\mathbb{R}^d)$-norm. Assume further that we have in hand not *one* principal shift-invariant space $S(\phi)$, but a collection of them, $\{S_h := S(\phi_h)\}_h$, where h varies over either the interval $(0, 1]$ or some discrete subset of this interval. A priori no connection between the generators $\{\phi_h\}$ of the various spaces is assumed. Each of the shift-invariant spaces S_h is then dilated to the h-level as follows:

$$S_h^h := S^h(\phi_h) := \{f(\cdot/h) : \ f \in S_h\}. \tag{1.3}$$

Note that the dilated space $S^h(\phi_h)$ is generated by the $h\mathbb{Z}^d$-shifts of the dilated function $\phi_h(\cdot/h)$. We say that $\{S_h\}_h$ (or $\{\phi_h\}_h$) *provides approximation order* $k > 0$ (in the 2-norm), if

$$E(f, S_h^h) = \mathcal{O}(h^k), \quad \forall f \in W, \tag{1.4}$$

where W is some smooth subspace of $L_2(\mathbb{R}^d)$ which depends on k, usually a Sobolev space.

The literature focuses primarily on the so-called *stationary case* in which only one function ϕ is employed, i.e., $\phi_h = \phi$, for all h. In this case $S_h = S(\phi)$, for all h, and the scaled spaces in (1.3) are all dilates of one basic space. The study of non-stationary settings was initiated in spline theory (exponential box splines [18], [9]), but *there are also very good reasons for considering non-stationary refinements in radial basis function theory*. This point is so important that we pause here momentarily to discuss the following example.

Example 1.5. Let ϕ be the Gaussian kernel, i.e., $\phi(x) = e^{-|x|^2}$. Despite the superior smoothness and decay properties of this function, it is a poor choice as far as approximation orders are concerned. The heuristic reason for this is that the dilated function $\phi(x/h) = e^{-h^{-2}|x|^2}$ converges to the δ functional faster than our linear refinement of the translates, and the task of approximating from this space becomes hopeless. But, a small change in the dilation process alters the picture dramatically: we obtain approximation orders as large as are wished by choosing $\lambda(h)$ to be a function that decays to 0 with h (e.g., $\lambda(h) = \mathcal{O}(1/|\log h|)$), and defining $\phi_h := e^{-\lambda(h)|\cdot|^2}$. (Note that $\phi_h(\cdot/h) = e^{-\lambda(h)h^{-2}|\cdot|^2}$, and so the $\lambda(h)$ parameter slows the convergence of $\phi(\cdot/h)$ to the δ functional.) Further, the choice $\lambda(h) := h^\nu$, $\nu > 0$, results in *spectral* approximation orders, i.e., approximation orders that depend only on the smoothness of the approximand.

Before [4] and [2] were written, the standard approach for the analysis of approximation orders went along the quasi-interpolation argument guidelines, which can roughly be divided into three steps. The first step is *localization* (referred to sometimes as "preconditioning"): since ϕ usually grows at ∞, one applies to ϕ a (finite/infinite) difference operator to obtain a function ψ with nice decay properties at ∞. Then, one tries to *reproduce polynomials*: if ψ decays fast enough at ∞, then, for some $k \geq 1$, the sum

$$\psi*'p := \sum_{\alpha \in \mathbb{Z}^d} p(\alpha)\psi(\cdot - \alpha)$$

converges uniformly on compact sets for every $p \in \Pi_{k-1}$, with Π_k the space of all polynomials of degree $\leq k$ (in d variables). Under certain conditions on ϕ and by a careful choice of the difference scheme employed, it is possible to prove that $\psi*'p = p$ for all $p \in \Pi_{k-1}$. This gives rise to the approximation scheme

$$f \approx \psi*'f, \quad f \in W. \tag{1.6}$$

The third step is the *error analysis* where the polynomial reproduction is shown to imply that the scheme (1.6) provides approximation order k. In case ψ is compactly supported, the conversion of polynomial reproduction to approximation orders provides no difficulty (cf. e.g., [1]), and the same holds in case it is known that ψ decays at ∞ like $\mathcal{O}(|\cdot|^{-k-d-\varepsilon})$ for some $\varepsilon > 0$ (cf. Proposition 1.1 and Corollary 1.2 of [8] and the arguments in [13]). However, things become more involved if the above decay holds only with $\varepsilon = 0$, and subtle information on ϕ and ψ is then required.

The focal point of our discussion here is that *we do not employ any step of the quasi-interpolation argument approach*; specifically, we do not reproduce polynomials (nor do we reproduce exponentials or any other "nice" functions). This results in a tremendous relaxation of the localization step, as the function ψ is no longer required to decay in a manner related to the desired approximation order k, but merely to lie in $L_2(\mathbb{R}^d)$.

Example 1.7. Let ϕ be the univariate inverse multiquadric, i.e., $d = 1$ and

$$\phi := (1 + |\cdot|^2)^{-1/2}.$$

In this case $\phi \in L_2(\mathbb{R})$, hence $S := S(\phi)$ is well-defined. We assume that S is refined by dilation, i.e., that $\phi_h = \phi$ for all h. It was conjectured by many (cf. e.g., [5]) that the above ϕ provides no positive approximation order; this, indeed, will be proved here. On the positive side, we show that

$$E(f, S^h) = o(1), \quad \forall f \in L_2(\mathbb{R}). \tag{1.8}$$

In the terminology of [2], ϕ *provides density order* 0. We will even determine the rate of decay of $E(f, S^h)$ for smooth f (e.g, for f in the Sobolev space $W_1^2(\mathbb{R})$). An L_∞-analogue of (1.8) has been recently established in [6].

We mentioned one drawback of the quasi-interpolation argument, i.e., that it requires high decay rates from the basis function ψ. There is, however, another significant deficiency in this argument: it provides only *lower bounds* on the approximation order in the sense that by quasi-interpolation one can only conclude that the approximation order is *at least* some k. General methods for the derivation of *upper bounds* on the approximation order were known only for the stationary case, and even there only for weaker versions of approximation orders (the so-called "controlled", "local" and "controlled-local" approximations), and only for basis functions which decay like $\mathcal{O}(|\cdot|^{-k-d-\varepsilon})$ at ∞ (cf. [20,3,14,12,10]). In contrast, [19,4,2], as well as this paper, employ methods that determine the *exact* approximation order.

It then becomes very interesting to compare the lower bounds on approximation orders provided by quasi-interpolation with the exact orders that will be described. For this purpose, we state (and prove in the next section) the following theorem.

Theorem 1.9. *Let ϕ be some function which grows no faster than polynomially at ∞, and let $\hat{\phi}$ be its Fourier transform. Assume that ψ_j, $j = 1, 2$, are two $L_2(\mathbb{R}^d)$-functions which satisfy the equations*

$$\widehat{\psi}_j = u_j \widehat{\phi}, \quad j = 1, 2,$$

where u_j, $j = 1, 2$ are some 2π-periodic functions each of which vanishes only on a set of measure 0. Then

$$S(\psi_1) = S(\psi_2).$$

How is this theorem connected to our discussion? As mentioned, whenever the basis function grows at ∞, a localization process precedes the construction of an approximation scheme. Whatever approach one chooses from the present literature, the connection between the original ϕ and its localized version ψ is given in the Fourier transform domain by an equation of the form $\widehat{\psi} = u\widehat{\phi}$, where u is some 2π-periodic function (the product $u\widehat{\phi}$ should be interpreted in a distributional sense). The above theorem then says that *the type of localization process used is immaterial to the approximation orders provided by the localized function. In particular, the decay rates at ∞ of the localized function ψ, as well as the zero that $\widehat{\psi}$ might or might not have at the origin, are irrelevant to approximation orders.* This is in stark contrast with the lower bounds on approximation order suggested by quasi-interpolation, which are improved together with the decay rates at ∞ of the localized function.

At first glance, the above remarks might seem surprising, since several authors (including myself, cf. [11,5,8]) proved that (for some specific basis functions) the approximation orders are improved together with the better decay rates of the function, and even proved that the approximation orders stated in their theorems are exact (i.e., best possible). This does not contradict the present statements: what is established in the above citations and other references is that for the given localization ψ, the approximation scheme (1.6) approximates to a certain (exact) order. The right conclusion is that the approximation scheme (1.6) fails to provide optimal approximation orders whenever ψ decays too slowly at ∞. As a matter of fact, for functions which decay slowly at ∞, *optimal approximation schemes* (i.e., those that realize the approximation order) *are not local.* For example, in case f happens to be compactly supported, (1.6) employs only finitely many shifts of f, and these shifts are determined by $\operatorname{supp} f$. In contrast, the approximation schemes used here employ infinitely many shifts, even if f is compactly supported, and the coefficients associated with these shifts decay sometimes at ∞ in a slow rate determined by the decay rate of the basis function ψ.

In the following, the symbol "const" stands for a generic positive constant; hence constants appearing in the same display need not to be the same. The notation "$f \sim g$ on Ω" means that $\operatorname{supp} f \cap \Omega$ differs from $\operatorname{supp} g \cap \Omega$ by a null-set, and $f/g, g/f \in L_\infty(\Omega)$.

§2. Approximation Orders in the L_2-Norm

The paper [2] provides a complete analysis of approximation orders from closed shift-invariant spaces of $L_2(\mathbb{R}^d)$. We could have applied those general results to the radial functions considered here, but prefer to derive our results more or less directly, since in this way we obtain finer statements and tighter bounds.

Given $\psi \in L_2(\mathbb{R}^d)$, the space $S(\psi)$ is defined as in the introduction. Throughout the paper, we assume that the generator $\widehat{\psi}$ is non-zero a.e. This assumption is not essential, but is satisfied by all examples in radial basis function theory.

Given the spaces $\{S_h := S(\psi_h)\}_h$, $\psi_h \in L_2(\mathbb{R}^d)$, our goal is to provide a realistic estimate for $\{E(f, S_h^h)\}_h$, $f \in L_2(\mathbb{R}^d)$. Since $E(f, S^h) = h^{d/2} E(f(h\cdot), S)$, as can be easily verified by scaling, we might study the identical quantities

$$h^{d/2} E(f(h\cdot), S_h),$$

as we occasionally do here. To make the analysis more concrete, we briefly discuss some of the possible choices for the sequence $\{\psi_h\}_h$.

Example 2.10. (a) The basis function ϕ is chosen to be a fundamental solution of a homogeneous elliptic differential operator (with constant coefficients) $P(D)$ of order $m > d/2$. In case $P(D)$ is the $m/2$-fold iterated Laplacian, (i.e., if $P(x) = |x|^m$), $\phi(x)$ can be chosen to be $c|x|^{m-d}$ or $c|x|^{m-d} \log|x|$, depending on the parity of d. The Fourier transform $\widehat{\phi}$ coincides with the reciprocal of $P(i\cdot)$ on $\mathbb{R}^d\backslash 0$. Since ϕ grows at ∞, we need to localize it before discussing approximation orders, and thus we assume that ψ is a localization of ϕ, which means that $\psi \in L_2(\mathbb{R}^d)$ and $\widehat{\psi} = u\widehat{\phi}$ for some 2π-periodic function (even a trigonometric polynomial may do, and recall from Theorem 1.9 that $S(\psi)$ is independent of the periodic u chosen). In the present example, we consider only the stationary case, i.e., defining $S := S(\psi)$, we study the decay rates of

$$h^{d/2} E(f(h\cdot), S),$$

for a smooth f. Since the localization ψ plays a dummy role, it is desirable to analyse the problem in terms of the basis function ϕ, or, if possible, in terms of the underlying polynomial P. We mention that in case $P(D)$ is the iterated Laplacian, the space S above is intimately related to the space of polyharmonic splines studied e.g., in [16].

(b) $\phi(x,c) = (|x|^2 + c^2)^{m/2}$, $(m \geq -d,\ m \notin 2\mathbb{Z}_+)$, or $\phi(x,c) = (|x|^2 + c^2)^{m/2} \log(|x|^2 + c^2)$ $(m \in 2\mathbb{Z}_+)$. This contains the multiquadrics and inverse multiquadrics which correspond to the values $m = +1, -1$ respectively. The present example has the following important advantage over the previous one: since ϕ here is infinitely smooth, its Fourier transform decays rapidly (as a matter of fact, exponentially) at ∞, and further this transform is known to vanish nowhere on $\mathbb{R}^d \backslash 0$. Because of these two properties, we will show that an appropriate change of the parameter c from one h-level to another results in an improvement of the approximation properties of the corresponding spaces. Thus, we have two alternatives to choose from:

(b1) The stationary case. As before, we might localize ϕ to get ψ, define $S := S(\psi)$, and do not change ψ with h. This case becomes very similar in its analysis to the one considered in (a). In both of them, the approximation orders are determined by the rate of growth of ϕ at ∞, or, more precisely, by the singularity order of $\hat{\phi}$ at the origin.

(b2) The non-stationary case. Here we change the parameter c with h, i.e., define $\phi_h := \phi(\cdot, c_h)$. Each ϕ_h is then localized to obtain a sequence $\{\psi_h\}_h$ of $L_2(\mathbb{R}^d)$-functions (again the type of localization used is insignificant, but it can be shown that the same periodic function u can be used for all ϕ_h). By letting $\{c_h\}_h$ grow to ∞, the fast decay of $\hat{\phi}$ would provide approximation orders that supersede the orders obtained in the stationary case (b1).

It should be observed that for the ϕ considered in (a), the trade-off between singularity order of $\hat{\phi}$ at 0 and its decay rate at ∞ provides no benefit. Because of the homogeneity of $\hat{\phi}$, the order of its pole at 0 is the same as its decay rate at ∞.

(c) Here we consider again a one-parametric family $\phi(\cdot, c)$ of very smooth functions, which, further, are in L_2. For example, $\phi(\cdot, c) = e^{-c|\cdot|^2}$, or $\phi = (|\cdot|^2 + c^2)^{-(d+1)/2}$. In such a case $\hat{\phi}$ admits no singularity at 0, and the only way to obtain positive approximation orders is as in (b2) above. In this regard, examples of the present type are advantageous over the examples in (b) since we do not need to localize our function.

In all of the above examples, the Fourier transform $\hat{\phi}$ of the basis function ϕ could have been identified on $\mathbb{R}^d \backslash 0$ with some smooth function. Since this is typical of radial basis functions, we adopt such an assumption from now on. In particular, in all subsequent analysis, the notation $\hat{\phi}$ also stands for the function which is defined on $\mathbb{R}^d \backslash 0$ and coincides there with the Fourier transform of ϕ.

§2.1. The PSI Space $S(\phi)$ and the Function Λ_ϕ

From the definition of $S(\psi)$ it is clear that $S(\psi)$ contains any finite linear combination of the shifts of ψ, and furthermore, any function $s \in S(\psi)$ can be arbitrarily closely approximated by these finite linear combinations. In terms of $\widehat{\psi}$ (which is known to be in $L_2(\mathbb{R}^d)$ since ψ is *assumed* to be so) we know that $\widehat{S(\psi)}$ contains all functions of the form $\tau\widehat{\psi}$, where τ is a trigonometric polynomial. The following characterization of *all* elements of $S(\psi)$, ($\psi \in L_2(\mathbb{R}^d)$) has been obtained in [2]:

$$f \in S(\psi) \quad \Longleftrightarrow \quad (f \in L_2(\mathbb{R}^d),\ \widehat{f} = \tau\widehat{\psi},\ \tau \text{ is } 2\pi - \text{periodic}). \qquad (2.11)$$

We want to emphasize that the 2π-periodic τ in the above characterization is not assumed to be integrable or square integrable or even measurable (although it can be proved to be measurable). Further, as for any L_2-function, the product $\tau\widehat{\psi}$ is defined almost everywhere, and consequently τ might be defined only a.e.

Proof of Theorem 1.9. To prove that $S(\psi_1) = S(\psi_2)$ it suffices to show that $\psi_1 \in S(\psi_2)$ and vice versa. Defining $\tau := u_1/u_2$, we know by the assumption on u_2 that τ is defined almost everywhere, and because the u_j's are 2π-periodic, so is τ. On the other hand,

$$\widehat{\psi_1} = \tau\widehat{\psi_2},$$

and $\psi_1 \in L_2(\mathbb{R}^d)$ by assumption, and therefore, by (2.11), $\psi_1 \in S(\psi_2)$, while the converse holds by symmetry. $\blacksquare$

The approximation properties of the space $S(\psi)$ are determined by the behaviour of the function

$$\Lambda_\psi := (1 - \frac{|\widehat{\psi}|^2}{\widetilde{\psi}^2})^{1/2}, \qquad (2.12)$$

where $\widetilde{\psi}$ is the following 2π-periodization of $\widehat{\psi}$:

$$\widetilde{\psi} := (\sum_{\beta \in 2\pi \mathbb{Z}^d} |\widehat{\psi}(\cdot + \beta)|^2)^{1/2}. \qquad (2.13)$$

The convergence of the sum in the last definition can be taken in the L_1-sense. It is easy to see that, with

$$C := [-\pi, \pi]^d,$$

$\widetilde{\psi} \in L_2(C)$ if and only if $\psi \in L_2(\mathbb{R}^d)$. Note that Λ_ψ is non-negative and bounded by 1.

We already know by Theorem 1.9, that at least from a theoretical point of view, the specific choice of the localization process is not important. This choice also lacks any significance in the practical computation of approximation orders: the approximation orders depend on the behaviour of Λ_ψ (see below), but we observe that, because $\widehat{\psi} = \tau\widehat{\phi}$ and τ is 2π-periodic, the function Λ_ϕ is also well-defined and coincides a.e. with Λ_ψ (subject to the assumption that τ is non-zero a.e.). Thus, to dispense entirely with ψ, we define

$$S(\phi) := S(\psi),$$

with ψ some (any) localization of ϕ. Note that, because of (2.11), the Fourier transform of every function f in $S(\phi)$ can be written in the form $\widehat{f} = \tau\widehat{\phi}$, for some 2π-periodic τ.

We could have defined $S(\phi)$ directly, without any recourse to localization, by an appropriate distributional interpretation of the product $\tau\widehat{\phi}$, τ 2π-periodic. There are two reasons for the indirect definition chosen above: first, the approximation map and its error analysis require the use of a localization ψ; second, there is no real loss in the indirect definition, since by our assumption on ϕ, $\widehat{\phi}$ has an isolated singularity at the origin, hence of finite order, and consequently this singularity can always be removed, e.g., by an application to ϕ of a finite difference operator which annihilates polynomials of sufficiently high degree.

§2.2. The Stationary Case

For the sake of clarity, we first consider the stationary case. Thus, the space $S := S(\phi)$ is fixed and, for the given smooth f, we need to study the quantities

$$E(f, S^h) = h^{d/2} E(f(h\cdot), S). \tag{2.14}$$

The space of smooth functions is chosen as the potential space $W_2^k(\mathbb{R}^d)$:

$$W_2^k(\mathbb{R}^d) := \{f \in L_2(\mathbb{R}^d) : \ \|f\|_{W_2^k(\mathbb{R}^d)} := (2\pi)^{-d/2}\|(1 + |\cdot|)^k\widehat{f}\|_{L_2(\mathbb{R}^d)} < \infty\}.$$

In case k is an integer, $W_2^k(\mathbb{R}^d)$ is the usual Sobolev space of the functions whose derivatives up to order k are in $L_2(\mathbb{R}^d)$.

Since the Fourier transform is an isometry on $L_2(\mathbb{R}^d)$, we might alternatively study the quantities $h^{d/2} E(\widehat{f(h\cdot)}, \widehat{S})$, with $\widehat{S}$ the range of $S(\phi)$ under the Fourier transform.

Our first step is *truncation*: instead of approximating $\widehat{f(h\cdot)}$, we approximate only the portion of it that is supported on some 0-neighborhood B, and add the rest of $\widehat{f(h\cdot)}$ to the error bound. Since $\widehat{f(h\cdot)} = h^{-d}\widehat{f}(\cdot/h)$, it is easy to prove (cf. Lemma 3.8 of [2]) that, for any fixed 0-neighborhood B and any $k \geq 0$,

$$h^{d/2}\|\widehat{f(h\cdot)}\|_{L_2(\mathbb{R}^d\setminus B)} \leq c_B h^k \varepsilon_f(h)\|f\|_{W_2^k(\mathbb{R}^d)}, \tag{2.15}$$

where $\varepsilon_f(h) \leq 1$ and decays to 0 with h. Hence, for $f \in W_2^k(\mathbb{R}^d)$,

$$|E(f, S^h) - (2\pi)^{-d/2}h^{-d/2}E(\chi_B\widehat{f}(\cdot/h), \widehat{S})| = o(1)h^k\|f\|_{W_2^k(\mathbb{R}^d)},$$

with the $o(1)$ factor bounded independently of f, and where χ_B is the characteristic function of any fixed neighborhood of the origin. Therefore, the truncation process is nondetrimental to the task of determining the approximation orders.

The function Λ_ϕ then enters the discussion because of the following result of [2]:

Result 2.16. *Let $f \in L_2(\mathbb{R}^d)$ and assume that supp $\widehat{f} \subset B \subset C$. Then*

$$E(\widehat{f}, \widehat{S}) = \|\widehat{f}\Lambda_\phi\|_{L_2(\mathbb{R}^d)} = \|\widehat{f}\Lambda_\phi\|_{L_2(B)}.$$

From the last result we conclude that, if $B \subset C$, then

$$E(\chi_B\widehat{f}(\cdot/h), \widehat{S}) = \|\widehat{f}(\cdot/h)\Lambda_\phi\|_{L_2(B)}.$$

Thus, since $h^{-d/2}\|\widehat{f}(\cdot/h)\Lambda_\phi\|_{L_2(B)} = \|\Lambda_\phi(h\cdot)\widehat{f}\|_{L_2(B/h)}$, we arrive at the following:

Corollary 2.17. *Let $B \subset C$ be a neighborhood of the origin, and let $f \in W_2^k(\mathbb{R}^d)$, $k \geq 0$. Then*

$$E(f, S^h) = (2\pi)^{-d/2}\|\Lambda_\phi(h\cdot)\widehat{f}\|_{L_2(B/h)} + o(1)h^k\|f\|_{W_2^k(\mathbb{R}^d)},$$

with the $o(1)$ factor bounded independently of f.

We will now make specific assumptions on the basis function ϕ which will allow us to replace Λ_ϕ in the last corollary by a simpler expression. Throughout the rest of this subsection we assume that ϕ satisfies the following conditions:

(a) Smoothness condition: The function M_ϕ which is defined by

$$M_\phi{}^2 := \sum_{\beta \in 2\pi \mathbb{Z}^d \backslash 0} |\widehat{\phi}(\cdot + \beta)|^2 \qquad (2.18)$$

is (essentially) bounded on some neighborhood B of the origin.

Condition (a) is satisfied by all the functions ϕ we consider, since they all enjoy the stronger property $|\widehat{\phi}(w)| = \mathcal{O}(|w|^{-(d/2+\varepsilon)})$ as $w \to \infty$. Without loss of generality, we assume that the set B appearing in the above condition is identical with B in Corollary 2.17.

(b) "Ellipticity" condition. The function M_ϕ is bounded below away from zero around the origin, while the function $|\widehat{\phi}|$ (which, by assumption, is defined on $\mathbb{R}^d \backslash 0$) converges to ∞ at 0.

In all of the examples considered here the boundedness of M_ϕ from below follows from the fact that $\widehat{\phi}$ is continuous on $\mathbb{R}^d \backslash 0$ and does not vanish either there or in some neighborhood of ∞.

Corollary 2.19. *Let* $f \in W_2^k(\mathbb{R}^d)$, $k \geq 0$. *If* ϕ *satisfies condition* (a) *above, then*

$$E(f, S^h) \leq \mathrm{const}\|\widehat{f}/(\widehat{\phi}(h\cdot))\|_{L_2(B/h)} + o(1)h^k\|f\|_{W_2^k(\mathbb{R}^d)}, \qquad (2.20)$$

with const independent of f *and* h, *and with* $o(1)$ *being bounded independently of* f. *Furthermore, if* ϕ *also satisfies condition* (b) *above, then the converse inequality holds as well (with, possibly, a different constant).*

Proof: With M_ϕ as in (2.18), we observe that

$$\Lambda_\phi^2 = \frac{M_\phi^2}{|\widehat{\phi}|^2 + M_\phi^2} \leq \frac{M_\phi^2}{|\widehat{\phi}|^2}.$$

Assuming (a), it thus follows that $\Lambda_\phi |\widehat{\phi}|$ is bounded on B, and an application of Corollary 2.17 yields (2.20).

Further, if we assume condition (b), then, assuming also (without loss of generality, since we can change B if necessary) that $1/|\widehat{\phi}|$ is bounded on B, we conclude that

$$M_\phi^2 + |\widehat{\phi}|^2 \leq c|\widehat{\phi}|^2 \quad \text{on } B.$$

Since we also know that $1/\mathrm{M}_\phi$ is bounded on some 0-neighborhood, it follows that around the origin

$$\Lambda_\phi^2 = \frac{\mathrm{M}_\phi^2}{\mathrm{M}_\phi^2 + |\widehat{\phi}|^2} \geq \mathrm{const}|\widehat{\phi}|^{-2}.$$

Again, an application of Corollary 2.17 proves that the converse inequality holds as well. ∎

Example 2.21. We proceed with case (a) of Example 2.10, i.e., assume that $|\widehat{\phi}| = 1/P$ (on $\mathbb{R}^d\backslash 0$) with $P(D)$ an elliptic operator, $\deg P > d/2$. It is then easy to verify that conditions (a) and (b) that were required in Corollary 2.19 hold, and therefore we obtain the following result.

Theorem 2.22. *Let ϕ be a fundamental solution of a constant coefficient homogeneous elliptic operator $P(D)$ of order $m > d/2$. Then ϕ provides approximation order m in the L_2-norm for every function $f \in W_2^m(\mathbb{R}^d)$. Further, for any such non-trivial f, $E(f, S^h) \neq o(h^m)$.*

Proof: By Corollary 2.19, the first statement of the theorem will be established as soon as we show that $\|\widehat{f}/\widehat{\phi}(h\cdot)\|_{L_2(B/h)} = \mathcal{O}(h^m)$ for every $f \in W_2^m(\mathbb{R}^d)$. Since $P = 1/\widehat{\phi} \sim |\cdot|^m$ on B (and, as a matter of fact, everywhere), due to the ellipticity of $P(D)$, we can replace $\|\widehat{f}/\widehat{\phi}(h\cdot)\|_{L_2(B/h)}$ by

$$\||h\cdot|^m\widehat{f}\|_{L_2(B/h)} = h^m\||\cdot|^m\widehat{f}\|_{L_2(B/h)} \leq (2\pi)^{d/2}h^m\|f\|_{W_2^m(\mathbb{R}^d)},$$

and the desired result follows.

If we assume that $E(f, S^h) = o(h^m)$, then, because of Corollary 2.19,

$$\|\widehat{f}/\widehat{\phi}(h\cdot)\|_{L_2(B/h)} = o(h^m),$$

which implies, as above, that

$$h^m\||\cdot|^m\widehat{f}\|_{L_2(B/h)} = o(h^m).$$

Consequently, $\||\cdot|^m\widehat{f}\|_{L_2(B/h)} = o(1)$, which can happen only if $|\cdot|^m\widehat{f} = 0$. Therefore, f is a polynomial, and hence null, since $L_2(\mathbb{R}^d)$ contains no non-trivial polynomials. ∎

The reader should observe that the assumption $m > d/2$ is essential: if $m \leq d/2$, ϕ is not locally in L_2.

An examination of the proof of Theorem 2.22 reveals that the actual choice of ϕ there played only a minor role. The properties of ϕ used were the satisfaction of conditions (a) and (b) and the fact that $\widehat{\phi} \sim |\cdot|^{-m}$ near the origin. Therefore, by arguments identical to those used in the last proof, we obtain the following:

Theorem 2.23. *Assume that ϕ satisfies the conditions stated before Corollary 2.19, and that $\widehat{\phi} \sim |\cdot|^{-k}$ on (say, the same) 0-neighborhood B, for some $k > 0$. Then ϕ provides approximation order k for all functions in the potential space $W_2^k(\mathbb{R}^d)$. Moreover, for every non-trivial $f \in W_2^k(\mathbb{R}^d)$, $E(f, S^h) \neq o(h^k)$.*

Example 2.24. We now revisit case (b1) of Example 2.10, and since c is fixed here, we denote $\phi := \phi(\cdot, c)$. The common feature of all of the basis functions considered in (b) of Example 2.10 is their Fourier transform (on $\mathbb{R}^d \backslash 0$):

$$\widehat{\phi}(w) = \text{const}(c, m, d)|w|^{-(m+d)/2} K_{(m+d)/2}(c|w|),$$

with K_ν being the modified Bessel function of third kind and order ν. The Bessel function is positive on $\mathbb{R}^d \backslash 0$ and decays exponentially at ∞, and from this we conclude that M_ϕ is bounded on C above and below by positive constants. Further, for $\nu > 0$ K_ν is known to a have a pole of order ν at the origin, and therefore, in case $m + d > 0$, we conclude that $\widehat{\phi} \sim |\cdot|^{-(m+d)}$ around the origin. Thus, we can apply Theorem 2.23 to the present ϕ with $k := m + d$ to obtain:

Corollary 2.25. *Let ϕ be as in Example 2.10 (b), and assume that $m + d > 0$. Then the results of Theorem 2.23 hold for this ϕ with $k = m + d$.*

Note that $\phi \in L_2(\mathbb{R}^d)$ whenever $-d < m < -d/2$, and hence for such a choice of m the definition of $S(\phi)$ does not require localization.

We now want to consider for the present example the extreme case when $m + d = 0$. Our analysis still applies to this case in the sense that conditions (a) and (b) required in Corollary 2.19 still hold here, and therefore this corollary reduces the study of $E(f, S^h)$ to the study of $\|\widehat{f}/\widehat{\phi}(h\cdot)\|_{L_2(B/h)}$. The difference between this case and the case $m + d > 0$ is that the singularity of the Bessel function is now of logarithmic type, i.e., $|\widehat{\phi}(w)| \sim |\log|w||$ around the origin, and thus the decay rates of $E(f, S^h)$ require the examination of

$$\|\widehat{f}(w)/\log|h|w|\| \|_{L_2(B/h)}.$$

Our result with respect to basis functions whose Fourier transform has a logarithmic singularity at the origin is as follows.

Theorem 2.26. *Assume that ϕ satisfies the following two conditions:*
(a) M_ϕ is essentially bounded below and above by positive constants on some 0-neighborhood.
(b) $\widehat{\phi}(w) \sim \log|w|$ around the origin.

Then:
(i) ϕ provides no positive approximation order k for any $f \in W_2^k(\mathbb{R}^d)$ and any $k > 0$.
(ii) $E(f, S^h) = o(1)$, for all $f \in L_2(\mathbb{R}^d)$.
(iii) For every $k > 0$, and every $f \in W_2^k(\mathbb{R}^d)$,

$$E(f, S^h) \leq \text{const}|\log h|^{-1}\|f\|_{W_2^k(\mathbb{R}^d)},$$

for all $h \leq h_0$, where const and h_0 depend on k but not on f.

Proof: Statement (ii) follows from Theorem 1.7 of [2]. That theorem says that the property $E(f, S^h) = o(1)$, $\forall f \in L_2(\mathbb{R}^d)$ (referred to as "the density property") is equivalent to Λ_ϕ^2 having a Lebesgue value 0 at the origin. The result applies here since, by the assumption made on ϕ, it is clear that Λ_ϕ is continuous at the origin and vanishes there.

Now fix $k > 0$. To prove (i) and (iii), we follow the remarks preceding this theorem and consider the quantities

$$\|\widehat{f}/\log|h\cdot|\,\|_{L_2(B/h)}.$$

We already know that, up to a term of order $o(h^k)$, these numbers determine the decay rates of $E(f, S^h)$ (as $h \to 0$). Without loss of generality we assume that B is the ball of radius $1/e$. For simplicity, we also assume that $h = e^{-l}$ for some integer l (other values of h are treated as below with some obvious modifications). We divide the ball $B/h = e^l B$ into annuli as follows:

$$R_0 := eB, \qquad R_j := \{w : e^{j-1} \leq |w| \leq e^j\}, \ j = 1,...,l-1. \tag{2.27}$$

On R_j, $j > 0$, we have the estimate $(\log(h|w|))^{-2} \leq (j-l)^{-2}$, and thus, for $f \in W_2^k(\mathbb{R}^d)$,

$$\|\widehat{f}/\log(h|\cdot|)\|_{L_2(R_j)}^2 \leq e^{-2k(j-1)}(j-l)^{-2}\||\cdot|^k|\widehat{f}|\|_{L_2(R_j)}^2$$
$$\leq \text{const}\, e^{-2kj}(j-l)^{-2}\|f\|_{W_2^k(\mathbb{R}^d)}^2.$$

Summing this last estimate for $j = 1,...,l-1$, we arrive at

$$\int_{(B/h)\backslash R_0} |\widehat{f}|^2(\log(h|\cdot|))^{-2} \leq \text{const}\|f\|_{W_2^k(\mathbb{R}^d)}^2 \sum_{j=1}^{l-1} e^{-2kj}(l-j)^{-2}$$

$$= \text{const}\|f\|_{W_2^k(\mathbb{R}^d)}^2 e^{-2kl} \sum_{m=1}^{l-1} e^{2km}/m^2.$$

Elementary integral tests show that the sum in the last expression is $\mathcal{O}(e^{2kl}/l^2) = \mathcal{O}(h^{2k}/\log^2 h)$, and therefore

$$\int_{(B/h)\setminus R_0} |\widehat{f}|^2 (\log(h|\cdot|))^{-2} \le \text{const}\|f\|^2_{W_2^k(\mathbb{R}^d)} |\log h|^{-2}.$$

Also, on R_0 we have

$$\int_{R_0} |\widehat{f}|^2 (\log(h|\cdot|))^{-2} \le (\log h)^{-2}\|\widehat{f}\|^2_{L_2(\mathbb{R}^d)} \le \text{const}(\log h)^{-2}\|f\|^2_{W_2^k(\mathbb{R}^d)}.$$

We conclude that for some f-independent h_0 and const, and for every $h \le h_0$,

$$\|\widehat{f}/\widehat{\phi}(h\cdot)\|_{L_2(B/h)} \le \text{const}\|f\|_{W_2^k(\mathbb{R}^d)}/|\log h|.$$

Substituting this into Corollary 2.19, we obtain (iii).

We now prove (i): let $f \in W_2^k(\mathbb{R}^d)$. Upon assuming that ϕ provides approximation order k to f, we conclude from Corollary 2.19 that

$$\|\widehat{f}/\log(h|\cdot|)\|_{L_2(B/h)} = \mathcal{O}(h^k). \tag{2.28}$$

Let $j \in \mathbb{Z}$ and let R_j be the annulus in (2.27). For sufficiently small h, $R_j \subset B/h$, hence $\|\widehat{f}/\log(h|\cdot|)\|_{L_2(R_j)} = \mathcal{O}(h^k)$, and since $(\log(h|\cdot|))^2 \le (\log h - j + 1)^2$ on R_j, we conclude that

$$\|\widehat{f}/\log(h|\cdot|)\|_{L_2(R_j)} \ge |\log h - j + 1|^{-1}\|\widehat{f}\|_{L_2(R_j)}.$$

Combining (2.28) with the last inequality, we arrive at

$$\limsup_{h\to 0} \frac{h^{-k}}{|\log h - j + 1|}\|\widehat{f}\|_{L_2(R_j)} < \infty,$$

which can happen only if $\widehat{f} = 0$ a.e. on R_j. Since j was arbitrary, $\widehat{f} = 0$ a.e. on $\mathbb{R}^d\setminus 0$, hence $f = 0$. $\blacksquare$

We have discussed cases (a) and (b) in Example 2.10. Let us briefly review case (c). In the two examples considered in (c), $\widehat{\phi}$ is a continuous positive function with exponential decay at ∞. Therefore, Λ_ϕ is a continuous positive function. If $f \in L_2(\mathbb{R}^d) = W_2^0(\mathbb{R}^d)$, then by Corollary 2.19, (with $S := S(\phi)$),

$$E(f, S^h) = \mathrm{const} \|\Lambda_\phi(h\cdot)\widehat{f}\|_{L_2(C/h)} + o(1)$$
$$\geq \mathrm{const} \|\widehat{f}\|_{L_2(C/h)} + o(1) \to \mathrm{const} \|\widehat{f}\|_{L_2(\mathbb{R}^d)}.$$

Therefore, unless $f = 0$,

$$E(f, S^h) \neq o(1),$$

i.e., ϕ does not provide even density order zero for any non-zero L_2-function. Here, the only information used is the fact that Λ_ϕ is non-zero on C (and actually C could have been replaced by any neighborhood of the origin). Therefore, we have

Corollary 2.29. *Assume that Λ_ϕ is bounded below by a positive constant in some neighborhood of the origin. Then, for every non-trivial $f \in L_2(\mathbb{R}^d)$,*

$$E(f, S^h) \neq o(1).$$

In particular, this holds for $\phi = e^{-c|\cdot|^2}$ and $\phi = (|\cdot|^2 + c^2)^{-(d+1)/2}$.

§2.3. The Non-stationary Case

In the context of radial basis functions, the notion of "non-stationary case" is connected with "spectral approximation orders". Here, we employ a sequence $\{\phi_h\}_h$ of basis functions each of which is some dilate $\phi(\lambda(h)\cdot)$ of one fixed function ϕ. In the analysis of the stationary case, the approximation orders provided by ϕ were determined by the behaviour of $\widehat{\phi}$ on small neighborhoods of the lattice $2\pi\mathbb{Z}^d$. This is no longer the case here, and our conditions on $\widehat{\phi}$ are of global nature.

Define $S_h := S(\phi_h)$. The approximation order provided by $\{\phi_h\}_h$ to the function f is determined by the rate of decay (as $h \to 0$) of the numbers

$$E(f, S_h^h), \quad h > 0. \tag{2.30}$$

For $f \in W_2^k(\mathbb{R}^d)$, Corollary 2.17 shows that

$$E(f, S_h^h) = (2\pi)^{-d/2} \|\Lambda_{\phi_h}(h\cdot)\widehat{f}\|_{L_2(B/h)} + o(1)h^k\|f\|_{W_2^k(\mathbb{R}^d)}.$$

It is important to note that the second term in the last equation, *the truncation error*, is independent of $\{\phi_h\}_h$.

The basic idea in the derivation of spectral orders is very simple: we estimate

$$\|\Lambda_{\phi_h}(h\cdot)\widehat{f}\|_{L_2(B/h)} \leq \|\widehat{f}\|_{L_2(\mathbb{R}^d)}\|\Lambda_{\phi_h}(h\cdot)\|_{L_\infty(B/h)}$$
$$= (2\pi)^{d/2}\|f\|_{L_2(\mathbb{R}^d)}\|\Lambda_{\phi_h}\|_{L_\infty(B)}.$$

We do not want to specify in advance the smoothness class from which f is selected, and therefore we are unable to bound the truncation error in the way we did in Corollary 2.17. Instead, we recall from (2.15) that this truncation error has the form

$$h^{d/2}\|\widehat{f(h\cdot)}\|_{L_2(\mathbb{R}^d\setminus B)} = \|\widehat{f}\|_{L_2(\mathbb{R}^d\setminus(B/h))}.$$

In summary, we obtain the following seemingly coarse estimate for $E(f, S_h^h)$:

$$E(f, S_h^h) \leq \|f\|_{L_2(\mathbb{R}^d)}\|\Lambda_{\phi_h}\|_{L_\infty(B)} + \text{const}\|\widehat{f}\|_{L_2(\mathbb{R}^d\setminus(B/h))}. \qquad (2.31)$$

Our objective is to make the first term above decay to zero so fast that, unless f is exceptionally smooth, the approximation rate provided by $\{\phi_h\}_h$ to f will be determined by the second term, i.e., by the smoothness of f. Recall that by (2.15) the second term here is $o(h^k)$ for every $f \in W_2^k(\mathbb{R}^d)$, hence in particular we have the following:

Proposition 2.32. *Assume that the sequence $\{\phi_h\}_h$ satisfies, for some 0-neighborhood B,*

$$\|\Lambda_{\phi_h}\|_{L_\infty(B)} = o(h^k), \quad \forall k \in \mathbb{R}_+.$$

Then, with $S_h := S(\phi_h)$,

$$E(f, S_h^h) = o(h^k)$$

for every $f \in W_2^k(\mathbb{R}^d)$.

In order to estimate $\|\Lambda_{\phi_h}\|_{L_\infty(B)}$, we write again

$$\Lambda_{\phi_h}^2 = \frac{\mathrm{M}_h^2}{|\widehat{\phi}_h|^2 + \mathrm{M}_h^2} \leq \frac{\mathrm{M}_h^2}{|\widehat{\phi}_h|^2},$$

with $\mathrm{M}_h^2 := \sum_{\beta \in 2\pi\mathbb{Z}^d\setminus 0}|\widehat{\phi}_h(\cdot + \beta)|^2$. In the stationary case we dispensed with M_h by assuming that in some small neighborhood of the origin M_h is bounded above; the desired properties of Λ_ϕ were then derived from the behaviour of $1/\widehat{\phi}$ at the origin. This is no longer the case: choosing

$$\phi_h := \lambda(h)^d\phi(\lambda(h)\cdot),$$

we have
$$\widehat{\phi}_h = \widehat{\phi}(\cdot/\lambda(h)),$$

and therefore, if $\lambda(h) \to 0$ with h and $\widehat{\phi}$ decays fast at ∞, the values M_h assumes around the origin tends to zero, with their decay rates being controlled by $\{\lambda(h)\}_h$. At the same time, $\widehat{\phi}_h$ also tends to 0, and thus an estimate of the form
$$\|\Lambda_{\phi_h}\|_{L_\infty(B)} \leq \text{const}\|M_h\|_{L_\infty(B)}$$

is not valid.

In order to focus our discussion, and in view of the examples that initiated this discussion, we assume that

$$|\widehat{\phi}(w)|^2 \sim \sigma(|w|)$$

for some univariate positive function σ which is non-increasing on $[0, \infty)$. Let $\rho < 2\pi$; then for w in
$$B_\rho := \{w : \ |w| < \rho\}$$

we have:

$$\begin{aligned}
M_h(w)^2 &= \sum_{\beta \in 2\pi \mathbb{Z}^d \backslash 0} |\widehat{\phi}((w+\beta)/\lambda(h))|^2 \\
&\leq \text{const} \sum_{\beta \in 2\pi \mathbb{Z}^d \backslash 0} \sigma((|\beta| - \rho)/\lambda(h)) \\
&\leq \text{const} \sum_{j=1}^{\infty} \sigma((2\pi j - \rho)/\lambda(h)) j^{d-1} \qquad (2.33) \\
&\leq \text{const} \int_{2\pi-\rho}^{\infty} \sigma(t/\lambda(h)) t^{d-1}\, dt \\
&= \text{const} \lambda(h)^d \int_{(2\pi-\rho)/\lambda(h)}^{\infty} \sigma(t) t^{d-1}\, dt.
\end{aligned}$$

On the other hand, we can also estimate

$$|\widehat{\phi}(w/\lambda(h))|^2 \geq \text{const}\,\sigma(\rho/\lambda(h)), \quad w \in B_\rho,$$

which, together with (2.33), yields the following bound for $\|\Lambda_{\phi_h}\|_{L_\infty(B_\rho)}$:

$$\|\Lambda_{\phi_h}\|_{L_\infty(B_\rho)}^2 \leq \text{const} \lambda(h)^d \frac{\int_{(2\pi-\rho)/\lambda(h)}^{\infty} \sigma(t) t^{d-1}\, dt}{\sigma(\rho/\lambda(h))}.$$

In view of (2.31) and Proposition 2.32, we arrive at the following:

Theorem 2.34. *Assume that $\widehat{\phi}(w)^2 \sim \sigma(|w|)$ on $\mathbb{R}^d$, where σ is a univariate non-increasing positive function defined on $[0, \infty)$. Let $S_h := S(\phi(\lambda(h)\cdot))$, $\lambda(h) > 0$. Then, for $0 < \rho < 2\pi$ and $f \in L_2(\mathbb{R}^d)$,*

$$E(f, S_h^h) \le c_\rho \|f\|_{L_2(\mathbb{R}^d)} \left(\frac{\lambda(h)^d \int_{(2\pi-\rho)/\lambda(h)}^\infty \sigma(t) t^{d-1}\, dt}{\sigma(\rho/\lambda(h))} \right)^{1/2} + \|\widehat{f}\|_{L_2(\mathbb{R}^d \setminus B_{\rho/h})}.$$

In particular, for $f \in W_2^k(\mathbb{R}^d)$,

$$E(f, S_h^h) = o(h^k)$$

if

$$\lambda(h)^d \frac{\int_{(2\pi-\rho)/\lambda(h)}^\infty \sigma(t) t^{d-1}\, dt}{\sigma(\rho/\lambda(h))} = o(h^{2k}).$$

The point in this theorem is to choose $\rho < \pi$ and to rely on the decay of σ (equivalently, $|\widehat{\phi}|$) at ∞. In order to capture Examples (b2) and (c) in Example 2.10, we assume $\widehat{\phi}$ decays exponentially with order $r > 0$ and type $n > 0$:

$$\sigma(t) := e^{-2n|t|^r}.$$

Assuming that $\lambda(h) < 1$, we have for this σ that

$$\lambda(h)^d \int_{(2\pi-\rho)/\lambda(h)}^\infty \sigma(t) t^{d-1}\, dt \le \text{const}_{n,r,d,\rho}\, \lambda(h)^r \sigma((2\pi-\rho)/\lambda(h)).$$

Consequently,

$$\lambda(h)^d \frac{\int_{(2\pi-\rho)/\lambda(h)}^\infty \sigma(t) t^{d-1}\, dt}{\sigma(\rho/\lambda(h))} \le \text{const}\, e^{-2nc\lambda(h)^{-r}},$$

with $c := (2\pi - \rho)^r - \rho^r$. Therefore, in this case, Theorem 2.34 reads as follows:

Corollary 2.35. *If $\widehat{\phi}(w) \sim e^{-n|w|^r}$ on $\mathbb{R}^d$, and $S_h := S(\phi(\lambda(h)\cdot))$, $0 < \lambda(h) < 1$, then, for $0 < \rho < \pi$ and $f \in L_2(\mathbb{R}^d)$,*

$$E(f, S_h^h) \le \text{const} \left(\|f\|_{L_2(\mathbb{R}^d)} e^{-nc\lambda(h)^{-r}} + \|\widehat{f}\|_{L_2(\mathbb{R}^d \setminus B_{\rho/h})} \right),$$

where $c = (2\pi - \rho)^r - \rho^r$. In particular, for $f \in W_2^k(\mathbb{R}^d)$,

$$E(f, S_h^h) = o(h^k)$$

if

$$e^{-nc\lambda(h)^{-r}} = o(h^k).$$

The last result indicates that the approximation properties of $\{S(\phi_h)\}_h$ are improved with the acceleration of the decay of $\lambda(h)$ to 0. However, when choosing $\{\lambda(h)\}_h$ it is good to keep in mind the effects this choice might have on the numerical stability of the approximation process: as $\lambda(h)$ becomes small, the function ϕ_h flattens, and approximation schemes from $S(\phi_h)$ become less and less stable.

Corollary 2.35 covers the examples of the Gaussian kernel $\phi = e^{-|\cdot|^2}$ ($r = 2$ $n = 1/4$), and $\phi = (|\cdot|^2 + 1)^{-(d+1)}$ ($r = n = 1$). Also, with some simple modifications, it can be used to cover the examples considered in Example 2.10(b). However, for the case $m + d > 0$ there, an improved version of Corollary 2.35 is available. This version takes a simultaneous account of the positive effect of the singularity of $\widehat{\phi}$ at the origin and its decay at ∞. We first state and prove a general result along these lines, and then apply it to Example 2.10(b2). In the theorem below, we use the notation

$$q_l(w) := \begin{cases} 1, & |w| \leq 1, \\ |w|^l, & |w| \geq 1. \end{cases}$$

Theorem 2.36. *Assume that $\widehat{\phi}(w) \sim |w|^{-j}e^{-n|w|^r}q_l(w)$ on $\mathrm{I\!R}^d$ for some positive j, n, r and real $l \leq j$. Let $\phi_h := \lambda(h)^d\phi(\lambda(h)\cdot)$, $S_h := S(\phi_h)$. Then, for $0 < c < (2\pi)^r$ and for every $f \in W_2^k(\mathrm{I\!R}^d)$, we have*

$$E(f, S_h^h) \leq o(h^k) + \mathrm{const}\lambda(h)^{-(j+l-r)}e^{-nc\lambda(h)^{-r}} \begin{cases} \|f\|_{W_2^k(\mathrm{I\!R}^d)}h^k, & k \leq j, \\ \|f\|_{W_2^j(\mathrm{I\!R}^d)}h^j, & k \geq j. \end{cases}$$

Proof: For $f \in W_2^k(\mathrm{I\!R}^d)$, $k \leq j$, we estimate $\|\Lambda_{\phi_h}(h\cdot)\widehat{f}\|_{L_2(B/h)}$ as follows:

$$\|\Lambda_{\phi_h}(h\cdot)\widehat{f}\|_{L_2(B/h)}$$
$$\leq h^k \| |\cdot|^k\widehat{f}\|_{L_2(B/h)} \| |\cdot|^{-k}\Lambda_{\phi_h}\|_{L_\infty(B)}$$
$$\leq \mathrm{const}\, h^k \|f\|_{W_2^k(\mathrm{I\!R}^d)} \|\frac{\mathrm{M}_{\phi_h}}{\widehat{\phi}_h}|\cdot|^{-k}\|_{L_\infty(B)}.$$

To estimate $\|\frac{\mathrm{M}_{\phi_h}}{\widehat{\phi}_h}|\cdot|^{-k}\|_{L_\infty(B)}$, we use the fact that for $\beta \in 2\pi\mathrm{Z\!\!Z}^d\backslash 0$ and w sufficiently small, since $l \leq j$,

$$\frac{q_l((w+\beta)/\lambda(h))}{|w+\beta|^j q_l(w/\lambda(h))} \leq \frac{(\frac{|w+\beta|}{\lambda(h)})^l}{|w+\beta|^j} \leq \mathrm{const}\,\lambda(h)^{-l}\,|\beta|^{l-j} \leq \mathrm{const}\lambda(h)^{-l}.$$

Together with the assumptions made in Theorem 2.36, this implies that

$$\frac{\widehat{\phi}_h(w+\beta)}{\widehat{\phi}_h(w)} = \frac{\widehat{\phi}((w+\beta)/\lambda(h))}{\widehat{\phi}(w/\lambda(h))} \leq \mathrm{const}\,|w|^j \lambda(h)^{-(j+l)} \frac{e^{-n|w+\beta|^r \lambda(h)^{-r}}}{e^{-n|w|^r \lambda(h)^{-r}}}.$$

Thus, a bound on $\frac{\mathrm{M}_{\phi_h}}{\widehat{\phi}_h}$ requires a bound on $\frac{\mathrm{M}_{\psi_h}}{\widehat{\psi}_h}$, with $\psi_h := \psi(\lambda(h)\cdot)$, and ψ the inverse transform of $e^{-n|\cdot|^r}$. Such a bound has been computed in the proofs of Theorem 2.34 and Corollary 2.35, where it was shown that

$$\|\frac{\mathrm{M}_{\psi_h}}{\widehat{\psi}_h}\|_{L_\infty(B_\rho)} \leq \mathrm{const}\,\lambda(h)^r e^{-nc\lambda(h)^{-r}},$$

with $c = (2\pi - \rho)^m - \rho^m$. Thus, our combined estimate is the following:

$$\frac{\mathrm{M}_{\phi_h}(w)}{\widehat{\phi}_h(w)}|w|^{-k} \leq \mathrm{const}\lambda(h)^{-(j+l-r)}|w|^{j-k}e^{-nc\lambda(h)^{-r}}, \quad w \in B_\rho.$$

Since $k \leq j$, $|w|^{j-k}$ is bounded on B_ρ, and our final estimate becomes

$$\|\Lambda_{\phi_h}(h\cdot)\widehat{f}\|_{L_2(B/h)} \leq \mathrm{const}h^k \lambda(h)^{-(j+l-r)}e^{-nc\lambda(h)^{-r}}\|f\|_{W_2^k(\mathbb{R}^d)}.$$

Invoking Corollary 2.17, we obtain the desired result for $k \leq j$.

If $k > j$, we alternatively use the bound

$$\|\Lambda_{\phi_h}(h\cdot)\widehat{f}\|_{L_2(B/h)}$$
$$\leq h^j \|\,|\cdot|^j \widehat{f}\|_{L_2(B/h)} \|\,|\cdot|^{-j}\Lambda_{\phi_h}\|_{L_\infty(B)}$$
$$\leq \mathrm{const}\,h^j \|f\|_{W_2^j(\mathbb{R}^d)} \|\frac{\mathrm{M}_{\phi_h}}{\widehat{\phi}_h}|\cdot|^{-j}\|_{L_\infty(B)},$$

and follow the proof of the first case. The only change is the disappearance of the factor $|w|^{k-j}$. $\blacksquare$

We can now revisit case (b2) in Example 2.10, where we choose the parameter c to be 1. Recall that for ϕ there,

$$\widehat{\phi} = \mathrm{const}|\cdot|^{-(m+d)/2}K_{(m+d)/2}(|\cdot|),$$

with K_ν the modified Bessel function of order ν. Since $K_\nu(w) \sim e^{-w}w^{-1/2}$ when $w \to \infty$, is positive on $\mathbb{R}_+$, and, in case $m + d > 0$, has a pole of order $(m+d)/2$ at the origin, we conclude that, for $m + d > 0$ and on $\mathbb{R}_+$,

$$w^{(m+d)/2}K_\nu(w) \sim e^{-w}q_{(m+d-1)/2}(w).$$

This shows that ϕ satisfies the conditions required in Theorem 2.36 for the choice $j = m + d$, $n = r = 1$, and $l = (m + d - 1)/2$, thereby proving the following:

Corollary 2.37. *Let ϕ be any of the functions considered in Example 2.10(b), with $m + d > 0$. Define $S_h := S(\phi(\lambda(h)\cdot))$, $\lambda(h) > 0$. If $f \in W_2^k(\mathbb{R}^d)$ then, for $0 < c < 2\pi$ and with $\nu := -3(m + d - 1)/2$,*

$$E(f, S_h^h) \le o(h^k) + \mathrm{const}\,\lambda(h)^\nu e^{-c/\lambda(h)} \begin{cases} \|f\|_{W_2^k(\mathbb{R}^d)} h^k, & k \le m + d, \\ \|f\|_{W_2^{m+d}(\mathbb{R}^d)} h^{m+d}, & k > m + d. \end{cases}$$

Acknowledgments. Supported in part by NSF Grant DMS - 9102857 and by ARO contract 89 - 0455.

References

1. de Boor, C., Quasiinterpolants and approximation power of multivariate splines, in *Computation of Curves and Surfaces*, M. Gasca and C. A. Micchelli (eds.), Dordrecht, Netherlands: Kluwer Academic Publishers, (1990), 313–345.

2. de Boor, C., R. A. DeVore and A. Ron, Approximation from shift-invariant subspaces of $L_2(\mathbb{R}^d)$, CMS TSR #92-2, University of Wisconsin-Madison, July 1991.

3. de Boor, C. and R.Q. Jia, Controlled approximation and a characterization of the local approximation order, Proc. Amer. Math. Soc. **95** (1985) 547-553.

4. de Boor, C. and A. Ron, Fourier analysis of approximation power of principal shift-invariant spaces, Constr. Approx., to appear.

5. Buhmann, M. D., Multivariate interpolation with radial basis functions, Constr. Approx. **6** (1990), 225–256.

6. Buhmann, M. D., On Quasi-Interpolation with Radial Basis Functions, ms., 1991.

7. Buhmann, M. D., and N. Dyn, Error estimates for multiquadric interpolation, in *Curves and Surfaces*, P.-J. Laurent, A. Le Méhauté, and L. L. Schumaker (eds.), Academic Press, New York, 1991, 51–58.

8. Dyn, N., I.R.H. Jackson, D. Levin, and A. Ron, On multivariate approximation by the integer translates of a basis function, Israel J. Math., to appear.

9. Dyn, N. and A. Ron, Local approximation by certain spaces of multivariate exponential-polynomials, approximation order of exponential box splines and related interpolation problems, Trans. Amer. Math. Soc. **319** (1990), 381-404.

10. Halton, E. J. and W. A. Light, On local and controlled approximation order, J. Approx. Theory , to appear.

11. Jackson, I .R .H., An order of convergence for some radial basis functions, IMA J. Numer. Anal. **9** (1989), 567–587.

12. Jia, R.-Q. and J. Lei, Approximation by multiinteger translates of functions having global support, J. Approx. Theory , to appear.

13. Lei, J. and R.-Q. Jia, Approximation by piecewise exponentials, SIAM J. Math. Anal. , to appear

14. Light, W. A. and E. W. Cheney, Quasi-interpolation with translates of a function having non-compact support, Constr. Approx. **8** (1992), 35–48.

15. Madych, W. R., Error estimates for interpolation by generalized splines, preprint.

16. Madych, W. R. and S. A. Nelson, Polyharmonic cardinal splines I, J. Approx. Theory **40** (1990), 141–156.

17. Powell, M.J.D., The theory of radial basis function approximation in 1990. in *Advances in Numerical Analysis II: Wavelets, Subdivision Algorithms and Radial Functions*, W. Light (ed.), Oxford University Press, Oxford, 1992, 105–210.

18. Ron, A., Exponential box splines, Constr. Approx. **4** (1988), 357-378.

19. Ron, A., A characterization of the approximation order of multivariate spline spaces, Studia Math. **98(1)** (1991), 73–90.

20. Strang, G. and G. Fix, A Fourier analysis of the finite element variational method. C.I.M.E. II Ciclo 1971, in *Constructive Aspects of Functional Analysis*, G. Geymonat (ed.), 1973, 793–840.

Amos Ron
Computer Sciences Department
University of Wisconsin-Madison
1210 W. Dayton St.
Madison, WI 53706, USA
amos @ cs.wisc.edu

Numerical Methods of Approximation Theory, Vol. 9
Dietrich Braess and Larry L. Schumaker (eds.), pp. 269–283.
International Series of Numerical Mathematics, Vol. 105
Copyright © 1992 by Birkhäuser Verlag, Basel
ISBN 3-7643-2746-4.

A Multi–Parameter Method for Nonlinear Least–Squares Approximation

R. Schaback

Dedicated to the memory of Lothar Collatz

Abstract. For discrete nonlinear least-squares approximation problems $\sum_{j=1}^{m} f_j^2(x) \to \min$ for m smooth functions $f_j : \mathbb{R}^n \to \mathbb{R}$, a numerical method is proposed which first minimizes each f_j separately, and then applies a penalty strategy to gradually force the different minimizers to coalesce. Though the auxiliary nonlinear least-squares method works on $\mathbb{R}^{n \cdot m}$, it is shown that the additional computational requirements consist of $m - 1$ gradient evaluations plus $\mathcal{O}(m \cdot n)$ operations. The application to discrete rational approximation is discussed, and numerical examples are given.

§1. Introduction

We consider the nonlinear unconstrained minimization problem of finding

$$\min_{x \in \mathbb{R}^n} f(x) \tag{1}$$

for a smooth function $f : \mathbb{R}^n \to \mathbb{R}$, and assume that it has several local minimizers. We want to construct an algorithm which enhances the probability of finding global minimizers. Motivated by the application to nonlinear least squares problems (which will be described in Section 5), we assume f to be additively decomposable in the form

$$f(x) = \sum_{j=1}^{m} f_j(x), \qquad f_j : \mathbb{R}^n \to \mathbb{R}^n \ \text{smooth,} \qquad (2)$$

where each f_j is easier to minimize than f itself.

The basic idea of the multi–parameter algorithm to be constructed is borrowed from multiple shooting methods in ordinary differential equations: each f_j gets a separate parameter vector $y_j \in \mathbb{R}^n$ to be optimized, and then the separate minimizing y_j will be gradually forced to coalesce, joining the separate minimizers gradually into a global solution. This strategy is not confined to additive decompositions, and proceeds as follows:

1) Starting with a given vector $x^{(0)} \in \mathbb{R}^n$, set $y_j^{(0)} := x^{(0)}$, $1 \le j \le m$, and calculate a (possibly only local) minimizer $y_j^{(1)} \in \mathbb{R}^n$ of f_j for $1 \le j \le m$.

2) Choosing a nonnegative penalty function P with control parameters Λ to be described later, perform an iterative minimization of the function

$$\sum_{j=1}^{m} f_j(y_j + y) + P(y_1, \ldots, y_m; \Lambda), \qquad y, y_j \in \mathbb{R}^n, \ 1 \le j \le m, \qquad (3)$$

on $(y_1, \ldots, y_m, y) \in \mathbb{R}^{n \cdot (m+1)}$, starting in $(y_j^{(1)}, \ldots, y_j^{(1)}, 0)$ and producing a minimizer $(y_1^{(2)}, \ldots, y_m^{(2)}, y^{(2)})$. Here we assume $P(y_1, \ldots, y_m, \Lambda) = 0$ iff $y_j = 0$ for all j, and the penalty parameters Λ should be suitably controlled during the optimization.

3) Starting with $x^{(2)} := y^{(2)} + \frac{1}{m} \sum_{j=1}^{m} y_j^{(2)}$, minimize (1) on $\mathbb{R}^n$ by an arbitrary unconstrained optimization procedure to get a minimizer $x^{(3)} \in \mathbb{R}^n$.

Note that the algorithm consists of a single execution of Steps 1–3. However, Steps 2 and 3 contain an inner iteration.

Depending on the penalty strategy in Step 2, we will often have $y_j^{(2)} = 0$ for all j, and $x^{(2)} = y^{(2)}$ will already be a local minimizer of f, so that Step 3 will often not be necessary at all. Nevertheless, the overall algorithm works from $\mathbb{R}^n$ to $\mathbb{R}^n$ as $x^{(0)} \mapsto x^{(3)}$, and the detour via $\mathbb{R}^{n \cdot (m+1)}$ in Step 2 can be concealed from the user.

We could do Step 2 without the additional vector y, and minimize

$$\sum_{j=1}^{m} f_j(y_j) + P(y_1 - \overline{y}, \ldots, y_m - \overline{y}; \Lambda)$$

for $(y_1, \ldots, y_m) \in \mathbb{R}^{m \cdot n}$, where $\overline{y} := \frac{1}{m} \sum_{j=1}^{m} y_j$. However, the form used in (3) is more easily handled.

The main problems to be addressed in the sequel are

1) Can the computational effort be reduced to a reasonable amount?
2) How should Λ in Step 2 be controlled?
3) Are there some classes of problems where the algorithm is superior to conventional methods?

For simplicity, we shall confine the majority of the rest of the paper to nonlinear least–squares calculations, and postpone the general nonlinear optimization problem to later investigations.

§2. Penalty Functions

Depending on how much control is required, one can consider different penalty functions of the general form $P(y_1, \ldots, y_m; \Lambda)$, e.g.:

$$\lambda^2 \sum_{j=1}^{m} \|y_j\|_2^2, \qquad \lambda \in \mathbb{R}, \tag{4}$$

$$\sum_{j=1}^{m} \lambda_j^2 \|y_j\|_2^2, \qquad \lambda_j \in \mathbb{R},\ 1 \le j \le m, \tag{5}$$

or

$$\sum_{j=1}^{m} \|\Lambda_j y_j\|_2^2, \tag{6}$$

with $n \times n$ nonsingular diagonal matrices

$$\Lambda_j = \begin{pmatrix} \lambda_{j1} & & 0 \\ & \ddots & \\ 0 & & \lambda_{jn} \end{pmatrix}, \qquad 1 \le j \le m. \tag{7}$$

We shall see later that good control of the parameters of the penalty function is a major problem which forces us to restrict ourselves to the simple case (4) when specific control strategies are treated later. However, the study of the computational complexity in the next section allows rather general controls of type (5) or (6).

§3. Computational Complexity

We consider a Gauss–Newton minimization step for the general composite objective function

$$\sum_{j=1}^{m} f_j^2(y_j + y) + \sum_{j=1}^{m} \|\Lambda_j y_j\|_2^2, \qquad y, y_j \in \mathbb{R}^n, \ 1 \le j \le m, \tag{8}$$

with (6) and (7). This arises in Step 2 of the algorithm and involves $n \cdot (m+1)$ variables instead of just n variables for the classical approach. Thus, we compare it with the corresponding step for minimization of the least–squares objective function

$$f(x) = \sum_{j=1}^{m} f_j^2(x), \qquad x \in \mathbb{R}^n \tag{9}$$

to find that not too much additional work is required.

Theorem 1. *A linear multi–parameter least–squares problem arising from linearization of the objective function (8) on the space $\mathbb{R}^{n \cdot (m+1)}$ can be solved by adding $\mathcal{O}(m \cdot n)$ operations to the solution of the corresponding problem arising from linearization of the least–squares function (9) on $\mathbb{R}^n$, which needs $\mathcal{O}(m \cdot n^2)$ operations. However, the number of gradient and function evaluations increases from 1 to m.*

Proof: We can write (8) as

$$\|H_\Lambda(y_1, \ldots, y_m, y)\|_2^2,$$

with

$$H_\Lambda(y_1, \ldots, y_m, y) := \begin{pmatrix} \Lambda_1 y_1 \\ y_m \\ f_1(y_1 + y) \\ \vdots \\ f_m(y_m + y) \end{pmatrix}, \qquad H_\Lambda : \mathbb{R}^{mn+n} \to \mathbb{R}^{mn+m}.$$

Writing gradients as row vectors, we calculate the Jacobian of H_Λ as

$$
J_\Lambda = \begin{pmatrix}
\Lambda_1 & 0 & \cdots & 0 & 0 \\
0 & \Lambda_2 & & 0 & 0 \\
\vdots & & \ddots & & \vdots \\
0 & 0 & \cdots & \Lambda_m & 0 \\
\nabla f_1 & 0 & \cdots & 0 & \nabla f_1 \\
0 & \nabla f_2 & \cdots & 0 & \nabla f_2 \\
\vdots & & \ddots & & \vdots \\
0 & 0 & \cdots & \nabla f_m & \nabla f_m
\end{pmatrix}
$$

where the argument of ∇f_j is $y_j + y$ throughout. This requires m evaluations of gradients.

We linearize H_Λ at $y_1, \ldots, y_m, y$ and consider the problem

$$
\left\| J_\Lambda \begin{pmatrix} z_1 \\ \vdots \\ z_m \\ z \end{pmatrix} + H_\Lambda(y_1, \ldots, y_m, y) \right\|_2^2 \to \min!
$$

for $z, z_j \in \mathbb{R}^n$, $1 \le j \le m$. This linear least–squares problem splits into the sum

$$
\sum_{j=1}^m \left\| \begin{pmatrix} \Lambda_j \\ \nabla f_j \end{pmatrix} z_j + \begin{pmatrix} \Lambda_j y_j \\ f_j + \nabla f_j \cdot z \end{pmatrix} \right\|_2^2 \tag{10}
$$

and we first minimize each single term with respect to z_j for arbitrary fixed z. With a positive definite $n \times n$ diagonal matrix D, vectors $u, v \in \mathbb{R}^n$, and a scalar α we want to find the minimizer $\tilde{x} \in \mathbb{R}^n$ of an expression of the form

$$
\left\| \begin{pmatrix} D \\ u^T \end{pmatrix} x + \begin{pmatrix} v \\ \alpha \end{pmatrix} \right\|_2^2 \to \min_x!
$$

A straightforward calculation yields

$$
\tilde{x} = -(D^2 + uu^T)^{-1}(Dv + \alpha u) = -D^{-1}v - \gamma D^{-2}u
$$

with

$$
\gamma = \frac{\alpha - u^T D^{-1} v}{1 + u^T D^{-2} u},
$$

and

$$\left\| \begin{pmatrix} D \\ u^T \end{pmatrix} \tilde{x} + \begin{pmatrix} v \\ \alpha \end{pmatrix} \right\|_2^2 = \gamma^2 (1 + \|D^{-1}u\|_2^2)$$

$$= (\alpha - u^T D^{-1} v)^2 / (1 + \|D^{-1}u\|_2^2).$$

Thus the minimizer of (10) for fixed z satisfies

$$z_j = -y_j - \frac{f_j - \nabla f_j \cdot (y_j - z)}{1 + \|\Lambda_j^{-1} \nabla f_j\|_2^2} \, \Lambda_j^{-2} (\nabla f_j)^T \tag{11}$$

as a function of z, and we insert (11) into (10) to get another linear least–squares problem for the vector $z \in \mathbb{R}^n$ as

$$\sum_{j=1}^m \frac{1}{1 + \|\Lambda_j^{-1} f_j\|_2^2} \, (f_j - \nabla f_j \cdot (y_j - z))^2 \to \min_z \tag{12}$$

which can be considered as a variation of the standard $m \times n$ least–squares calculation required after linearization of (2). The additional work needed for (12) arises in forming the $m \times n$ coefficient matrix and the m right–hand–side entries, and is of order $\mathcal{O}(m \cdot n)$. After solving (12) by orthogonalization or singular–value decomposition, we need only substitute z into (11), which again will introduce $\mathcal{O}(m \cdot n)$ operations. ∎

For the penalty function (4) and a Levenberg–Marquardt strategy based on normal equations, a similar result was obtained by K. Nottbohm in [4].

§4. Control of Penalty Parameters

Discrete least–squares approximation of a function $\varphi : T \to \mathbb{R}$ on a finite set $T = \{t_1, \ldots, t_m\}$ of m distinct points by a parametrized family of functions

$$\Phi(x, \cdot) : T \to \mathbb{R}, \quad x \in \mathbb{R}^n$$

takes the form (9) with

$$f_j(x) = \Phi(x, t_j) - \varphi(t_j), \quad 1 \le j \le m. \tag{13}$$

Section 5 will treat rational approximation as an example for this approach. In most applications the sets

$$X_j := \{x \in \mathbb{R}^n : \ \Phi(x, t_j) = \varphi(t_j)\}, \quad 1 \le j \le m,$$

are non–empty and easy to reach by minimization of $f_j^2(x)$ on $\mathbb{R}^n$. Consequently, the direct separate minimization of each f_j^2 in Step 1 will normally produce parameters $y_j^{(1)} \in X_j$. In most cases the interpolation equation

$$f_j(y_j^{(1)}) = \Phi(y_j^{(1)}, t_j) - \varphi(t_j) = 0 \tag{14}$$

still has $n - 1$ degrees of freedom available for variation of $y_j^{(1)}$. This freedom should be used to find parameters $y_j^{(1)}$ which are "as equal as possible" without violating the interpolation condition (14).

Thus, it is advisable to begin Step 2 of the algorithm by keeping all penalty parameters at very small positive and fixed values, and to iterate on H_Λ within Step 2 until no reasonable progress is possible.

The overall control strategy can be illustrated as in Figure 1, where the values $f = \sum_{j=1}^{m} f_j^2(y_j + y)$ of iterates are plotted against g, where g is the value of the penalty function (4) for $\lambda = 1$.

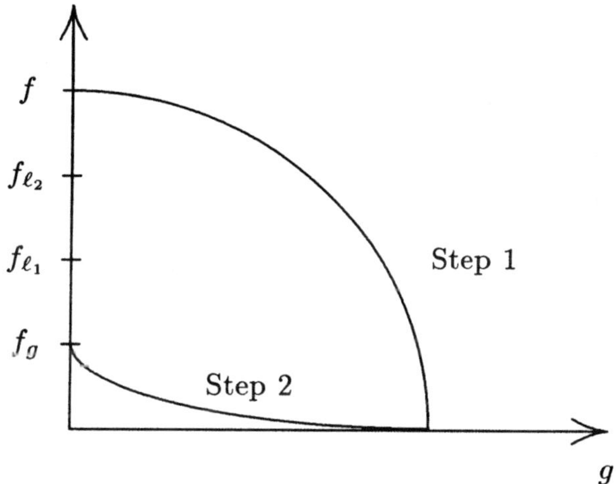

Figure 1. The f - g diagram

The global minimizer of $f = \sum_{j=1}^{m} f_j^2(x)$ on $\mathbb{R}^n$ yields a minimal value f_g on the f axis, while the values of local best approximations are denoted by $f_{\ell_1}, f_{\ell_2}, \ldots$ in Figure 1. A typical application of Step 1 will dash down to the g axis from a high starting point on the f axis. Then our suggestion to start Step 2 with small values of penalty parameters will cause the method to creep along the g axis towards the origin, avoiding large values of f. This can be put on a more rigorous basis by using small values of ρ in the following theorem.

Theorem 2. *Let the given starting values* $y_j := x^{(1)}$, $1 \le j \le m$, $y := 0$ *and* $\rho > 0$ *satisfy*

$$\rho > \sum_{j=1}^{m} f_j^2(x^{(1)}).$$

Then minimizing

$$\sum_{j=1}^{m} f_j^2(y_j + y) + \lambda^2 \sum_{j=1}^{m} \|y_j\|^2 \tag{15}$$

for the penalty function (4) with λ *bounded by*

$$0 < \lambda^2 \le \frac{\rho - \sum\limits_{j=1}^{m} f_j^2(x^{(1)})}{m\|x^{(1)}\|^2}$$

yields parameters $y^{(2)}, y_j^{(2)}, 1 \le j \le m$, *with*

$$\sum_{j=1}^{m} f_j^2(y_j^{(2)} + y^{(2)}) \le \rho,$$

provided that an iteration with guaranteed descent of (15) is used.

Proof: For each iterate,

$$
\begin{aligned}
\sum_{j=1}^{m} f_j^2(y_j + y) \; &\le \; \sum_{j=1}^{m} f_j^2(y_j + y) + \lambda^2 \sum_{j=1}^{m} \|y_j\|_2^2 \\
&\le \; \sum_{j=1}^{m} f_j^2(x^{(1)}) + \lambda^2 \sum_{j=1}^{m} \|x^{(1)}\|_2^2 \\
&\le \; \rho. \quad \blacksquare
\end{aligned}
$$

Similar results can easily be obtained for the other penalty functions. The main purpose of Theorem 2 is to keep the principal error term within reasonable bounds while trying to coalesce the parameters as far as possible.

Another easy result follows for large λ in the penalty function (4):

Theorem 3. *Assume y_j^μ, y^μ and y_j^λ, y^λ, $1 \le j \le m$, in $\mathbb{R}^n$ satisfy*

$$
\begin{aligned}
\sum_{j=1}^m f_j^2(y_j^\mu + y^\mu) + \mu^2 \sum_{j=1}^m \|y_j^\mu\|_2^2 &\le \sum_{j=1}^m f_j^2(y_j^\lambda + y^\lambda) + \mu^2 \sum_{j=1}^m \|y_j^\lambda\|_2^2 \\
\sum_{j=1}^m f_j^2(y_j^\lambda + y^\lambda) + \lambda^2 \sum_{j=1}^m \|y_j^\lambda\|_2^2 &\le \sum_{j=1}^m f_j^2(y_j^\mu + y^\mu) + \lambda^2 \sum_{j=1}^m \|y_j^\mu\|_2^2.
\end{aligned}
\tag{16}
$$

Then for $\mu < \lambda$ we have

$$
\begin{aligned}
\sum_{j=1}^m \|y_j^\lambda\|_2^2 &\le \sum_{j=1}^m \|y_j^\mu\|_2^2, \\
\sum_{j=1}^m f_j^2(y_j^\mu + y^\mu) &\le \sum_{j=1}^m f_j^2(y_j^\lambda + y^\lambda).
\end{aligned}
\tag{17}
$$

If the monotonic function

$$
\sum_{j=1}^m f_j^2(y_j^\lambda + y^\lambda) + \lambda^2 \sum_{j=1}^m \|y_j^\lambda\|_2^2
$$

stays bounded for $\lambda \to \infty$, then $y_j^\lambda \to 0$ for $\lambda \to \infty$ and all j, $1 \le j \le m$.

Proof: The theorem is a consequence of the theory of penalty function methods (see e.g.: Fletcher [2]). ∎

Inequalities (16) will typically hold if (15) is minimized by a method that ensures (weak) descent, and if y_j^μ, y^μ and y_j^λ, y^λ are optimal for minimization for μ and λ fixed, respectively, where y_j^μ, y^μ are admissible for minimization with fixed λ, and conversely. Theorem 3 allows us to force the y_j^λ to coalesce for large λ, if the penalty function (15) remains bounded during the iteration. This can be used as the final strategy of the internal iteration within Step 2 of the basic algorithm. Standard arguments of nonlinear optimization show that minimization of (15) for large λ is equivalent to minimization of (2) under the rather strict constraint

$$
\sum_{j=1}^m \|y_j\|^2 \le \varepsilon
$$

on the y_j. Thus, if the limits

$$
\lim_{\lambda \to \infty} y_j^\lambda = 0, \qquad \lim_{\lambda \to \infty} y^\lambda = y
$$

exist, the point y is critical for (2). This shows that Step 3 is not necessary, if the penalized objective function remains bounded when the penalty parameters are driven to infinity in Step 2.

The two extreme cases described above leave the strategy of the internal iteration within Step 2 for "moderate" values of λ open. As described by (17), pushing λ up will normally

$$\text{increase} \quad \sum_{j=1}^{m} f_j^2(y_j^\lambda + y^\lambda) \quad =: \quad f^\lambda,$$

$$\text{decrease} \quad \sum_{j=1}^{m} \|y_j^\lambda\|_2^2 \quad =: \quad g^\lambda,$$

while lowering λ has the opposite effect. If the value of f^λ gets much larger than expected or wanted when pushing up λ, the user may prefer to go back to $\lambda = 0$ for a couple of iterations, hoping that this restart will give a better starting point for Step 2. Monitoring the values (f^λ, g^λ) in a diagram like Figure 1 may help to detect progress.

For all penalty functions of Section 2 the critical points $(y_1^\Lambda, \ldots, y_m^\Lambda, y^\Lambda)$ of $\|H_\Lambda\|^2$ for Λ fixed correspond to points $z_j^\Lambda = y_j^\Lambda + y^\Lambda$ of the set

$$M := \left\{ \begin{pmatrix} z_1 \\ \vdots \\ z_m \end{pmatrix} \in \mathbb{R}^{mn} : \sum_{j=1}^{m} f_j(z_j) \cdot \nabla f_j(z_j) = 0 \right\}.$$

This follows from

$$\frac{1}{2} \nabla_y \|H_\Lambda\|^2 = \frac{1}{2} \nabla_y \sum_{j=1}^{m} f_j^2(y_j + y) = \sum_{j=1}^{m} (f_j \cdot \nabla_y f_j)(y_j + y)$$

for arguments $(y_1, \ldots, y_m, y)$ of H_Λ, independent of the penalty parameters. Control of Λ will steer $y_j^\Lambda + y^\Lambda$ along M, possibly jumping between connected components of M.

Penalty function (4), if fully optimized for each value of λ, will generically produce pieces of curves on M, which may be visualized as curves in Figure 1 if we plot (f^λ, g^λ) as a function of λ. However, we do not recommend following these curves closely, because this would spoil the main advantage of the method: its high space dimension allows a lot of freedom to manoeuver

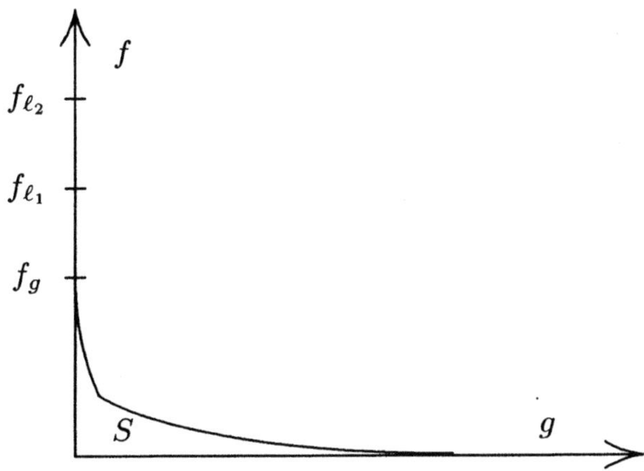

Figure 2. The unattainable set S

around. Performing just one linearization step for each instance of penalty parameters Λ is quite enough and allows much more freedom, jumping between many trajectories on M.

Theoretically, the best strategy would be to follow the boundary of the "unattainable set"

$$S = \{(f,g) \in \mathbb{R}^2_{\geq 0} : \text{ if } g = \sum_{j=1}^m \|y_j\|^2, \ y_j \in \mathbb{R}^n, \text{ then}$$
$$\sum_{j=1}^m f_j^2(y_j + y) > f \text{ for all } y \in \mathbb{R}^n\}$$

in Figure 2, which always links the global minimum of (2) with the global minimum of $\sum_{j=1}^m f_j^2(y_j)$ having least value of $\sum_{j=1}^m \|y_j\|^2$. However, this is not easier to solve than the original problem, but it indicates that one should try to keep as "southwest" in Figures 1 and 2 as possible.

Since the set M is described by n equations on $\mathbb{R}^{n \cdot m}$, the penalty function (6) has enough degrees of freedom to allow appropriate "steering" along M in a sophisticated way that may involve a lot of information about $f_j(y_j)$ or $\nabla f_j(y_j)$. Thus, the user may assign large penalty parameters to "good" y_j (or components thereof) to make sure that they will not be moved around too much. Details can easily be provided for $m = 2$ and $n = 1$, but we will leave further analysis to future work.

Ute Jäger [3] treated the penalty function (4) with $\lambda^2 \sim (g^\lambda)^{-1}$ and showed that nondegenerate attractors $\tilde{x}$ for (9) on $\mathbb{R}^n$ yield nondegenerate attractors $(0, \ldots, 0, \tilde{x})$ for (15) on $\mathbb{R}^{n(m+1)}$, provided that a well–controlled trust–region method is used.

§5. Application to Discrete Rational Approximation

On a finite subset $T = \{t_1, \ldots, t_m\} \subset [-1, +1]$ we consider the approximation of a function $\varphi : T \to \mathbb{R}$ by rational functions

$$\Phi(x, t) = \frac{p(x, t)}{q(x, t)} \qquad t \in [-1, +1] \tag{18}$$

where

$$p(x, t) = \sum_{i=1}^{k} x_i t^{i-1}, \quad q(x, t) = 1 + \sum_{i=k+1}^{n} x_i t^{i-k}.$$

We use (13) and obtain a nonlinear least–squares problem of type (9).

To study Step 1 of the algorithm, we note that $f_j(x)$ vanishes on the affine subspace of $\mathbb{R}^n$ consisting of all $x \in \mathbb{R}^n$ with

$$p(x, t_j) - \varphi(t_j) \cdot q(x, t_j) = 0 \tag{19}$$

for $1 \le j \le m$, except for points where

$$p(x, t_j) = 0 = q(x, t_j)$$

holds, and at which l'Hospital's rule for evaluating $\Phi(x, t_j)$ does not yield $\varphi(t_j)$. If we ignore these exceptional x for a moment, we find that minimizers $y_j^{(1)}$ of f_j^2 with $f_j(y_j^{(1)}) = 0$ exist and form an affine subspace. Thus, there normally are no problems with Step 1 of the algorithm.

If Step 2 is carried out with very small penalty parameters, it is comparable to a minimization of the positive definite penalty term under the affine constraints (19). Thus, Step 2 yields a unique minimizer $(y_1^{(2)}, \ldots, y_m^{(2)}, y^{(2)})$ depending on Λ, but not on common factors of the various possible penalty variables in Λ. In particular, the minimizer does not depend on the result of Step 1. It is a multiple parameter set that satisfies all linearized one–point interpolation problems

$$p(y_j^{(2)}, t_j) - \varphi(t_j)q(y_j^{(2)}, t_j) = 0, \qquad 1 \le j \le m,$$

and Step 2 has tried to make the parameters $y_j^{(2)}$ as equal as possible. Thus, for all discrete rational approximation problems, a fixed implementation of the multi–parameter method will almost surely produce the same output for each possible starting parameter vector, since the exceptional points form a set of measure zero. Of course, this feature does not generalize to other nonlinear families of approximating functions.

§6. Examples

We take the example

$$f^2(x,y) = \sum_{i=1}^{m} f_i^2(x,y), \quad \begin{pmatrix} f_1(x,y) \\ f_2(x,y) \\ f_3(x,y) \\ f_4(x,y) \end{pmatrix} = \begin{pmatrix} 1 - x + 25xy \\ 1 + x \\ 1 - y \\ 1 + y \end{pmatrix}$$

from [4] for $n = 2$, $m = 4$ with three critical points:

$$\text{at } (x,y) = (0,0) \quad \text{is a saddle point with } f(x,y) = 2$$
$$\text{at } (x,y) = (0.12, -0.24) \quad \text{is a local minimum with } f(x,y) = 1.843$$
$$\text{at } (x,y) = (-1.006, 0.079) \quad \text{is a global minimum with } f(x,y) = 1.419.$$

For various possibilities of controlling λ in Step 3, we always found the multi–parameter method to converge to the global minimum (see also [3]). Figure 3 contains a contour plot of the approximation error for this example, where the saddle point and the global minimum are clearly visible. The local minimum is overlaid by its surrounding (black) basin of attraction of a regularized Gauss–Newton method. Regularization was done by a Levenberg–Marquardt strategy to prevent rank loss, and damping by a stepsize strategy was used to enforce descent even for small or zero Levenberg–Marquardt parameters. The same Gauss–Newton routine was used in Step 3 of the multi–parameter algorithm. The corresponding plot for the multi–parameter method does not contain the black blob, because the basin of attraction of the local minimum with respect to the multi–parameter method was empty.

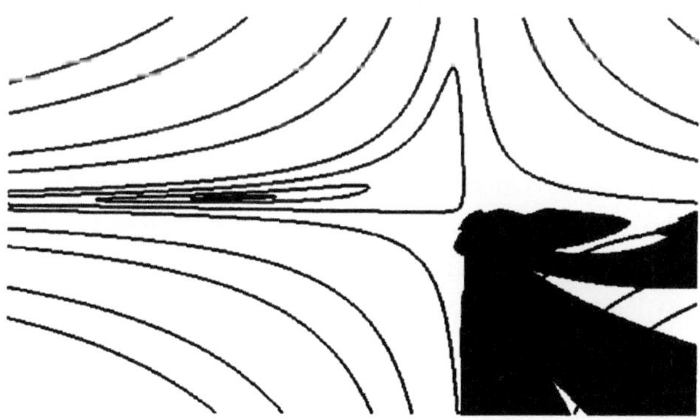

Figure 3. Contours and attractors

The next example is taken from [1] and approximates $\varphi(t) = t^2 - 0.6$ on 11 equidistant points of $[-1, 1]$ by rational functions (18) with constant numerator and linear denominator. The problem is symmetric, and we plot the error in the (x_1, x_2)–plane in Figure 4. Note that there are singularities for $x_2 = -5/j$ for $1 \le |j| \le 5$, and the horizontal axis $x_2 = 0$ extends through the global minimum at $x_1 = -0.2$, $x_2 = 0$ near the lower margin of Figure 4. The local minima and error values ϵ are

$$
\begin{array}{lllllll}
x_1 & = & -0.200000 & x_2 & = & 0.000000 & \epsilon & = 1.172 \\
x_1 & = & -0.035051 & x_2 & = & \pm1.075847 & \epsilon & = 1.232 \\
x_1 & = & -0.053199 & x_2 & = & \pm1.459828 & \epsilon & = 1.219 \\
x_1 & = & -0.062364 & x_2 & = & \pm2.204753 & \epsilon & = 1.211 \\
x_1 & = & -0.067159 & x_2 & = & \pm4.423534 & \epsilon & = 1.207
\end{array}
$$

the latter differing only by about 3 %, making the problem hard to solve globally, if not started near the global minimum. Figure 5 shows the basins of attraction of the stabilized Gauss–Newton method, where we used black and white alternatively for the attractors.

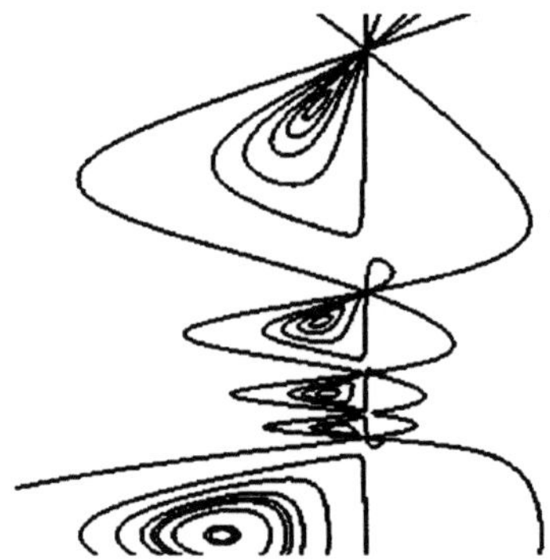

Figure 4. Contours of approximation error

The multi–parameter method always converged to the global minimum, though the global minimum does not have the "nicest" basin of attraction. Other discrete rational approximation problems we tested were less hazardous. In all cases the results were qualitatively the same. Tests on discrete exponential approximation problems, however, did not show such a good behaviour.

Acknowledgments. Special thanks go to Dr. Immo Diener for a series of useful discussions and some valuable help preparing the final version of the paper.

Figure 5. Gauss–Newton attractors

References

1. Diener, I., On Nonuniqueness in Nonlinear L_2-Approximation, J. Approx. Th. **51** (1987), 54–67.
2. Fletcher, R., *Practical Methods of Optimization Vol. 1: Unconstrained Optimization*, Wiley, Chichester, 1980.
3. Jäger, U., Das Mehrfach–Parameter–Verfahren unter Benutzung der trust region–Methode zur Lösung nichtlinearer, diskreter Approximationsprobleme in der L_2–Norm, Diplomarbeit, Göttingen 1989.
4. Nottbohm, K., Das Mehrfach–Parameter–Verfahren zur Lösung nichtlinearer diskreter Approximationsprobleme in der L_2–Norm, Ph.D. Dissertation, Göttingen 1986.

R. Schaback
Georg–August–Universität Göttingen
Lotzestrasse 16–18
D–3400 Göttingen, Germany
schaback@namu01.gwdg.de

Numerical Methods of Approximation Theory, Vol. 9

Dietrich Braess and Larry L. Schumaker (eds.), pp. 285–300.

International Series of Numerical Mathematics, Vol. 105

ISBN 3-7643-2746-4.

285

Analog VLSI Networks

Walter Schempp

Dedicated to the memory of Lothar Collatz

Abstract. The retina forms a multilayered precortical structure which collects and preprocesses the information that reaches the visual cortex. To simulate neural network computation by the analog VLSI implementation technology, an analog model of the first stages of retinal processing has been constructed on a single silicon chip by CMOS VLSI circuitry and applied to machine vision. The purpose of this paper is to study the exactly solvable hexagonal resistive networks which model the horizontal cell layer of the retina. An overview of free-space multilayer architectures of hybrid optoelectronic interconnection networks is also given. It is shown that the shift register stage used in the horizontal and vertical scanner of the silicon retina implementation is in correspondence to the S-SEED technology in the hybrid optoelectronic implementation of interconnection networks.

Formalism often is viewed as standing on its own, without regard for applications insight that breathes life into it.

Carver A. Mead (1989)

§1. Introduction

The main function of the visual sense is to identify objects and to evaluate relationships between them. A major part ($> 75\%$) of the cortical activities is concerned with visual processing. The segmentation of the visual scenes into distinct objects and shapes, and the ability of visual perception to segregate visual scenes into distinct figures and background, is by itself a major achievement of the visual processing system. The complex functional architecture for mammalian primary visual cortex and precortical structures has been experimentally elucidated in impressive detail over the past 20-25 years. Although visual processes are complex, it was found that a surprisingly simple set of physiologically and anatomically plausible rules suffices to generate many of the salient features of this functional architecture. To understand the basic rules, it is of fundamental importance to study the vertebrate retina as an outpost of the cortex cerebri. It forms not only a pixel array of detectors sensitive to single photons but it also acts as a neural preprocessor for computing weighted spatial and temporal averages and segregating objects of visual scenes with overlapping contours from each other and from background. Higher levels of retinal functions include the detection of oriented light-intensity edges, edge enhancement, adaption and gain control, spatial-frequency matched filtering, time series analysis, and statistical optimization.

Unlike conventional computer hardware, neural circuitry is not hardwired. The organization is not specified as an explicit net of point-to-point connections. The tangential network in the retina is a flat mesh of dense processes that are highly interconnected by resistive gap junctions. Any given cell is locally and globally connected with many others, and there is a great deal of overlap among interconnected cells. Therefore the horizontal layer of the retinal architecture may serve as a biological paradigm of the parallel computations that can be performed by modular self-adaptive Artificial Neural Networks (ANNs) and their various analog and digital implementations.

Most perceptual information processing in the cerebral cortex may be considered as pattern recognition. The basic principle of ANNs designed for pattern recognition and completion can be formulated as follows: Define a hierarchical multilayer geometry to compute the response of the ANN as weighted spatial and temporal averages over the layers. In this way the difficulties that originate from using only the levels of activity in individual neurons to encode the information will be avoided and a cooperative organization form of neuronal collectives arises. Presently there are three main categories of architectures available for these neural computations:

1) Analog electronic architectures computing averages by CMOS VLSI circuitry,

2) Analog photonic architectures recording interference patterns of wavelets,

3) Hybrid optoelectronic architectures based on free-space digital photonics.

ANNs are fascinating objects involving many different structures, and having highly interesting applications in machine vision and robotics. The mathematical modelling of cortical encoding procedures ([17,18,21]) and their practical implementations by interconnection networks, however, is just beginning. It is the purpose of this paper to focus on the analog VLSI implementations of ANNs and to give a short overview of the hybrid optoelectronic implementation technology. The hybrid implementations use photonics for interconnections and electronics for routing and control. A hybrid optoelectronic recording of stationary interference patterns of neuronal wavelets which reflect cerebral cortex activities has been described in the paper [9].

Although a cell's response function is in general nonlinear, visual neurophysiologists have found that for many cells, a linear summation approximation is appropriate.

Ralph Linsker (1988)

§2. Linearity

The single most important principle in the analysis of electrical circuits with components of linear characteristics is the principle of linear superposition. Provided there exists a reference value for voltages, the ground to which all node potentials revert when all sources are reduced to zero, the principle states that for any network containing resistors and voltage sources, the solution for the network is the solution for the network in response to each voltage source individually, with all the other voltage sources reduced to zero. Thus the solution for the network, determining the voltage at every node and the current through every resistor and including the effects of all voltage sources, is the sum of the solutions for the individual voltage sources.

To maintain the principle of linear superposition for certain categories of ANNs with components of nonlinear characteristics, linearization can be performed by introducing an infinite number of variables. In particular, an application of the method of covering linearization provides exactly solvable interconnection network models ([21]). Because the procedure of covering linearization preserves the relative phase of coherent state wavelet responses, it avoids the superposition catastrophe of cortical encoding ([3,23]). The recently developed symmetric self electro-optic-effect device (S-SEED) technology ([15,2]) for ANN implementation, which integrates photodetectors and quantum-well modulators with transistor circuits, has the capability to provide both optical inputs and outputs for VLSI circuitry. Most of the work on S-SEEDs has concentrated on devices for digital applications. The devices, however, also have analog modes to implement two-port model neurons. Their greatest strength is that changing states is a function of the relative phase and not of the absolute intensity of the two input beams, so that one light beam forms the reference for the other one. The optical inputs and outputs are based on quantum-well absorptive structures consisting of many very thin layers of two different semiconductor materials. The hybrid optoelectronic architectures based on processing arrays of S-SEEDs offer a solution for high-performance free-space digital photonic systems that are highly connection-intensive. The routing within the interconnection network architecture can all be optically controlled by dynamically combining S-SEED two-port model neuron arrays via an amacrine beam-steering module with planar arrays of individual addressed micro-lasers to generate focal-plane structures that are photonically and competitively coupled with local electronics processing cells.

Signal transmission with its attendant gain control necessitates the introduction of nonlinearities. But pulse to wave conversion at synapses once more linearizes the system. Thus the unconstrained dendritic computational microprocess is essentially linear.

Karl H. Pribram (1991)

§3. Nonlinearity

One may well ask why execute vision algorithms on programmable digital computers when the signals themselves are self-adaptive wavelets and therefore analog signals which allow resonance by phase coherence? Indeed, coherent wavelets have recently come to the forefront of attention with the experimental discovery of cooperatively synchronized neural activity in the visual cortex ([6,24]). Why not exploit the mature electronics technology presently available to implement analog special purpose vision systems?

In analog VLSI implementation technology, discrete two-dimensional resistive networks are arranged in a regular array by interconnection of nearest neighbors. The retina is modelled by a triangular network. Each node is connected with its 6 neighbors by a resistance R. Moreover, each node is connected to ground which acts as a reference, by a conductance G. Figure 1 displays the hexagonal resistive network. Note that the hexagonal lattice is dual to the triangular lattice.

The silicon retina implemented by complementary metal-oxide-silicon (CMOS) VLSI technology ([1,12,13,14,11]) yields results remarkably similar to those obtained from the biological paradigm. A photoreceptor whose output

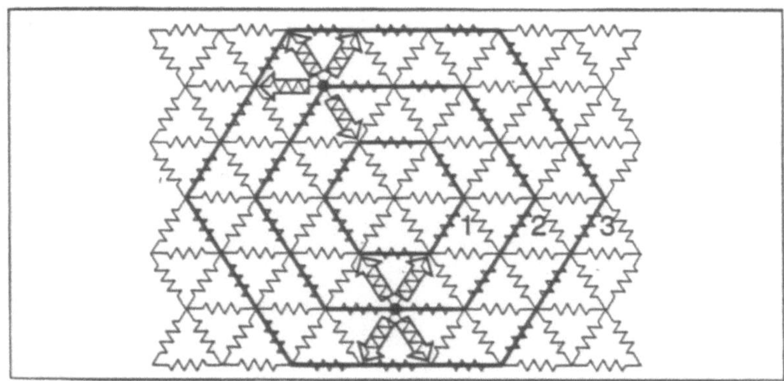

Figure 1. Concentric hexagons in a triangular resistive network.

voltage is proportional to the logarithm of the incoming light intensity is coupled via the conductance G to drive the resistive network specified above. The output of the silicon chip is proportional to the voltage difference between the logarithmic photoreceptor circuit and the network. Figure 2 displays an idealized retina which represents in the radial direction a logarithmically deformed version of the hexagonal resistive network. The deformation is performed by a log-polar transformation which improves pattern-recognition systems by making them less sensitive to the rotation of their targets.

It is often argued that conversion from photonics to electronics and back to photonics is an inefficient procedure. It usually is. The inefficiencies often result, however, from the fact that the hybrid system is not integrated. It is therefore very important that synthetic perceptual systems can be implemented by analog VLSI technology. The retinal image of a visual scene consists of a two-dimensional continuous distribution of gray levels, whereas the retina chip consists of an array of pixels, and a scanning arrangement for reading-out the results of retinal computing. The output of any pixel can be accessed through the scanner, which is made up of a vertical scan register and a horizontal scan register along the sides of the chip. Each scan register stage has 1-bit of shift register, with the associated signal-selection circuits. The main pixel array is made up of alternating rows of rectangular tiles arranged to form a hexagonal pattern. The scanner along the vertical side allows any row of pixels to be selected, the scanner along the horizontal side allows the output current of any selected pixel to be gated onto the output line where it is sensed by the offchip current-sensing amplifier.

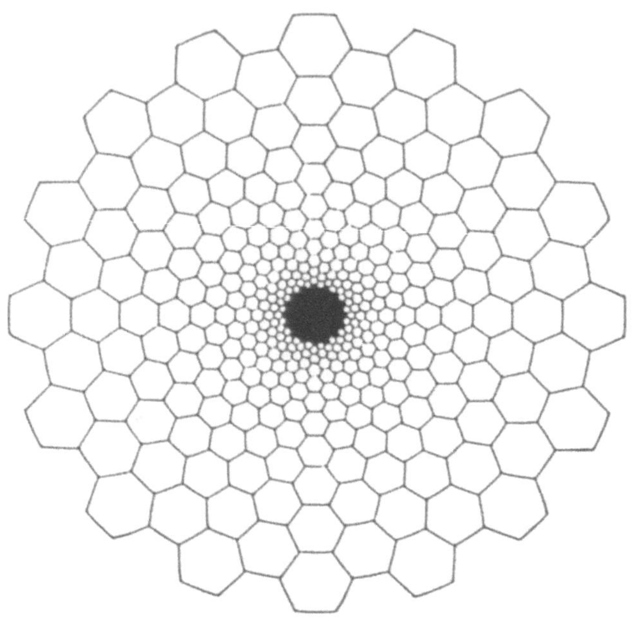

Figure 2. An idealized retina.

With respect to sensory driven aspects of perception the models point to the importance of successive iterations of the process.

Karl H. Pribram (1991)

§4. Circular Approximation

When considering the circular approximation of the retinal ANN, we fix a center in the network specified in §3 and displayed in Figure 1 above. It is assumed that the voltage is constant around the perimeters of concentric hexagons [4]. The concentric hexagon n contains $6n$ nodes from the pixel array on its perimeter. Six of these nodes are vertices of the hexagon n, and the remaining $6(n-1)$ are located along its edges. Each of the 6 vertex nodes makes 3 outside connections, i.e., 3 connections to hexagon $n+1$, while each of the other $6(n-1)$ nodes makes only 2 connections.

The total number of connections between the concentric hexagons n and $n+1$ is therefore $12n + 6$. Since each link in the triangular mesh corresponds to a resistance R, the impedance between the two neighboring hexagons n and $n+1$ is

$$\frac{R}{12n + 6}.$$

Similarly, the impedance connecting hexagon n to hexagon $n-1$ is

$$\frac{R}{12n - 6}.$$

Moreover, there are equal conductances from each node to ground, making a net admittance of

$$6nG$$

from hexagon n to ground.

Introduce the parameter $a = RG$. By hypothesis, the potential, say V_n, is constant along hexagon n. Kirchhoff's current law then yields the finite-difference equation for the steady-state node voltage:

$$(2n + 1)V_{n+1} - n(4 + a)V_n + (2n - 1)V_{n-1} = 0.$$

The boundary conditions of the difference equation are

$$\begin{cases} V_0 \text{ is given,} \\ V_n \text{ is bounded as } n \to \infty. \end{cases}$$

A global analysis of the voltages V_n for the circular approximation is performed by the generating function

$$F(z) = \sum_{n \geq 1} V_n z^n, \qquad z \in \mathbf{C}.$$

It follows from elementary complex function theory that F is holomorphic in the open unit disc $|z| < 1$ of the complex plane $\mathbf{C}$. In the engineering terminology, the function F forms the z-transform of the node voltages $(V_n)_{n \geq 1}$. It satisfies the first-order inhomogeneous linear differential equation

$$2\left(z^2 - (2 + \frac{a}{2})z + 1\right) F'(z) + (z - \frac{1}{z})F(z) = V_1 - zV_0.$$

The right-hand side includes the unknown voltage V_1. The problem is to compute V_1 in terms of V_0 in order to apply the three-term recurrence for the circular approximation to calculate the node voltages $(V_n)_{n \geq 1}$ step by step.

The factorization of the quadratic term reads

$$Q(z) = z^2 - (2 + \frac{a}{2})z + 1 = (z - r_+)(z - r_-)$$

where

$$r_- = 1 - \frac{2}{1 + \sqrt{1 + 8L^2}}, \qquad L = \frac{1}{\sqrt{a}} \geq 0$$

so that

$$0 \leq r_- \leq 1, \qquad r_+ r_- = 1, \qquad r_+ \geq 1.$$

The homogenous linear differential equation takes the form

$$2Q(z)\Phi'(z) + (z - \frac{1}{z})\Phi(z) = 0.$$

The solution of the homogeneous equation

$$\Phi(z) = \sqrt{\frac{z}{|Q(z)|}}$$

is singular at $z = r_-$. Assuming

$$F(z) = \Phi(z)f(z)$$

gives

$$2Q(z)\Phi(z)f'(z) = V_1 - zV_0$$

and

$$\begin{cases} f(z) = \dfrac{1}{2} \displaystyle\int\limits_0^z \mathrm{sgn}Q(\zeta)\dfrac{V_1 - \zeta V_0}{\sqrt{\zeta|Q(\zeta)|}}\, d\zeta \\[2ex] f(r_-) = 0 \end{cases}$$

for real values of z. Thus

$$V_1 \int\limits_0^{r_-} \frac{1}{\sqrt{\xi Q(\xi)}}\, d\xi = V_0 \int\limits_0^{r_-} \frac{\xi}{\sqrt{\xi Q(\xi)}}\, d\xi$$

with singularities at $z = 0$ and $z = r_-$. In the next step we perform the singularities absorbing substitution

$$\xi = r_- \sin^2 \theta.$$

Hence,

$$\frac{V_1}{V_0} = r_+ \left(1 - \frac{E(r_-^2)}{K(r_-^2)} \right),$$

where K and E are the complete elliptic integrals of the first and second kind, respectively,

$$\begin{cases} K(r_-^2) = \int_0^{\pi/2} \frac{1}{\sqrt{1 - r_-^2 \sin^2 \theta}} \, d\theta, \\[2ex] E(r_-^2) = \int_0^{\pi/2} \sqrt{1 - r_-^2 \sin^2 \theta} \, d\theta. \end{cases}$$

The elliptic integrals for the parameter value r_-^2 can be evaluated by the algorithm of the arithmetic-geometric mean. Set

$$a_0 = 1 \qquad\qquad b_0 = \cos r_-^2 \qquad\qquad c_0 = \sin r_-^2$$

$$a_j = \tfrac{1}{2}(a_{j-1} + b_{j-1}) \qquad b_j = \sqrt{a_{j-1} b_{j-1}} \qquad c_j = \tfrac{1}{2}(a_{j-1} - b_{j-1}).$$

Starting from the product $a_{j-1} b_{j-1}$, the number b_j of the jth iteration loop can be computed by an iteration subloop of Newton linearizations:

$$b_{jk} = \frac{1}{2}\left(b_{jk-1} + \frac{a_{j-1} b_{j-1}}{b_{jk-1}} \right) \qquad k \geq 1.$$

The algorithm of successive linearizations corresponds to the retinal strategy of superposing average computations in a multilayer architecture. If we have $c_N \approx 0$ in the Nth step of the algorithm, $N \geq 1$, then

$$\begin{cases} K(r_-^2) = \frac{\pi}{2 a_N}, \\[2ex] E(r_-^2) = K(r_-^2) \left(1 - \tfrac{1}{2} \sum_{0 \leq j \leq N} 2^j c_j^2 \right). \end{cases}$$

Thus, in the circular approximation, the node voltages $(V_n)_{n \geq 1}$ are obtained by an algorithm of successive averaging. The asymptotic analysis of the node voltages is based on the assumption that

$$V_n \sim V \frac{r_-^n}{\sqrt{n}} \qquad \text{as } n \to \infty$$

with a constant V. Now V can be expressed in terms of elliptic integrals by a procedure of singularities absorbing substitution similar to the treatment of V_1 above:

$$V = \frac{1}{\sqrt{\pi(1 - r^2)}} \left(V_0 E(1 - r_-^2) - r_- V_1 K(1 - r_-^2) \right).$$

An application of the Legendre relation yields

$$V = \frac{V_0}{2K(r_-^2)} \sqrt{\frac{\pi}{1 - r_-^2}}$$

where, as before, r_- denotes the zero of Q in the interval $[0, 1]$. Recall that r_- depends on the free parameter $a = RG$ in the three-term recurrence.

Two-dimensional networks compute an average that is a nearly ideal way to derive a reference with which local signals can be compared. If currents are injected at many nodes, the network performs an automatic weighted average: the farther away the inputs are, the less weight they are given. The interconnection density of the silicon system, however, is limited by the total amount of wire required to accomplish the average computation. The hardwiring aspect contributes to the lack of flexibility, which forms one of the principal drawbacks of analog CMOS VLSI circuitry.

Brain function is neither analog nor digital, as these terms are defined for computer usage. Pulse trains that appear to be digital are in fact analog, and sums of dendritic current that appear to be analog are in fact time-segmented by bifurcations into discrete wave packets.

Walter J. Freeman (1991)

§5. Temporal Coding by Coherent State Wavelets

In neurobiology, if enough current is injected into the dendritic network, then the neuron will release neurotransmitter from any output synapses it has on its dendrites. Due to the quantal nature of neurotransmitter release and the role of phase coherency in the cortical encoding procedure ([3,5,6,7,8,10,19,22,23,24,25,26]), a quantum theoretic approach to self-adaptive wavelets and their resonant interactions is appropriate. This idea leads us to consider L^2 harmonic analysis on the Heisenberg nilpotent Lie group N as the mathematical fundament of cortical self-organization by dynamical functional connectivity. In particular, the dendritic membrane can be identified with a linear metaplectic manifold to explain the basic cortical self-organization principles by the multilayer geometry of the unitary dual of N ([16,17,19,20]). The coadjoint action of N on the planar layers of this multilayer architecture solves the binding problem of synthetic perceptual systems ([22,23]) by phase coherence. Indeed, each of these planar coadjoint orbits can be interpreted as an N-homogeneous data page in a stack of layers, where N admits a representation by real unipotent matrices ([19]):

$$\begin{pmatrix} 1 & x & z \\ 0 & 1 & \xi \\ 0 & 0 & 1 \end{pmatrix} \qquad (x \in \mathbf{R}, \xi \in \mathbf{R}, z \in \mathbf{R}).$$

In the matrices representing the Heisenberg group N, the entry ξ denotes the phase shift of the cooperative coherent state wavelets, and the entry x is the dual variable associated with ξ in the phase plane that performs a shift of the temporal coding of neural activity ([21,20]). Due to the Fourier duality, the real variables (x, ξ) always appear in pairs. In an associative memory implemented by a cross-linking interconnection ANN, the phase shift ξ encodes the degree of association between coherent stimulus components, whereas the amplitude of self-adaptive wavelets encodes the significance of synchronized signal components ([19]).

Because the one-dimensional center C of N is given by the matrices

$$\begin{pmatrix} 1 & 0 & z \\ 0 & 1 & 0 \\ 0 & 0 & 1 \end{pmatrix} \qquad (z \in \mathbf{R}),$$

the third entry z denotes the central extension of the phase plane N/C with real coordinates (x, ξ). Thus N forms a principal line bundle over the phase plane, and the vector space dual C^* of C parametrizes the stack of transverse layers which represent the affine planar coadjoint orbits of the unitary dual of N.

In the SEED technology, the concept of diode-biased SEED (D-SEED) allows us to photonically shift the energy levels associated with the planar layers. Moreover, if L now denotes the three-dimensional uniform lattice of unipotent matrices with integer entries

$$\begin{pmatrix} 1 & x & z \\ 0 & 1 & \xi \\ 0 & 0 & 1 \end{pmatrix} \qquad (x \in \mathbf{Z}, \xi \in \mathbf{Z}, z \in \mathbf{Z}),$$

embedded as a principal subgroup into N, the Heisenberg group N projects onto the compact Heisenberg nilmanifold $N\backslash L$ which forms a principal circle bundle over the compact two-dimensional torus $\mathbf{T}^2$ ([20]). Due to the periodization, the projection $N \to N\backslash L$ performs a time averaging and induces a shift register action by N on $N\backslash L$. The shift register stage used in the horizontal and vertical scanner of the silicon retina implementation is in correspondence to the S-SEED technology in the hybrid free-space digital photonics implementation of ANNs. Though a young technology, the architecture of planar two-dimensional processing arrays of S-SEEDs offers a large-scale connectivity that cannot be achieved in the more mature electronics technology of hardwired VLSI circuitry ([21]).

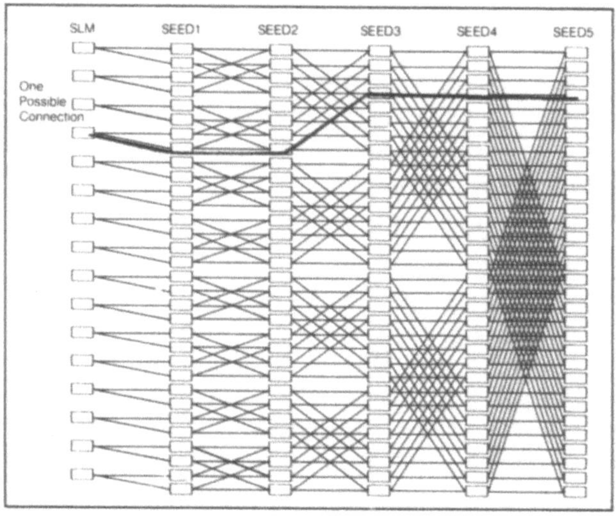

Figure 3. An extended generalised shuffle network.

At the time of writing, S-SEED arrays are at the point of early commercial availability. Figure 3 displays S-SEED arrays stacked in planar layers. The strengths of the S-SEED technology include the ability to provide lower signal energy per interconnection, lower onchip power dissipation per connection, lower crosstalk between different connections, lower interconnection skew, high connection density, and massive parallelism. This is quite different from conventional integrated-system silicon chips. Because the dividing line between photonics and electronics becomes indistinct, the SEED technology poses challenges as well as opportunities.

An analog parallel model of computation is especially interesting from the point of view of the present understanding of the biophysics of neurones, membranes and synapses.

Tomaso Poggio (1985)

§6. Conclusions

The ability of the visual system to recognize objects in spite of the enormous number of possible shape variations cannot be explained in terms of template matching or object-filtering; the number of templates or filters required would be too large. Adopting a more general point of view, biological organization processes are far too complex to hope that a relatively complete understanding of how perceptual systems like the retina and the visual cortex function will soon emerge. But the basic principles of organization of ANNs can be understood without one's knowing in detail how the components actually work. Furthermore, the same principles can be used to implement ANNs in any of several different technologies. Among these technologies, free-space photonics has many attractive features, such as the large bandwidths of optical signals and the high connectivity of holographic image processing. The use of theoretical neural networks that embody biologically-motivated rules and constraints is a powerful tool in the development and study of synthetic perceptual systems that demonstrate high data capacity, operate in a massively parallel way, and require no explicit programming. The resulting process is non-local with respect to spatial distance, resistant to ablation, and adaptable, which are the characteristics that suggest the neural holographic model.

Acknowledgments. The author is grateful to the neurophysiologist and neurosurgeon Professor Karl H. Pribram, M.D., Eminent Scholar of the Commonwealth of Virginia, for providing encouragement as well as vision. Technical discussions with him at his Laboratory of Brain Research and Informational Sciences were of enormous help from the neurobiology side. Moreover, the author wishes to thank the neurobiologists Professor Walter J. Freeman, University of California at Berkeley, Professor E. Roy John, New York University Medical Center, and Professor Hillel Pratt, Technion - Israel Institute of Technology, for their valuable comments from the experimentalists' point of view. Finally, the author would like to thank Professor Dietrich Braess, Ruhr-Universität Bochum, for painstaking efforts to improve the readability of this paper.

References

1. Allen, T., C. Mead, F. Faggin, and G. Gribble, Orientation selective VLSI retina, in *Visual Communications and Image Processing '88*, T. Russell Hsing (ed.), Proc. SPIE 101, 1988, 1040–1046.

2. Chirovsky, L. M. F., M. W. Focht, J. M. Freund, G. D. Guth, R. E. Leibenguth, G. J. Przybylek, L. E. Smith, L. A. D'Asaro, A. L. Lentine, R. A. Novotny, and D. B. Buchholz, Large arrays of symmetric self-electro-optic effect devices, *OSA Proc. on Photonic Switching, Vol. 8*, H. S. Hinton and J. W. Goodman (eds.), 1991, 56–59.

3. Engel, A. K., P. König, A. K. Kreiter, C. M. Gray, and W. Singer, Temporal coding by coherent oscillations as a potential solution to the binding problem, in *Physiological evidence. Nonlinear Dynamics and Neuronal Networks*, H. G. Schuster (ed.), VCH Publishers, Weinheim, New York, Basel, Cambridge, 1991, 3–25.

4. Feinstein, D. I., The hexagonal resistive network and the circular approximation, Caltech Computer Science Technical Report Caltech-CS-TR-88-7. Computer Science Department, California Institute of Technology, 1988.

5. Eckhorn, R., P. Dicke, W. Kruse, and H. J. Reitboeck, Stimulus-related facilitation and synchronization among visual cortical areas, in *Experiments and models. Nonlinear Dynamics and Neuronal Networks*, H. G. Schuster (ed.), VCH Publishers, Weinheim, New York, Basel, Cambridge, 1991, 57–75.

6. Eckhorn, R., H. J. Reitboeck, M. Arndt, and P. Dicke, Feature linking via stimulus - Evoked oscillations: Experimental results from cat visual cortex and functional implications from a network model, in *International Joint Conference on Neural Networks, Vol. I*, IEEE/INNS Washington D.C., 1989, 723–730.

7. Freeman, W. J., What are the state variables for modeling brain dynamics with neural networks?, in *Nonlinear Dynamics and Neuronal Networks*, H. G. Schuster (ed.), VCH Publishers, Weinheim, New York, Basel, Cambridge, 1991, 243–255.

8. Gray, C. M., A. K. Engel, P. König, and W. Singer, Temporal properties of synchronous oscillatory neuronal interactions in cat striate cortex, in

Nonlinear Dynamics and Neuronal Networks, H. G. Schuster (ed.), VCH Publishers, Weinheim, New York, Basel, Cambridge, 1991, 27–55.

9. Grinvald, A., Real-time optical mapping of neuronal activity, Ann Rev. Neurosci. **8** (1985), 263–305.

10. Grossberg, S., and D. Somers, Synchronized oscillations during cooperative feature linking in a cortical model of visual perception, Neural Networks **4** (1991), 453–466.

11. Harris, J., C. Koch, J. Luo, and J. Wyatt, Resistive fuses: Analog hardware for detecting discontinuities in early vision, in *Analog VLSI Implementation of Neural Systems*, C. Mead and M. Ismail (eds.), Kluwer Academic Publishers, Boston, Dordrecht, London, 1989.

12. Mead, C. A., and M. A. Mahowald, A silicon model of early visual processing, Neural Networks **1** (1988), 91–97.

13. Mead, C., *Analog VLSI and Neural Systems*, Addison-Wesley, Reading, MA, 1989.

14. Mead, C., Adaptive retina, in *Analog VLSI Implementation of Neural Systems*, C. Mead and M. Ismail (eds.), Kluwer Academic Publishers, Boston, Dordrecht, London, 1989, 239–246.

15. Miller, D. A. B., Quantum-well self-electro-optic effect devices, Optical and Quantum Electronics **22** (1990), S61–S98.

16. Pribram, K. H., Prolegomenon for a holonomic brain theory, in *Synergetics of Cognition*, H. Haken and M. Stadler (eds.), Springer-Verlag, Berlin, Heidelberg, New York, 1990, 150–184.

17. Pribram, K. H., *Brain and Perception: Holonomy and Structure in Figural Processing*, Lawrence Erlbaum, Publishers, Hillsdale, NJ, 1991.

18. Pribram, K. H., and W. Schempp, The dendritic membrane as a linear metaplectic manifold, to appear.

19. Reitboeck, H. J., R. Eckhorn, M. Arndt, and P. Dicke, A model for feature linking via correlated neural activity, in *Synergetics of Cognition*, H. Haken and M. Stadler (eds.), Springer-Verlag, Berlin, Heidelberg, New York, 1990, 112–125.

20. Schempp, W., *Harmonic analysis on the Heisenberg nilpotent Lie group, with applications to signal theory*, Pitman Research Notes in Math., Vol. 147, Longman Scientific and Technical, Harlow, Essex, 1986.

21. Schempp, W., Bohr's indeterminacy principle in quantum holography, self-adaptive neural networks, cortical self-organization, molecular computers, magnetic resonance imaging, and solitonic nanotechnology, in *Nonlinear Image Processing III*, E. R. Dougherty, J. Astola, and C. G. Boncelet, Jr., (eds.), Proc. SPIE 1658, 1992, 297–343.

22. Schillen, T. B., and P. König, Temporal coding by coherent oscillation as a potential solution to the binding problem, in *Nonlinear Dynamics and Neuronal Networks*, H. G. Schuster (ed.), VCH Publishers, Weinheim, New York, Basel, Cambridge, 1991, 153–171.
23. Singer, W., Search for coherence: A basic principle of cortical self-organization, Concepts Neurosci. **1** (1990), 1–26.
24. Singer, W., C. Gray, A. Engel, P. König, A. Artola, and S. Bröcher, Formation of cortical cell assemblies, in *Cold Spring Harbor Symposium on Quantitative Biology, Vol. 55*, 1990, 939–952.
25. Sompolinsky, H., D. Golomb, and D. Kleinfeld, Phase coherence and computation in a neural network of coupled oscillators, in *Nonlinear Dynamics and Neuronal Networks*, H. G. Schuster (ed.), VCH Publishers, Weinheim, New York, Basel, Cambridge, 1991, 113-130.
26. Sporns, O., G. Tononi, and G. M. Edelmann, Dynamic interactions of neuronal groups and the problem of cortical integration, in *Nonlinear Dynamics and Neuronal Networks*, H. G. Schuster (ed.), VCH Publishers, Weinheim, New York, Basel, Cambridge, 1991, 205–240.

Walter Schempp
Lehrstuhl für Mathematik I
Hölderlinstrasse 3
D-5900 Siegen
GERMANY
schempp@hrz.uni-siegen.dbp.de

Numerical Methods of Approximation Theory, Vol. 9
Dietrich Braess and Larry L. Schumaker (eds.), pp. 301–316.
International Series of Numerical Mathematics, Vol. 105
Copyright © 1992 by Birkhäuser Verlag, Basel
ISBN 3-7643-2746-4.

Converse Theorems for Approximation
on Discrete Sets II

G. Schmeisser

Dedicated to the memory of Lothar Collatz

Abstract. Converse theorems describe regularity properties of a function in terms of the speed of convergence of its best approximations on an interval. Here we consider approximation in the L^p norm ($p \in [1, +\infty)$) by polynomials, trigonometric polynomials and entire functions of exponential type. We extend some of the known results to approximation on a sequence of discrete sets instead of an interval. The main tools are upper estimates of an L^p norm in terms of an l^p norm for the three classes of approximating functions in consideration.

§1. Introduction

Denote by $E_n^*(f)_p$ the error of best approximation of a 2π-periodic function f by trigonometric polynomials of degree at most n in the L^p norm on $[0, 2\pi]$. As usual, we write $\omega(f, h)_{L^p[0,2\pi]} := \sup_{|t| \le h} \| f(\cdot + t) - f(\cdot) \|_{L^p[0,2\pi]}$ for the modulus of continuity in L^p. A fascinating consequence of results of Jackson and Bernstein (and analogues due to Akhiezer, E. Quade, Stechkin, A.F. & M.F. Timan and others, see [7, Sec. 5.1.4, 5.5.5, 6.1.4]) is the following

Theorem A. *Let f be a 2π-periodic function belonging to $L^p[0, 2\pi]$, where $p \in [1, +\infty]$; let $k \in \mathbb{N}_0$ and $\alpha \in (0, 1)$. Then*

$$E_n^*(f)_p = \mathcal{O}\left(n^{-k-\alpha}\right) \quad \text{as} \quad n \to \infty, \tag{1}$$

if and only if f has an absolutely continuous derivative of order $k - 1$ such that $f^{(k)} \in L^p[0, 2\pi]$ and $\omega(f^{(k)}, h)_{L^p[0,2\pi]} = \mathcal{O}(h^\alpha)$ as $h \to 0$.

Trigonometric polynomials are entire functions of exponential type (see [2] for more information about this class). If $A_\sigma(f)_p$ denotes the error of best approximation of a function $f \in L^p(\mathbb{R})$ by entire functions of exponential type σ in the L^p norm on $\mathbb{R}$, then the following perfect analogue (and extension in the case $p = +\infty$) of Theorem A holds.

Theorem B [7, Sec. 5.1.4, 6.1.6]. *Let $f \in L^p(\mathbb{R})$, where $p \in [1, +\infty]$; let $k \in \mathbb{N}_0$ and $\alpha \in (0, 1)$. Then*

$$A_\sigma(f)_p = \mathcal{O}(\sigma^{-k-\alpha}) \quad \text{as} \quad \sigma \to \infty, \tag{2}$$

if and only if f has an absolutely continuous derivative of order $k - 1$ such that $f^{(k)} \in L^p(\mathbb{R})$ and $\omega(f^{(k)}, h)_{L^p(\mathbb{R})} = \mathcal{O}(h^\alpha)$ as $h \to 0$.

In the important case of approximation by polynomials, there is no naive analogue of Theorems A and B. The various attempts to find the right setting for polynomials could become an exciting topic in the history of approximation theory. One remarkable approach is due to Ditzian and Totik. Introducing $\varphi(x) := \sqrt{1 - x^2}$, they designed a modified central difference on $[-1, 1]$ by defining

$$\Delta_{h\varphi(x)} f(x) := \Delta_{h\varphi(x)}^1 f(x) := f\left(x + \tfrac{1}{2}h\varphi(x)\right) - f\left(x - \tfrac{1}{2}h\varphi(x)\right)$$

and as the jth power

$$\Delta_{h\varphi(x)}^j f(x) := \sum_{i=0}^{j} (-1)^i \binom{j}{i} f\left(x + \tfrac{1}{2}(j - 2i)h\varphi(x)\right)$$

with the agreement that $\Delta_{h\varphi(x)}^j f(x) = 0$ if $x \pm \tfrac{1}{2}jh\varphi(x) \notin [-1, 1]$. Denoting by $E_n(f)_p$ the error of best approximation of a function $f \in L^p[-1, 1]$ by polynomials of degree at most n in the L^p norm on $[-1, 1]$, they obtained

Theorem C [3, Corollary 7.2.5]. *Let* $f \in L^p[-1,1]$, *where* $p \in [1,+\infty]$; *let* $k \in \mathbb{N}_0$ *and* $\alpha \in (0,1)$. *Then*

$$E_n(f)_p = \mathcal{O}(n^{-k-\alpha}) \quad \text{as} \quad n \to \infty, \tag{3}$$

if and only if $\left\| \Delta_{h\varphi}^{k+1} f \right\|_{L^p[-1,1]} = \mathcal{O}(h^{k+\alpha})$ *as* $h \to 0$, *where* $\varphi(x) := \sqrt{1-x^2}$.

By computational methods the best approximation can be achieved on a discrete set of points only. We may therefore ask for sequences of discrete sets on which a prescribed rate of convergence of the form (1), (2), or (3) implies already the same regularity properties of f as in the above theorems. Such point sets may be considered as rich enough for replacing the original approximation problem by a discrete one.

Discrete versions of the converse parts of Theorems A–C have been established in [6] for $p = +\infty$. The purpose of this paper is to treat the case $p \in [1,+\infty)$.

§2. Statement of Results

We start with approximation by entire functions of exponential type, since this is in some sense the most general case under consideration, and provides access to the trigonometric and polynomial case.

Definition 1. *For positive real numbers* δ, L *and* τ, *we denote by* $\Lambda(\tau) := \Lambda(\delta, L, \tau)$ *a set* $\{x_\nu \in \mathbb{R} : \nu \in \mathbb{Z}\}$ *of points such that*

$$x_{\nu+1} - x_\nu \geq 2\pi\frac{\delta}{\tau} \quad \text{and} \quad \left| x_\nu - \frac{\pi\nu}{\tau} \right| \leq \pi\frac{L}{\tau}.$$

For fixed δ *and* L, *we write* $\Lambda_\mu(\tau) := \Lambda(\delta, L, \tau 2^\mu)$ *if we have a sequence of such sets with the property that*

$$\Lambda(\delta, L, \tau 2^\mu) \subset \Lambda(\delta, L, \tau 2^{\mu+1})$$

for $\mu \in \mathbb{N}_0$.

An example is given by the sets $\Lambda_\mu(\tau) := \{\frac{\pi\nu}{\tau 2^\mu} : \nu \in \mathbb{Z}\}$ with $\delta \in (0, \frac{1}{2}]$ and $L > 0$. Concerning Theorem B, we are now ready to state

Theorem 1. *Let $f \in C(\mathbb{R}) \cap L^p(\mathbb{R})$, where $p \in [1, +\infty)$; let $k \in \mathbb{N}_0$, $\alpha \in (0,1)$ and $\sigma > 0$. Suppose there exists a sequence of entire functions $g_{\sigma 2^\mu}$ of exponential type $\sigma 2^\mu$ such that*

$$\frac{1}{\tau 2^{\mu+1}} \sum_{x \in \Lambda_{\mu+1}(\tau)} |f(x) - g_{\sigma 2^\mu}(x)|^p \le c(\sigma 2^\mu)^{-(k+\alpha)p}, \quad \mu = 0, 1, \dots \quad (4)$$

with constants $c > 0$ and $\tau > \sigma$. Then f has an absolutely continuous derivative of order $k - 1$ such that $f^{(k)} \in L^p(\mathbb{R})$ and $\omega(f^{(k)}, h)_{L^p(\mathbb{R})} = \mathcal{O}(h^\alpha)$ as $h \to 0$.

Now we turn to trigonometric approximation.

Definition 2. *For positive real numbers δ, L and $n \in \mathbb{N}$, we denote by $\Omega(n) := \Omega(\delta, L, n)$ a set*

$$\{x_\nu \in [-\pi, \pi] : \nu = -n+1, -n+2, \dots, n\}$$

of $2n$ points such that

$$x_{\nu+1} - x_\nu \ge 2\pi \frac{\delta}{n} \quad \text{and} \quad \left| x_\nu - \frac{\pi\nu}{n} \right| \le \pi \frac{L}{n}$$

for $\nu = -n+1, -n+2, \dots, n$, where $x_{n+1} := x_{-n+1} + 2\pi$. For fixed δ and L, we write $\Omega_\mu(n) := \Omega(\delta, L, n2^\mu)$ if we have a sequence of such sets with the property that

$$\Omega(\delta, L, n2^\mu) \subset \Omega(\delta, L, n2^{\mu+1})$$

for $\mu \in \mathbb{N}_0$.

An example is given by the sets

$$\Omega_\mu(n) := \left\{ \frac{\pi\nu}{n2^\mu} : \nu = -n2^\mu + 1, -n2^\mu + 2, \dots, n2^\mu \right\}$$

with $\delta \in (0, \frac{1}{2}]$ and $L > 0$. Concerning Theorem A, we have

Theorem 2. *Let $f \in C(\mathbb{R})$ be a 2π-periodic function; let $k \in \mathbb{N}_0$, $\alpha \in (0,1)$ and $n \in \mathbb{N}$. Suppose there exists a sequence of trigonometric polynomials t_{n2^μ} of degree at most $n2^\mu$ such that for some $p \in [1, +\infty)$ we have*

$$\frac{1}{n2^{\mu+1}} \sum_{x \in \Omega_{\mu+1}(n+1)} |f(x) - t_{n2^\mu}(x)|^p \le c(n2^\mu)^{-(k+\alpha)p}, \quad \mu = 0, 1, \dots$$

with a constant $c > 0$. Then f has an absolutely continuous derivative of order $k - 1$ such that $f^{(k)} \in L^p[0, 2\pi]$ and $\omega(f^{(k)}, h)_{L^p[0,2\pi]} = \mathcal{O}(h^\alpha)$ as $h \to 0$.

Finally, we turn to approximation by polynomials.

Definition 3. *For positive real numbers δ, L and an integer $n \geq 2$, we denote by $\Xi(n) := \Xi(\delta, L, n)$ a set of $n-1$ points $x_\nu = \cos\theta_\nu$ $(\nu = 1, 2, ..., n-1)$ such that $0 < \theta_1 < ... < \theta_{n-1} < \pi$,*

$$\theta_{\nu+1} - \theta_\nu \geq 2\pi \frac{\delta}{n-1} \quad \text{and} \quad \left| \theta_\nu - \frac{\pi\nu}{n-1} \right| \leq \frac{\pi L}{n-1}$$

for $\nu = 0, 1, ..., n-1$, where $\theta_0 := -\theta_1$ and $\theta_n := 2\pi - \theta_{n-1}$. For fixed δ and L, we write $\Xi_\mu(n) := \Xi(\delta, L, n2^\mu)$ if we have a sequence of such sets with the property that

$$\Xi(\delta, L, n2^\mu) \subset \Xi(\delta, L, n2^{\mu+1})$$

for $\mu \in \mathbb{N}_0$.

Obviously, for $\delta \in (0, 1/2]$ and $L > 0$ the zeros of the Chebyshev polynomials of the second kind $U_{n2^\mu - 1}(x)$ form such sets $\Xi_\mu(n)$. Concerning Theorem C, we state

Theorem 3. *Let $f \in C[-1, 1]$; let $k \in \mathbb{N}_0$, $\alpha \in (0, 1)$ and $n \in \mathbb{N}$. Suppose there exists a sequence of polynomials P_{n2^μ} of degree at most $n2^\mu$ such that for some $p \in [1, +\infty)$ we have*

$$\frac{1}{n2^{\mu+1}} \sum_{x \in \Xi_{\mu+1}(n+2)} \varphi(x) \, | \, f(x) - P_{n2^\mu}(x) \, |^p \leq c \, (n2^\mu)^{-(k+\alpha)p}, \qquad \mu = 0, 1, \ldots$$

with a constant $c > 0$. Then $\left\| \Delta_{h\varphi}^{k+1} f \right\|_{L^p[-1,1]} = \mathcal{O}(h^{k+\alpha})$ as $h \to 0$, where $\varphi(x) = \sqrt{1 - x^2}$.

For the sharpness or possible improvements of these theorems, we refer to the discussion in the section "Sharpness of the Results" in [6].

§3. Lemmas

The crucial tools in our proofs are upper estimates of L^p norms in terms of l^p norms. Such estimates exist for entire functions of exponential type. They allow us to deduce analogous estimates for trigonometric and algebraic polynomials, as we shall do below for $p \in [1, +\infty)$. The case of polynomials and $p = +\infty$ has been settled by Erdős [4]. For $p \in [1, +\infty)$ and trigonometric polynomials, inequalities of the desired form, but with l^p norms on discrete sets of *equidistant* points, have been obtained by Marcinkiewicz & Zygmund [5].

In what follows, we shall denote by K_1, K_2, K_3 and c_1, c_2, etc. appropriate positive constants which may depend on δ, L, λ and p, but do *not* depend on x, θ, σ, m, n, and the functions in consideration.

Lemma 1. *Let g be an entire function of exponential type $\sigma > 0$; let $\lambda > 1$ and $p \in [1, +\infty)$. Suppose that $\sum_{x \in \Lambda(\lambda\sigma)} |g(x)|^p$ converges (cf. Definition 1). Then $g \in L^p(\mathbb{R})$ and*

$$\int_{-\infty}^{+\infty} |g(x)|^p \, dx \leq K_1 \frac{1}{\sigma} \sum_{x \in \Lambda(\lambda\sigma)} |g(x)|^p .$$

Proof: This result is obtained by applying a theorem of Boas [1] to $f(z) := g(\frac{\pi}{\lambda\sigma} z + x_0)$, choosing $\Phi(x) \equiv x^p$ and $\beta(x) \equiv x$. ∎

Lemma 2. *Let t be a trigonometric polynomial of degree at most n and let m be an integer with $m \geq \lambda n$, where $\lambda > 1$. For a set $\Omega(m)$ as specified in Definition 2 with $L \geq 1$ and*

$$-\pi \leq x_{-m+1} < x_{-m+2} < \cdots < x_m \leq \pi,$$

define $x_{\mu+2\nu m} := x_\mu + 2\nu\pi$ for $\nu \in \mathbb{Z}$ and $\mu = -m+1, -m+2, \ldots, m$. Then

$$\max_{|x| \leq 4\pi L/m} |t(x + x_j)| \leq c_1 \sum_{\mu=-m+1}^{m} \beta_\mu |t(x_{j+\mu})|, \tag{5}$$

where

$$\beta_\mu := \begin{cases} 1 & \text{if } |\mu| \leq 4L, \\ |\mu|^{-3} & \text{otherwise.} \end{cases} \tag{6}$$

Proof: It is shown in [1, p. 195–196] that for an entire function f of exponential type $\tau < \pi$ and points $\lambda_\mu \in \mathbb{R}$ ($\mu \in \mathbb{Z}$) with $\lambda_0 = 0$, $\lambda_{\mu+1} - \lambda_\mu \geq 2\delta > 0$ and $|\lambda_\mu - \mu| \leq L$ there exist constants N and b_μ ($\mu \in \mathbb{Z}$), depending only on L and δ, such that

$$\max_{|x| \leq 2L} |f(x + \lambda_j)| \leq N \sum_{\mu=-\infty}^{+\infty} b_\mu |f(\lambda_{j+\mu})|. \tag{7}$$

Furthermore, if $L > 1$, then

$$0 \leq b_\mu \leq \begin{cases} (1 + |\mu| + 2L)^{8L+1}(|\mu| - 2L)^{-8L-4} & \text{if } |\mu| > 2L, \\ (1 + 4L)^{8L+1}\delta^{-8L-4} & \text{if } |\mu| \leq 2L. \end{cases}$$

A trigonometric polynomial of degree at most n is an entire function of exponential type n. We may therefore apply the just described result to $f(z) := t(\frac{\pi}{m} z + x_0)$ with $\lambda_\mu = \frac{m}{\pi}(x_\mu - x_0)$ and L replaced by $2L$. Because of the periodicity of t, the series on the right-hand side of (7) can be reduced to the finite sum in (5), writing

$$N \sum_{\nu=-\infty}^{+\infty} b_{\mu+2m\nu} =: c_1 \beta_\mu \quad \text{for} \quad \mu = -m+1, -m+2, \ldots, m.$$

It is not difficult to see that this is possible with the β_μ's defined in (6), if we choose c_1 appropriately (depending on L and δ). ∎

In the last two lemmas, we denote by Z a set of points

$$-\pi = \xi_0 < \xi_1 < \cdots < \xi_\ell = \pi$$

and by $S(f, Z)$ the Riemann sum

$$S(f, Z) := \sum_{\nu=1}^{\ell} f(\xi_\nu)(\xi_\nu - \xi_{\nu-1}). \tag{8}$$

Lemma 3. *Let t be a trigonometric polynomial of degree at most n; let $\lambda > 1$, $m \in \mathbb{N}$, $m \geq \lambda(n+1)$ and $p \in [1, +\infty)$. Then, for every set $\Omega(m)$ as specified in Definition 2 and every Riemann sum (8) such that $\Omega(m) \subset Z$, we have*

$$S(|t|^p, Z) \leq K_2 \frac{1}{m} \sum_{x \in \Omega(m)} |t(x)|^p. \tag{9}$$

In particular,

$$\int_{-\pi}^{\pi} |t(x)|^p \, dx \leq K_2 \frac{1}{m} \sum_{x \in \Omega(m)} |t(x)|^p. \tag{10}$$

Proof: We may assume that $L \geq 1$, since according to Definition 2 every set $\Omega(m) := \Omega(\delta, L, m)$ may also be conceived as $\Omega(\delta, L', m)$ if $L' \geq L$. For the points x_j of $\Omega(m)$, we readily find that

$$|x_{-m+1} + \pi| \leq 2\pi \frac{L}{m}, \qquad |x_m - \pi| \leq \pi \frac{L}{m} \tag{11}$$

and

$$|x_{j+1} - x_j| \le \left| x_{j+1} - \frac{(j+1)\pi}{m} \right| + \left| x_j - \frac{j\pi}{m} \right| + \frac{\pi}{m} \le 3\pi\frac{L}{m}. \tag{12}$$

Since by assumption $\Omega(m) \subset Z$, the Riemann sum (8) for $f(x) :\equiv |t(x)|^p$ on $[-\pi, \pi]$ may be split into $2m+1$ sections $S_{-m}, S_{-m+1}, \ldots, S_{m-1}, S_m$, which are themselves Riemann sums on the subintervals

$$[-\pi, x_{-m+1}], [x_{-m+1}, x_{-m+2}], \ldots \ldots, [x_{m-1}, x_m], [x_m, \pi],$$

respectively, such that $S(|t|^p, Z) = \sum_{j=-m}^{m} S_j$. Of course, $S_{-m} = 0$ if $x_{-m+1} = -\pi$ and $S_m = 0$ if $x_m = \pi$. Using (12), we may estimate

$$S_j \le |x_{j+1} - x_j| \max_{x_j \le x \le x_{j+1}} |t(x)|^p \le 3\pi\frac{L}{m} \max_{0 \le x \le 3\pi L/m} |t(x + x_j)|^p$$

for $j = -m + 1, \ldots, m - 1$; further, taking into account the periodicity of t, we find with the help of (11) that

$$S_{-m} + S_m \le 3\pi\frac{L}{m} \max_{0 \le x \le 3\pi L/m} |t(x + x_m)|^p.$$

Hence

$$S(|t|^p, Z) \le 3\pi\frac{L}{m} \sum_{j=-m+1}^{m} \max_{|x| \le 3\pi L/m} |t(x + x_j)|^p. \tag{13}$$

Now, we use Lemma 2, raise both sides of (5) to the power p, write the sum on the right-hand side as a convex combination and utilize the convexity of the function $x \mapsto |x|^p$. Introducing $c_2 := \sum_{\mu=-m+1}^{m} \beta_\mu$, we obtain this way

$$\max_{|x| \le 3\pi L/m} |t(x + x_j)|^p \le (c_1 c_2)^p \left(\sum_{\mu=-m+1}^{m} \frac{\beta_\mu}{c_2} |t(x_{j+\mu})| \right)^p$$

$$\le (c_1 c_2)^p \sum_{\mu=-m+1}^{m} \frac{\beta_\mu}{c_2} |t(x_{j+\mu})|^p.$$

Substituting this bound in (13) and exchanging the order of summation, we get

$$S(|t|^p, Z) \le 3\pi\frac{L}{m} c_1^p c_2^{p-1} \sum_{\mu=-m+1}^{m} \beta_\mu \sum_{j=-m+1}^{m} |t(x_{j+\mu})|^p.$$

Finally, because of the periodicity of t and the definition of $x_{j+\mu}$ for $j + \mu \notin \{-m+1, -m+2, \ldots, m\}$ given in Lemma 2, we find that $\sum_{j=-m+1}^{m} |t(x_{j+\mu})|^p$ does not depend on μ. Thus,

$$S(|t|^p, Z) \leq 3\pi \frac{L}{m} (c_1 c_2)^p \sum_{j=-m+1}^{m} |t(x_j)|^p,$$

which is (9) with $K_2 := 3\pi L (c_1 c_2)^p$. Inequality (10) is an immediate consequence of (9). $\blacksquare$

Lemma 4. *Let Q be a polynomial of degree at most n; let $\lambda > 1$, $m \in \mathbb{N}$, $m \geq \lambda n$ and $p \in [1, +\infty)$. Then, for every set $\Xi(m+1)$ with associated angles $\theta_\mu \in (0, \pi)$ as specified in Definition 3 and every Riemann sum (8) such that $\pm \theta_\mu \in Z$ $(\mu = 1, \ldots, m)$, we have*

$$\frac{1}{2} S\Big((\varphi |Q|^p) \circ \cos(.), Z\Big) \leq K_3 \frac{1}{m} \sum_{x \in \Xi(m+1)} \varphi(x) |Q(x)|^p, \qquad (14)$$

where $\varphi(x) := \sqrt{1 - x^2}$. In particular,

$$\int_{-1}^{1} |Q(x)|^p \, dx \leq K_3 \frac{1}{m} \sum_{x \in \Xi(m+1)} \sqrt{1 - x^2} \, |Q(x)|^p. \qquad (15)$$

Proof: First, we briefly indicate a proof of (14) in the case that m is restricted from above by a constant c. By Lagrange interpolation with nodes from $\Xi(m+1)$, the polynomial Q may be represented as

$$Q(x) = \sum_{\mu=1}^{m} L_\mu(x) Q(\cos \theta_\mu).$$

Since $\theta_{\mu+1} - \theta_\mu \geq 2\pi \delta / m \geq 2\pi \delta / c$, the fundamental polynomials $L_\mu(x)$ are bounded by a constant c_3 so that we may write

$$|Q(x)| \leq m c_3 \sum_{\mu=1}^{m} \frac{1}{m} |Q(\cos \theta_\mu)| \quad \text{for} \quad x \in [-1, 1].$$

Raising both sides of this inequality to the power p and using a convexity argument, we obtain

$$|Q(x)|^p \leq (m c_3)^p \sum_{\mu=1}^{m} \frac{1}{m} |Q(\cos \theta_\mu)|^p \quad \text{for} \quad x \in [-1, 1].$$

Further, $|\sin\theta_\mu| \geq \min\{|\sin\theta_1|, |\sin\theta_m|\} \geq \sin(\pi\delta/m) \geq \sin(\pi\delta/c) =: c_4$; therefore,

$$\max_{-1\leq x\leq 1} \sqrt{1-x^2}\,|Q(x)|^p \leq \frac{(cc_3)^p}{c_4}\frac{1}{m}\sum_{\mu=1}^{m} |\sin\theta_\mu|\,|Q(\cos\theta_\mu)|^p,$$

which yields (14).

For "large" m, we want to imitate the proof of Lemma 3. Defining $\theta_{-\mu+1} := -\theta_\mu$ for $\mu = 1, 2, \ldots, m$ and noting that

$$\left|\theta_{-\mu+1} - \frac{(-\mu+1)\pi}{m}\right| \leq \left|-\theta_\mu + \frac{\mu\pi}{m}\right| + \frac{\pi}{m} \leq \frac{\pi}{m}(L+1),$$

the points $\theta_{-m+1}, \theta_{-m+2}, \ldots, \theta_m$ form a set $\Omega(m) = \Omega(\delta, L', m)$ as specified in Definition 2 with $L' := L+1$. Now the proof can be carried out completely analogously to the previous one, if we succeed in showing that

$$\max_{|\theta|\leq 3\pi L'/m} |\sin(\theta+\theta_j)|\,|Q(\cos(\theta+\theta_j))|^p$$
$$\leq c_5 \sum_{\mu=-m+1}^{m} \beta_\mu |\sin\theta_{j+\mu}|\,|Q(\cos\theta_{j+\mu})|^p, \tag{16}$$

where the β_μ's, defined by (6), are with respect to L' and $\theta_{\mu+2\nu m} := \theta_\mu + 2\nu\pi$ for $\nu \in \mathbb{Z}$, $\mu = -m+1, -m+2, \ldots, m$.

In view of the above observation, we may assume that

$$\frac{m}{2} > \left[\frac{(3\pi+2)L'+3}{2}\right] =: j_0, \tag{17}$$

where $[\,.\,]$ produces the integer part. We shall distinguish two cases.

Case 1: $(0 \leq |j| < j_0$ or $m - j_0 < |j| \leq m)$. Applying Lemma 2 to the trigonometric polynomial $t(\theta) := Q(\cos\theta)$, raising both sides of the corresponding inequality (5) to the power p and utilizing the convexity of the function $x \mapsto |x|^p$, we obtain

$$\max_{|\theta|\leq 3\pi L'/m} |Q(\cos(\theta+\theta_j))|^p \leq c_6 \sum_{\mu=-m+1}^{m} \beta_\mu |Q(\cos\theta_{j+\mu})|^p. \tag{18}$$

Since for the j's of Case 1,

$$\max_{|\theta|\leq 3\pi L'/m} |\sin(\theta+\theta_j)| \leq \frac{c_7}{m},$$

and as a consequence of $\theta_\mu \in \Omega(\delta, L', m)$ for $\mu = -m+1, -m+2, \ldots, m$,

$$|\sin\theta_{j+\mu}| \geq \sin(\pi\delta/m) \geq \frac{c_8}{m}\,,$$

we see that (16) follows readily from (18).

Case 2: $(j_0 \leq |j| \leq m - j_0)$ Applying Lemma 2 to the trigonometric polynomial $t(\theta) := (\sin\theta)Q(\cos\theta)$, which is of degree $n+1$, we obtain

$$\max_{|\theta|\leq 3\pi L'/m} |t(\theta + \theta_j)| \leq c_9 \sum_{\mu=-m+1}^{m} \beta_\mu |\sin\theta_{j+\mu}|\,|Q(\cos\theta_{j+\mu})| \tag{19}$$

$$= c_9 b_j \sum_{\mu=-m+1}^{m} \gamma_{j\mu} |Q(\cos\theta_{j+\mu})|\,,$$

where

$$b_j := \sum_{\mu=-m+1}^{m} \beta_\mu |\sin\theta_{j+\mu}| \quad \text{and} \quad \gamma_{j\mu} := \frac{\beta_\mu |\sin\theta_{j+\mu}|}{b_j}. \tag{20}$$

Since the last sum on the right-hand side of (19) is a convex combination, we conclude that

$$\max_{|\theta|\leq 3\pi L'/m} |t(\theta + \theta_j)|^p \leq (c_9 b_j)^p \sum_{\mu=-m+1}^{m} \gamma_{j\mu} |Q(\cos\theta_{j+\mu})|^p\,.$$

Consequently, defining

$$B_j := b_j / \min_{|\theta|\leq 3\pi L'/m} |\sin(\theta + \theta_j)|\,, \tag{21}$$

and using (20), we get

$$\max_{|\theta|\leq 3\pi L'/m} |\sin(\theta + \theta_j)|\,|Q(\cos(\theta + \theta_j))|^p \leq$$

$$c_9^p B_j^{p-1} \sum_{\mu=-m+1}^{m} \beta_\mu |\sin\theta_{j+\mu}|\,|Q(\cos\theta_{j+\mu})|^p\,.$$

In order to complete the proof of (16), it remains to show that, for the j's of Case 2, the B_j's remain bounded by a constant, say c_{10}.

First, since

$$|\theta_{j+\mu} - \theta_j| \le \left|\theta_{j+\mu} - \frac{(j+\mu)\pi}{m}\right| + \left|\theta_j - \frac{j\pi}{m}\right| + \left|\frac{\mu\pi}{m}\right| \le (2L' + |\mu|)\frac{\pi}{m},$$

we have

$$|\sin\theta_{j+\mu}| = |\sin\theta_j \cos(\theta_{j+\mu} - \theta_j) + \cos\theta_j \sin(\theta_{j+\mu} - \theta_j)|$$
$$\le |\sin\theta_j| + |\sin(\theta_{j+\mu} - \theta_j)| \le |\sin\theta_j| + (2L' + |\mu|)\frac{\pi}{m}.$$

Hence, taking into account (6), the quantities b_j from (20) may be estimated as

$$|b_j| \le |\sin\theta_j| \sum_{\mu=-m+1}^{m} \beta_\mu + \frac{\pi}{m} \sum_{\mu=-m+1}^{m} (2L' + |\mu|)\beta_\mu \tag{22}$$
$$\le c_{11}|\sin\theta_j| + \frac{c_{12}}{m}.$$

Secondly, for the j's of Case 2, we have

$$|\sin\theta_j| \ge \min\{|\sin\theta_{j_0}|, |\sin\theta_{m-j_0}|\} \ge \frac{2}{\pi}\min\{|\theta_{j_0}|, |\pi - \theta_{m-j_0}|\},$$

where the inequality $\sin\theta \ge 2\theta/\pi$ for $\theta \in [0, \pi/2]$ has been used in the last step. Since each θ_j may be written as $\theta_j = \frac{\pi}{m}(j + \eta_j L')$ with an appropriate $\eta_j \in [-1, 1]$, we conclude, using (17), that

$$|\sin\theta_j| \ge \frac{2}{m}(j_0 - L') \ge \frac{1}{m}(3\pi L' + 1). \tag{23}$$

Next, by the mean-value theorem

$$\sin(\theta_j + \theta) = \sin\theta_j + \theta \cos(\theta_j + \eta\theta) \quad \text{with} \quad \eta \in [0, 1].$$

Hence, together with (23), we get

$$\min_{|\theta|\le 3\pi L'/m} |\sin(\theta_j + \theta)| \ge |\sin\theta_j| - 3\pi\frac{L'}{m} \ge \frac{1}{m}. \tag{24}$$

Finally, going back to (21), the inequalities (22) and (24) yield

$$B_j \le \frac{m|\sin\theta_j|c_{11} + c_{12}}{m|\sin\theta_j| - 3\pi L'}.$$

The right-hand side becomes largest when $m \,|\sin\theta_j\,|$ attains its smallest value. Thus, by (23), we obtain $B_j \leq (3\pi L' + 1)c_{11} + c_{12}$. This completes the proof of (16) and so (14) is settled.

Since

$$\frac{1}{2}\int_{-\pi}^{\pi} \sqrt{1 - \cos^2\theta}\,|Q(\cos\theta)|^p\,d\theta = \int_{-1}^{1}|Q(x)|^p\,dx,$$

(15) is an obvious consequence of (14). ∎

§4. Proofs of the Theorems

We now denote by γ_1, γ_2, etc. appropriate positive constants which do *not* depend on i, j and μ.

Proof of Theorem 1: By (4) and the property $\Lambda_\mu(\tau) \subset \Lambda_{\mu+1}(\tau)$, we have

$$\frac{1}{\tau 2^\mu}\sum_{x\in\Lambda_\mu(\tau)}|f(x) - g_{\sigma 2^\mu}(x)|^p \leq \frac{1}{\tau 2^\mu}\sum_{x\in\Lambda_{\mu+1}(\tau)}|f(x) - g_{\sigma 2^\mu}(x)|^p$$

$$\leq 2c\,(\sigma 2^\mu)^{-(k+\alpha)p}.$$

Hence, defining

$$G_\mu(x) := g_{\sigma 2^\mu}(x) - g_{\sigma 2^{\mu-1}}(x),$$

we find, using Minkowski's inequality, that

$$\left(\frac{1}{\tau 2^\mu}\sum_{x\in\Lambda_\mu(\tau)}|G_\mu(x)|^p\right)^{1/p} \leq \left(\frac{1}{\tau 2^\mu}\sum_{x\in\Lambda_\mu(\tau)}|f(x) - g_{\sigma 2^\mu}(x)|^p\right)^{1/p} +$$

$$\left(\frac{1}{\tau 2^\mu}\sum_{x\in\Lambda_\mu(\tau)}|f(x) - g_{\sigma 2^{\mu-1}}(x)|^p\right)^{1/p} \leq \gamma_1\,(\sigma 2^\mu)^{-k-\alpha}.$$

Since G_μ is an entire function of exponential type $\sigma 2^\mu$, we obtain with the help of Lemma 1 that $G_\mu \in L^p(\mathbb{R})$ and

$$\|G_\mu\|_p := \left(\int_{-\infty}^{+\infty}|G_\mu(x)|^p\,dx\right)^{1/p} \leq \gamma_2\,(\sigma 2^\mu)^{-k-\alpha}.$$

Now, employing a theorem of Plancherel & Pólya [2, Theorem 6.7.15], we find that for every $j \in \mathbb{N}$ with $j \geq \mu$

$$\frac{1}{\tau 2^j} \sum_{x \in \Lambda_j(\tau)} |G_\mu(x)|^p \leq \frac{2}{\pi^2 \delta} e^{\pi \delta p} \|G_\mu\|_p^p \leq \gamma_3 \left(\sigma 2^\mu\right)^{-(k+\alpha)p}.$$

Thus, for any two integers $j > i > 0$, repeated application of Minkowski's inequality yields

$$\left(\frac{1}{\tau 2^j} \sum_{x \in \Lambda_j(\tau)} |f(x) - g_{\sigma 2^i}(x)|^p\right)^{1/p} \leq \sum_{\mu=i+1}^{j} \left(\frac{1}{\tau 2^j} \sum_{x \in \Lambda_j(\tau)} |G_\mu(x)|^p\right)^{1/p} +$$

$$\left(\frac{1}{\tau 2^j} \sum_{x \in \Lambda_j(\tau)} |f(x) - g_{\sigma 2^j}(x)|^p\right)^{1/p} \leq \gamma_4 \left(\sigma 2^i\right)^{-k-\alpha}. \tag{25}$$

Denoting the points of $\Lambda_j(\tau)$ by $x_{j\nu}$ $(\nu \in \mathbb{Z})$, arranged in increasing order, we know (cf. Definition 1) that

$$\frac{2\pi\delta}{\tau 2^j} \leq x_{j,\nu+1} - x_{j\nu} \leq \frac{(2L+1)\pi}{\tau 2^j}.$$

Hence, if $[a, b]$ is any subinterval of $\mathbb{R}$, we get from (25)

$$\left(\sum_{x_{j\nu} \in \Lambda_j(\tau) \cap [a,b]} |f(x_{j\nu}) - g_{\sigma 2^i}(x_{j\nu})|^p (x_{j,\nu+1} - x_{j\nu})\right)^{1/p} \leq$$

$$\left(\frac{(2L+1)\pi}{\tau 2^j} \sum_{x \in \Lambda_j(\tau)} |f(x) - g_{\sigma 2^i}(x)|^p\right)^{1/p} \leq \gamma_5 \left(\sigma 2^i\right)^{-k-\alpha}.$$

Since $f - g_{\sigma 2^i}$ is a continuous function, the left-hand side converges to

$$\left(\int_a^b |f(x) - g_{\sigma 2^i}(x)|^p \, dx\right)^{1/p}$$

as $j \to \infty$, while the right-hand side remains bounded and does not depend on a and b. Consequently, $f - g_{\sigma 2^i} \in L^p(\mathbb{R})$ for every $i \in \mathbb{N}$ and

$$\|f - g_{\sigma 2^i}\|_p \leq \gamma_5 \left(\sigma 2^i\right)^{-k-\alpha}.$$

A fortiori,

$$A_s(f)_p \leq A_{\sigma 2^i}(f)_p \leq \gamma_5 \left(\sigma 2^i\right)^{-k-\alpha} \leq \gamma_6 s^{-k-\alpha}$$

for $s \in [\sigma 2^i, \sigma 2^{i+1})$ and $i \in \mathbb{N}$. Thus (2) holds, and the proof is completed by applying the converse part of Theorem B. ∎

Proof of Theorems 2 and 3: We proceed as in the previous proof but use Lemmas 3 and 4, respectively. Only at the place where we employed the theorem of Plancherel & Pólya we argue differently. It may be enough to give some indications in case of Theorem 2.

Defining $T_\mu(x) := t_{n2^\mu}(x) - t_{n2^{\mu-1}}(x)$, it is not difficult to see (cf. Definition 2) that for $j \geq \mu$

$$\frac{1}{(n+1)2^j} \sum_{x \in \Omega_j(n+1)} |T_\mu(x)|^p \leq \gamma_7 \, S\left(|T_\mu|^p, \Omega_j(n+1)\right).$$

Further, from (9) we get

$$S\left(|T_\mu|^p, \Omega_j(n+1)\right) \leq K_2 \frac{1}{(n+1)2^\mu} \sum_{x \in \Omega_\mu(n+1)} |T_\mu(x)|^p.$$

Consequently,

$$\frac{1}{(n+1)2^j} \sum_{x \in \Omega_j(n+1)} |T_\mu(x)|^p \leq \gamma_7 K_2 \frac{1}{(n+1)2^\mu} \sum_{x \in \Omega_\mu(n+1)} |T_\mu(x)|^p$$

for all $j \geq \mu$. This enables us to establish an inequality analogous to (25). ∎

References

1. Boas, R.P., Integrability along a line for a class of entire functions, Trans. Amer. Math. Soc. **73** (1952), 191–197.
2. Boas, R.P., *Entire Functions*, Academic Press, New York, 1954.
3. Ditzian, Z., and V. Totik, *Moduli of Smoothness*, Springer, New York, 1987.
4. Erdös, P., On the boundedness and unboundedness of polynomials, J. d'Analyse Math. **19** (1967), 135–148.
5. Marcinkiewicz, J., and A. Zygmund, Mean values of trigonometric polynomials, Fund. Math. **28** (1937), 131–166.

6. Schmeisser, G., Converse theorems for approximation on discrete sets, in *Approximation, Optimization and Computing*, A.G. Law and C.L. Wang (eds.), North-Holland, Amsterdam, 1990, 159–162.
7. Timan, A.F., *Theory of Approximation of Functions of a Real Variable*, Pergamon Press, Oxford, 1963.

Mathematisches Inst.
Univ. Erlangen-Nürnberg
Bismarckstrasse $1\frac{1}{2}$
D-85 Erlangen, GERMANY
mi @ cnve.rrze.uni-erlangen.dbp.de

Numerical Methods of Approximation Theory, Vol. 9
Dietrich Braess and Larry L. Schumaker (eds.), pp. 317–329.
International Series of Numerical Mathematics, Vol. 105
Copyright © 1992 by Birkhäuser Verlag, Basel
ISBN 3-7643-2746-4.

A Dual Method for Smoothing Histograms
Using Nonnegative C^1-Splines

Jochen W. Schmidt

Dedicated to the memory of Lothar Collatz

Abstract. For smoothing histograms under positivity constraints, we use the objective function K_2 proposed in [6]. The feasible functions are assumed to be quadratic C^1-splines. This leads to finite dimensional programming problems with a partially separable structure which can be solved efficiently via dualization.

§1. Introduction

Let $F = \{f_1, \ldots, f_n\}$ be a histogram given on a mesh $\Delta = \{x_0 < x_1 < \cdots < x_n\}$, and let $h_i = x_i - x_{i-1}$, $i = 1(1)n$ be the corresponding step sizes. For smoothing F under monotonicity or convexity constraints, it is recommended in [6] to minimize the functional

$$K_2(s) = \ell \int_{x_o}^{x_n} s'(x)^2 dx + \sum_{i=1}^{n} p_i (f_i - \frac{1}{h_i} \int_{x_{i-1}}^{x_i} s(x)dx)^2, \qquad (1.1)$$

where the feasible functions s are supposed to be quadratic C^1-splines, e.g. on the mesh Δ. The global weight $\ell > 0$ and the local weights $p_1 > 0, \ldots, p_n > 0$ are assumed to be given.

In [6] this approach is shown to lead to so-called *partially separable optimization problems* which can be solved effectively by dual algorithms. In this paper the approach will be continued by treating positivity constraints which are also of importance in applications. If we discretize the positivity constraint by a linear but only sufficient condition, we are in a position to develop the dual algorithm up to the same level as was done in [6] for monotonicity and convexity.

Besides well-known applications of constrained histogram smoothing in statistics, here we refer to the following model occurring in the one-dimensional motion of a material point. Find the velocity $s = s(x)$ as a function of the time x such that the point is at the position g_i at the time x_i, $i = 0(1)n$. Because of

$$\int_{x_{i-1}}^{x_i} s(x)dx = g_i - g_{i-1}, \quad i = 1(1)n,$$

in F we now have to set

$$f_i = (g_i - g_{i-1})/h_i, \quad i = 1(1)n. \tag{1.2}$$

We are led to the same model in controlling the flow of a production. Here g_i denotes the required output at the time x_i, $i = 0(1)n$. Let us remark that in most problems of this type, the velocity has to be nonnegative.

Earlier papers in unconstrained area matching are [10,11], while for area matching under constraints we refer to the recent papers [4,5,8,9], as well as to the book [12]. As mentioned above, monotone and convex smoothing of histograms are considered in [6]. Positive volume matching problems are treated, e.g., in [1,3,13].

§2. Discretization of Positive Smoothing Problems

First we recall well-known spline properties which we need below. For $x \in [x_{i-1}, x_i]$, a quadratic spline s on the mesh Δ can be defined by

$$s(x) = y_{i-1} + h_i m_{i-1} t + (y_i - y_{i-1} - h_i m_{i-1})t^2, \quad 0 \le t \le 1, \tag{2.1}$$

with the local variable $t = (x - x_{i-1})/h_i$, $i = 1(1)n$. We have $s \in C^1[x_0, x_n]$ if and only if

$$h_i(m_{i-1} + m_i) = 2(y_i - y_{i-1}), \quad i = 1(1)(n-1), \tag{2.2}$$

and for the parameters in (2.1) it follows
$$y_i = s(x_i), \quad m_i = s'(x_i), \quad i = 0(1)n. \tag{2.3}$$
In view of the Bernstein-Bézier representation
$$s(x) = y_{i-1}(1 - t)^2 + (2y_{i-1} + h_i m_{i-1})t(1 - t) + y_i t^2, \quad 0 \le t \le 1,$$
a sufficient condition for the positivity of s on $[x_0, x_n]$ reads
$$y_i \ge 0, \quad i = 0(1)n, \quad 2y_{i-1} + h_i m_{i-1} \ge 0, \quad i = 1(1)n \tag{2.4}$$
and it can be shown that (2.4) is the sharpest affine linear positivity condition not enlarging the number of inequalities.

We give two further examples. The quadratic spline s is positive as well as monotone increasing if and only if
$$y_0 \ge 0, \quad m_i \ge 0, \quad i = 0(1)n, \tag{2.5}$$
while the inequalities
$$y_0 \ge 0, \quad m_0 \ge 0, \quad m_{i-1} \le m_i, \quad i = 1(1)n \tag{2.6}$$
are equivalent for s to be positive, monotone, and convex.

Further, the functional K_2 defined by (1.1) is computed to be
$$K_2(s) = \sum_{i=1}^{n} F_i(y_{i-1}, m_{i-1}, y_i, m_i) \tag{2.7}$$
with
$$F_i(f, x, g, y) = \frac{\ell h_i}{3}(x^2 + xy + y^2) + \frac{p_i}{36}(6f_i - 4f - 2g - h_i x)^2. \tag{2.8}$$
Thus, the problem of positive histogram smoothing with quadratic C^1-splines leads to the partially separable program

$$\text{minimize} \quad \Big\{ \sum_{i=1}^{n} F_i(y_{i-1}, m_{i-1}, y_i, m_i) : \ (y_{i-1}, m_{i-1}, y_i, m_i) \in W_i, \tag{2.9}$$
$$i = 1(1)n, \quad y_0 = \alpha \Big\},$$
where, in addition to (2.8), we have to set
$$W_i = \{(f, x, g, y) : \ h_i(x + y) = 2(g - f), \ g \ge 0, \ 2f + h_i x \ge 0\}. \tag{2.10}$$
The constant α must be nonnegative.

Notice that the feasible domain in program (2.9) is nonempty; e.g. the vector with the components $y_i = \alpha$, $m_i = 0$, $i = 0(1)n$ satisfies the constraints. The objective function is quadratic, and the Hessian is positive definite on the (convex) feasible domain. Thus, we obtain the following proposition:

Proposition 1. *The program (2.9) with (2.8), (2.10) arising in positive histogram smoothing is uniquely solvable.*

Analogously, if the smoothing spline is required to be positive and monotone we get the program

$$\text{minimize} \quad \Big\{ \sum_{i=1}^{n} F_i(y_{i-1}, m_{i-1}, y_i, m_i) : \ (y_{i-1}, m_{i-1}, y_i, m_i) \in W_i,$$

$$i = 1(1)n, \quad y_0 = \alpha, \ m_0 = \beta \big\}. \tag{2.11}$$

Since F_i is defined by (2.8), we now have to set

$$W_i = \{(f, x, g, y) : \ h_i(x + y) = 2(g - f), \ y \ge 0\}, \tag{2.12}$$

and the constants α and β have to be nonnegative.

Since there exist feasible vectors, e.g. the vector with the components $y_0 = \alpha$, $m_0 = \beta$, $y_i = \alpha + (h_1 + \ldots + h_i)\beta$, $m_i = \beta$, $i = 1(1)n$, *program (2.11) with (2.8), (2.12) turns out to be uniquely solvable.*

Smoothing under positivity, monotonicity, and convexity again yields the same program (2.11), with the exception that now

$$W_i = \{(f, x, g, y) : \ h_i(x + y) = 2(g - f), \ x \le y\}. \tag{2.13}$$

This program (2.11) with (2.8),(2.13) is uniquely solvable; the vector from above is seen to be feasible for this program as well.

§3. Dualization of the Programs for Positive Histogram Smoothing

For solving the programs (2.9) and (2.11) numerically, their very special structure can be taken into account via dualization; see e.g. the earlier papers [2,7]. If $F_1, \ldots, F_n$ are convex functions and $W_1, \ldots, W_n$ are closed convex sets, a program dual to (2.9) reads

$$\text{maximize} \quad \Big\{ -\sum_{i=1}^{n} H_i^*(y_{i-1}^*, m_{i-1}^*, -y_i^*, -m_i^*) + \alpha y_0^* :$$

$$m_0^* = y_n^* = m_n^* = 0 \big\}. \tag{3.1}$$

Here H_i^* denotes the Fenchel conjugate to F_i and W_i defined by

$$H_i^*(\rho, \xi, \sigma, \eta) = \sup\{\rho f + \xi x + \sigma g + \eta y - F_i(f, x, g, y) : (f, x, g, y) \in W_i\}. \tag{3.2}$$

Analogously, a dual program to (2.11) is

$$\text{maximize} \quad \Big\{ -\sum_{i=1}^{n} H_i^*(y_{i-1}^*, m_{i-1}^*, -y_i^*, -m_i^*) + \alpha y_0^* + \beta m_0^* :$$

$$y_n^* = m_n^* = 0\}. \tag{3.3}$$

From [7] we recall the following duality statements.

Theorem 2. *In the primal programs (2.9) and (2.11), let $F_1, \ldots, F_n$ be convex functions and $W_1, \ldots, W_n$ be closed polyhedral sets. Assume that the programs (3.2) for computing the conjugate functions have unique maximizers for any $(\rho, \xi, \sigma, \eta) \in \mathbb{R}^4$, $i = 1(1)n$. If the primal program is feasible, then both the primal program and the corresponding dual one are solvable, and the optimal values are equal. Further, if a solution $(y_0^*, m_0^*, \ldots, y_n^*, m_n^*)$ of the dual program is known, then the unique solution of the primal one is explicitly given by the formula*

$$(y_{i-1}, m_{i-1}, y_i, m_i)^T = \text{grad } H_i^*(y_{i-1}^*, m_{i-1}^*, -y_i^*, -m_i^*), \qquad i = 1(1)n. \tag{3.4}$$

In the present cases (2.8), (2.10) as well as (2.8), (2.12) and (2.8), (2.13), the conjugates (3.2) are everywhere finite. This is of importance because then the dual programs become unconstrained. They are much easier to handle than the constrained primal programs. Thus, for solving the histogram problems in question, we can again recommend the following general procedure:

Step 1: Test whether the feasible domain of the primal program is nonempty. In the present cases this holds true.

Step 2: Compute a solution of the dual program numerically, e.g. by Newton's method combined with the steepest descent method for supplying sufficiently good initial vectors.

Step 3: Determine the solution of the primal program by means of the formula (3.4).

Further, we have the advantage that the Fenchel conjugates, of interest here, are explicitly computed.

Proposition 3. *The conjugate to (2.8), (2.10) is*

$$H_i^*(\rho, \xi, \sigma, \eta) = \begin{cases} E_i^* & \text{for } A_i \geq 0, \ B_i \geq 0 \\[2mm] E_i^* - \dfrac{3\ell p_i}{3\ell + h_i p_i} A_i^2 & \text{for } A_i \leq 0, \ C_i \geq 0 \\[2mm] E_i^* - \dfrac{3\ell p_i}{3\ell + h_i p_i} B_i^2 & \text{for } B_i \leq 0, \ D_i \geq 0 \\[2mm] F_i^* & \text{for } C_i \leq 0, \ D_i \leq 0, \end{cases} \tag{3.5}$$

where

$$E_i^* = E_i^*(\rho, \xi, \sigma, \eta) =$$
$$f_i(\rho + \sigma) + \frac{h_i}{12\ell}(\rho^2 - \rho\sigma + \sigma^2) + \frac{(\rho + \sigma)^2}{4p_i} + \frac{\eta\sigma - \xi\rho}{2\ell} + \frac{\xi^2 - \xi\eta + \eta^2}{\ell h_i},$$

$$F_i^* = F_i^*(\rho, \xi, \sigma, \eta) =$$
$$\frac{1}{12\ell + h_i p_i}\left(12\ell p_i f_i^2 + 3h_i p_i f_i(\rho - \frac{2\xi}{h_i}) + \frac{9h_i}{4}(\rho - \frac{2\xi}{h_i})^2\right),$$

$$A_i = A_i(\rho, \xi, \sigma, \eta) = f_i + \frac{h_i}{12\ell}(2\sigma - \rho) + \frac{\rho + \sigma}{2p_i} + \frac{\eta}{2\ell},$$

$$B_i = B_i(\rho, \xi, \sigma, \eta) = f_i - \frac{h_i}{12\ell}(\rho + \sigma) + \frac{\rho + \sigma}{2p_i} + \frac{\xi - \eta}{2\ell},$$

$$C_i = C_i(\rho, \xi, \sigma, \eta) = f_i - \frac{h_i \rho}{12\ell} + \frac{\rho}{2p_i} + \frac{2\xi - \eta}{6\ell} + \frac{\xi - 2\eta}{h_i p_i},$$

$$D_i = D_i(\rho, \xi, \sigma, \eta) = f_i + \frac{h_i(\sigma - \rho)}{12\ell} + \frac{\rho + 2\sigma}{2p_i} + \frac{\xi + \eta}{6\ell} + \frac{2\eta - \xi}{h_i p_i}.$$

Proposition 4. *The conjugate to (2.8), (2.12) is*

$$H_i^*(\rho, \xi, \sigma, \eta) = E_i^* - \frac{\ell h_i}{4}\left(-\frac{\sigma}{2\ell} + \frac{\xi - 2\eta}{\ell h_i}\right)_+^2. \tag{3.6}$$

Proposition 5. *The conjugate to (2.8), (2.13) is*

$$H_i^*(\rho, \xi, \sigma, \eta) = E_i^* - \frac{\ell h_i}{12}\left(-\frac{\rho + \sigma}{2\ell} + \frac{3(\xi - \eta)}{\ell h_i}\right)_+^2. \tag{3.7}$$

§4. Proof of Proposition 3

For the proofs of Propositions 4 and 5, we can refer to [6]. We now give details of the proof of Proposition 3. The Lagrangian to program (3.2) with F_i and W_i according to (2.8) and (2.10) reads

$$
\begin{aligned}
\phi(f, x, g, y, \lambda, \mu, \nu) = {} & -\rho f - \xi x - \sigma g - \eta y + \frac{\ell h_i}{3}(x^2 + xy + y^2) \\
& + \frac{p_i}{36}(6f_i - 4f - 2g - h_i x)^2 + \lambda(-2f - h_i x) + \mu(-g) \\
& + \nu(h_i(x + y) - 2(g - f)),
\end{aligned}
\tag{4.1}
$$

and the (necessary and sufficient) Kuhn-Tucker conditions are

$$
\begin{aligned}
& \phi_f = \phi_g = \phi_x = \phi_y = 0, \\
& \phi_\lambda \le 0, \ \lambda \ge 0, \ \lambda\phi_\lambda = 0, \ \phi_\mu \le 0, \ \mu \ge 0, \ \mu\phi_\mu = 0, \ \phi_\nu = 0.
\end{aligned}
\tag{4.2}
$$

From $\phi_f = 0, \phi_g = 0$ it follows that

$$
6\nu = \rho - 2\sigma + 2\lambda - 2\mu,
\tag{4.3}
$$

$\phi_x = 0, \phi_f = 0$ implies

$$
-\xi + \frac{h_i \rho}{4} + \frac{\ell h_i}{3}(2x + y) + \frac{h_i}{2}(\nu - \lambda) = 0,
\tag{4.4}
$$

and $\phi_y = 0$ implies

$$
-\eta + \frac{\ell h_i}{3}(x + 2y) + h_i \nu = 0.
\tag{4.5}
$$

Substituting ν from (4.3) and solving the linear system of equations (4.4), (4.5), we get

$$
x = -\frac{\rho}{2\ell} + \frac{2\xi - \eta}{\ell h_i} + \frac{\lambda}{\ell},
\tag{4.6}
$$

$$
y = \frac{\sigma}{2\ell} + \frac{2\eta - \xi}{\ell h_i} - \frac{\lambda}{\ell} + \frac{\mu}{2\ell}.
\tag{4.7}
$$

Next, using (4.6), (4.7) we obtain from $\phi_\nu = 0$

$$
g - f = \frac{h_i}{4\ell}(\sigma - \rho) + \frac{\xi + \eta}{2\ell} + \frac{h_i \mu}{4\ell},
\tag{4.8}
$$

and from $\phi_g = 0$ we get

$$2f + g = 3f_i + \frac{h_i\rho}{4\ell} + \frac{3}{2p_i}(\rho + \sigma) + \frac{\eta - 2\xi}{2\ell} - \frac{h_i\lambda}{2\ell} + \frac{3}{2p_i}(2\lambda + \mu). \tag{4.9}$$

This linear system of equations (4.8), (4.9) yields

$$f = f_i + \frac{h_i}{12\ell}(2\rho - \sigma) + \frac{\rho + \sigma}{2p_i} - \frac{\xi}{2\ell} - \frac{h_i}{12\ell}(2\lambda + \mu) + \frac{2\lambda + \mu}{2p_i}, \tag{4.10}$$

$$\begin{aligned} g &= f_i + \frac{h_i}{12\ell}(2\sigma - \rho) + \frac{\rho + \sigma}{2p_i} + \frac{\eta}{2\ell} + \frac{h_i}{6\ell}(\mu - \lambda) + \frac{2\lambda + \mu}{2p_i} \\ &= A_i(\rho, \xi, \sigma, \eta) + \frac{h_i}{6\ell}(\mu - \lambda) + \frac{2\lambda + \mu}{2p_i}. \end{aligned} \tag{4.11}$$

Furthermore, (4.6) and (4.10) imply

$$f + \frac{h_i x}{2} = B_i(\rho, \xi, \sigma, \eta) + \frac{h_i}{12\ell}(4\lambda - \mu) + \frac{2\lambda + \mu}{2p_i}. \tag{4.12}$$

In view of the complementarity conditions in (4.2), four different cases need to be considered.

Case 1: Let $\lambda = 0, \mu = 0$. Because of (4.11)–(4.12), using $\phi_\lambda \leq 0$ and $\phi_\mu \leq 0$, the domain is given by $A_i \geq 0, B_i \geq 0$. The maximizer *(f,x,g,y)* given by (4.6), (4.7), (4.10) and (4.11) with $\lambda = 0$ and $\mu = 0$ is unique. If the property of the maximizer

$$(f, x, g, y)^T = \text{grad} H_i^*(\rho, \xi, \sigma, \eta) \tag{4.13}$$

is taken into account, the first part of formula (3.5) is immediately verified.

Case 2: Let $\lambda = 0, \phi_\mu = 0$. Now (4.11) leads to

$$\mu = -\frac{6\ell p_i A_i}{3\ell + h_i p_i}, \tag{4.14}$$

and using (4.12), in view of $\mu \geq 0, \phi_\lambda \leq 0$, the domain is straightforwardly found to be defined by $A_i \leq 0, C_i \geq 0$. Again, the maximizer is unique and is given by (4.6), (4.7), (4.10) and (4.11) with $\lambda = 0$ and μ according to (4.14). By means of (4.13), after some computation, the second part of formula (3.5) is seen to be valid.

Case 3: Let $\phi_\lambda = 0, \mu = 0$. From (4.12) we now get

$$\lambda = -\frac{3\ell p_i B_i}{3\ell + h_i p_i}, \tag{4.15}$$

and for the remainder of this case we can proceed as in Case 2.

Case 4: Let $\phi_\lambda = 0, \phi_\mu = 0$. The linear system resulting from (4.11), (4.12) is solved by

$$\lambda = -\frac{6\ell p_i}{12\ell + h_i p_i}\left(\frac{2\ell}{h_i p_i}(B_i - A_i) + \frac{A_i + 2B_i}{3}\right) = \frac{-6\ell p_i C_i}{12\ell + h_i p_i}, \tag{4.16}$$

$$\mu = -\frac{12\ell p_i}{12\ell + h_i p_i}\left(\frac{2\ell}{h_i p_i}(A_i - B_i) + \frac{2A_i + B_i}{3}\right) = \frac{-12\ell p_i D_i}{12\ell + h_i p_i}. \tag{4.17}$$

Because of $\lambda \geq 0, \mu \geq 0$, the domain is described by $C_i \leq 0, D_i \leq 0$. Further, $g = 0$ and $2f + h_i x = 0$ imply $y = 0$, and we get the one-dimensional program

$$\text{maximize}\quad \left\{\rho f - \frac{2\xi}{h_i}f - \frac{4\ell}{3h_i}f^2 - \frac{p_i}{36}(6f_i - 2f)^2\right\}$$

having the maximizer

$$f = \frac{1}{12\ell + h_i p_i}\left(3h_i p_i f_i + \frac{9}{2}h_i \rho - 9\xi\right).$$

Thus, the fourth part of formula (3.5) follows immediately, and the proof of Proposition 3 is complete. ∎

§5. Numerical Examples

Several numerical tests have been performed on the basis of the described results. For solving the unconstrained dual program (3.1), (3.5), Newton's method is used with the rough starting vector $y_i^* = m_i^* = 0, i = 0(1)n$. In general, we have convergence after a small number NS of steps. Of course, Newton's method without modifications may also fail to converge. For the present problems, this frequently occurs, e.g., for small values of ℓ. Better results are obtained by a combination of Newton's method with the steepest descent method.

The following figures illustrate the influence of the parameters ℓ and $p_1, \ldots, p_n$ on the approximating splines. The splines in Figure 1 result by setting

	ℓ	α	p_i	NS
spline 1	1	0.5	$p_i = 1, i = 1(1)10$	3
spline 2	1	0.5	$p_5 = p_6 = p_7 = 10$ $p_i = 1$ elsewhere	3
spline 3	1	0.5	$p_5 = p_6 = p_7 = 50$ $p_i = 1$ elsewhere	4

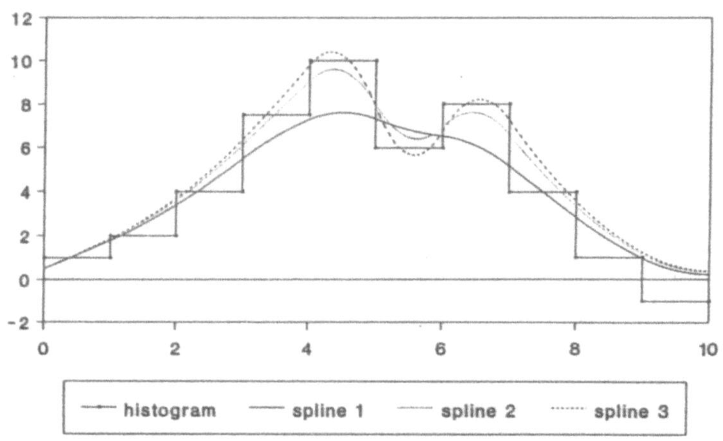

Figure 1. Approximating splines for varying the local parameters p_i.

In Figure 2 the splines are obtained by choosing

	ℓ	α	p_i	NS
spline 1	1	1	$p_i = 1, i = 1(1)10$	1
spline 2	0.01	1	$p_i = 1, i = 1(1)10$	2

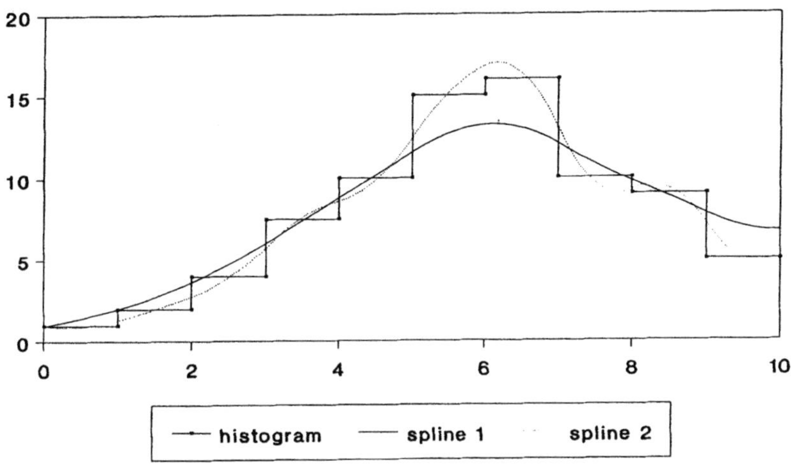

Figure 2. Approximating splines for varying the global parameter ℓ.

Finally, the splines in Figure 3 belong to the data

	ℓ	p_i	NS
spline 1	1	$p_i = 1, i = 1(1)10$	1
spline 2	1	$p_5 = p_6 = 10,$ $p_i = 1$ elsewhere	1
spline 3	1	$p_5 = p_6 = 50,$ $p_i = 1$ elsewhere	4

We remark that in most of the examples some of the positivity constraints used in program (2.9) are active. E.g., for the splines 2 and 3 in Figure 3, this happens in the subintervals in the middle.

Acknowledgement. The author is indebted to cand. math. C. Gemmer for performing the numerical tests.

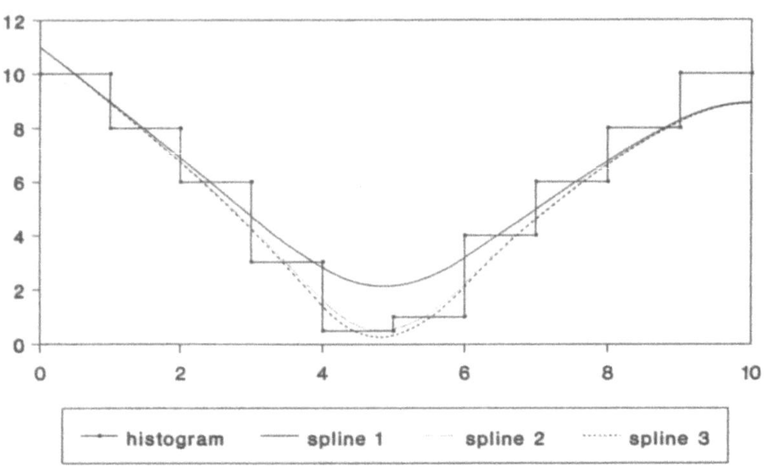

Figure 3. Approximating splines for varying the local parameters p_i.

References

1. Barrett, J.W., and R. Chakrabarti, Finite element approximation of the volume-matching problem, Numer. Math. **60** (1991), 291–313.
2. Dietze, S., and J.W. Schmidt, Determination of shape preserving spline interpolants with minimal curvature via dual programs, J. Approx. Theory **52** (1988), 43–57
3. Dyn, N., and W. H. Wong, On the characterization of non-negative volume-matching surface splines, J. Approx. Theory **51** (1987), 1–10.
4. Morandi, R., and P. Costantini, Piecewise monotone quadratic histosplines, SIAM J. Sci. Stat. Comput. **10** (1989), 397–406.
5. Sakai, M., and R. A. Usmani, A shape preserving area true approximation of histograms by rational splines, BIT **28** (1989), 329–339.
6. Schmidt, J.W., Constrained smoothing of histograms by quadratic splines, Computing **48** (1992), 97–107.
7. Schmidt, J.W., Dual algorithms for solving convex partially separable optimization problems, Jahresber. Deutsch. Math.-Verein. **94** (1992), 40–62.
8. Schmidt, J.W., and W. Heß, Shape preserving C^2-spline histopolation, Hamburger Beitr. Angew. Math., Preprint A41, 1991.

9. Schmidt, J.W., W. Heß, and T. Nordheim, Shape preserving histopolation using rational quadratic splines, Computing **44** (1990), 245–258.
10. Schoenberg, I.J., Splines and histograms, in *Spline Functions and Approx. Theory*, A. Meir and A. Sharma (eds.), Birkhäuser Verlag, Basel und Stuttgart, ISNM 21, 1973, 277–327.
11. Späth, H., Zur Glättung empirischer Häufigkeitsverteilungen, Computing **10** (1972), 353–357.
12. Späth, H., *Eindimensionale Spline–Interpolations–Algorithmen*, Oldenbourg-Verlag, München, Wien, 1990.
13. Tobler, W.R., Smooth pycnophylactic interpolation for geographical regions, J. Amer. Statist. Assoc. **74** (1979), 519–530.

Jochen W. Schmidt
Institut für Numerische Mathematik
Technische Universität
D-O-8027 Dresden, Germany
numath@ urzdfn.mathematik.tu-dresden.dbp.de

Numerical Methods of Approximation Theory, Vol. 9

Dietrich Braess and Larry L. Schumaker (eds.), pp. 331–346.

International Series of Numerical Mathematics, Vol. 105

Copyright © 1992 by Birkhäuser Verlag, Basel

ISBN 3-7643-2746-4.

Segment Approximation Using Linear Functionals

Manfred Sommer

*Dedicated to the memory of Lothar Collatz
and to Günter Meinardus on his 65th birthday*

Abstract. Properties of linear functionals which annihilate finite dimensional Haar or weak Chebyshev subspaces of $C[a, b]$ are studied. It is shown that such functionals can be used to compute optimal sets of knots for piecewise approximation of certain functions in $C[a, b]$ with respect to best L_∞- and L_1-approximation.

§0. Introduction

In Meinardus, Nürnberger, Sommer and Strauss [4], a special functional is applied to simplify the computation of a set of approximate knots for best uniform piecewise polynomial approximation of continuous functions. The numerical results show that, at least for an important subclass of continuous functions, such knot sets are nearly optimal. Hence, starting with these knots, the algorithm developed in Nürnberger, Sommer and Strauss [5] rapidly converges to an optimal set of knots.

In this paper we extend this method by defining certain classes of functionals which annihilate finite dimensional Haar or weak Chebyshev subspaces of $C[a, b]$. We show that they can be applied to approximately compute best uniform piecewise polynomial approximations and best piecewise L_1-approximations, respectively.

§1. L_∞-Approximation

Given $n \in \mathbb{N}$ and a set $T = \{\tilde{t}_0, \ldots, \tilde{t}_n\}$ such that

$$-1 = \tilde{t}_0 < \tilde{t}_1 < \cdots < \tilde{t}_n = 1,$$

we transform this point set to any real interval $[a, b]$ by

$$t_i = ((b - a)\tilde{t}_i + (b + a))/2, \quad 0 \le i \le n.$$

Let U_n denote an n-dimensional Haar subspace of $C[a, b]$ and $f \in C[a, b]$. We define a linear functional

$$L(f, U_n, [a, b], T) := \sum_{i=0}^{n} \lambda_i f(t_i), \quad \lambda_i \in \mathbb{R}, \quad 0 \le i \le n, \qquad (1.1)$$

satisfying $\lambda_0 > 0$, $\sum_{i=0}^{n} |\lambda_i| = 1$ and $L(u, U_n, [a, b], T) = 0$ for every $u \in U_n$. Then the λ_i are uniquely determined and can be computed as follows: Let $U_n = \mathrm{span}\,\{u_1, \ldots, u_n\}$ such that

$$D\binom{u_1 \quad \cdots \quad u_n}{y_1 \quad \cdots \quad y_n} := \det(u_i(y_j))_{i,j=1}^{n} > 0 \quad \text{for all } a \le y_1 < \cdots < y_n \le b$$

(it is well known that this property is equivalent to the Haar property of U_n). Then setting

$$\tilde{\lambda}_i := D \begin{pmatrix} u_1 & \cdots & u_n \\ t_0 & \ldots t_{i-1} t_{i+1} \ldots & t_n \end{pmatrix}, \quad 0 \le i \le n,$$

$$\tilde{\lambda} := \sum_{i=0}^{n} \tilde{\lambda}_i,$$

we have that

$$\lambda_i := (-1)^i \tilde{\lambda}_i / \tilde{\lambda}, \quad 0 \le i \le n.$$

Moreover, it is well known that

$$|L(f, U_n, [a, b], T)| \le d(f, U_n, [a, b]) := \min_{u \in U_n} ||f - u||_{[a,b]},$$

where $|| \cdot ||_{[a,b]}$ denotes the supremum norm on $[a, b]$.

If the function u is the best uniform approximation of f from U_n on $\{t_o, \ldots t_n\}$, then there exists $\lambda \in \mathbb{R}$ such that

$$(-1)^i (f(t_i) - u(t_i)) = \lambda, \quad 0 \le i \le n.$$

Hence, it follows from (1.1) that

$$|\lambda| = |L(f, U_n, [a, b], T)|.$$

Therefore, at least for certain point sets $(t_i)_{i=0}^n$, $|L(f, U_n, [a, b], T)|$ is a good approximation of $d(f, U_n, [a, b])$.

In [4] a special functional of type (1.1) was applied to simplify the computation of best piecewise polynomial approximations with free knots. The idea was the following: Let $k \in \mathbb{N}$ and $D := \{(x, y) \in \mathbb{R}^2 : a \le x \le y \le b\}$. Define a function $d : D \to \mathbb{R}$ by

$$d \text{ is continuous,} \tag{1.2}$$

$$d(x, x) = 0 \quad \text{for every} \quad x \in [a, b], \tag{1.3}$$

$$d(x, y) \le d(\tilde{x}, \tilde{y}), \text{ if } [x, y] \subset [\tilde{x}, \tilde{y}] \subset [a, b]. \tag{1.4}$$

(For instance, the minimal deviation $d(f, U_n, [x, y])$ satisfies these properties for any $(x, y) \in D$.) A partition $a = x_o \le x_1 \le \cdots \le x_{k+1} = b$ is called a *leveled set* of knots if

$$d(x_{i-1}, x_i) = d(x_i, x_{i+1}), \quad 1 \le i \le k, \tag{1.5}$$

and it is called *optimal* if

$$\max_{0 \le i \le k} d(x_i, x_{i+1}) \le \max_{0 \le i \le k} d(y_i, y_{i+1}) \tag{1.6}$$

for every partition $a = y_o \le y_1 \le \cdots \le y_{k+1} = b$. It was proved in Lawson [2] and Meinardus [3] that for a function d satisfying (1.2)-(1.4), the following statements hold:

(i) Every leveled set of knots is optimal.

(ii) There exists a leveled set of knots.

(iii) If

$$d(x, y) < d(\tilde{x}, \tilde{y}) \quad \text{for all } [x, y] \subsetneq [\tilde{x}, \tilde{y}] \subset [a, b], \tag{1.7}$$

then there exists a unique optimal set of knots. This set contains k distinct knots.

In [4, 5] an algorithm is developed to compute a sequence of knot sets converging to a leveled (and therefore to an optimal) set w.r.t. d. This algorithm is applied to compute best uniform approximations of a function $f \in C[a, b]$ by piecewise polynomials with free knots. In phase one of this method, the function d is defined by

$$d(x, y) := |L(f, \Pi_{n-1}, [x, y], T)|, \qquad [x, y] \subset [a, b], \qquad (1.8)$$

where L denotes a functional of type (1.1) with $U_n = \Pi_{n-1}$, the polynomials of degree at most $n - 1$, and $T = \{\tilde{t}_i\}_{i=0}^n = \{\cos((n - i)\pi/n)\}_{i=0}^n$. For this simple function d, a leveled set of knots is computed. Starting with this set of knots, in phase two of the algorithm the function d is defined by

$$d(x, y) := \min_{p \in \Pi_{n-1}} \|f - p\|_{[x,y]}, \qquad [x, y] \subset [a, b],$$

and an optimal set of knots for best piecewise polynomial approximation of f is determined.

Unfortunately, the function d in (1.8) does not satisfy (1.2)-(1.4) for every $f \in C[a, b]$. Therefore in [4] a characterization of those differentiable functions is given for which the function d does satisfy (1.2)-(1.4) and (1.7).

Theorem 1.1 ([4]). *Let d be defined as in (1.8). Then for $f \in C[a, b] \cap C^n(a, b)$ the following statements are equivalent:*

$$\text{The function } d \text{ satisfies (1.2)–(1.4) and (1.7).} \qquad (1.9)$$

The function f satisfies $\sigma f^{(n)}(t) \geq 0$, $t \in (a, b)$, $\sigma \in \{-1, 1\}$,

and $f^{(n)}$ does not vanish identically on a subinterval of (a, b). $\qquad (1.10)$

Remark 1.2. To prove this theorem, it is important that for Π_{n-1} and $\tilde{t}_i = \cos((n - i)\pi/n)$, $0 \leq i \leq n$, the coefficients in (1.1) are independent of the subintervals $[x, y]$ of $[a, b]$. Actually,

$$\lambda_0 = 1/(2n), \quad \lambda_i = (-1)^i/n, \quad 1 \leq i \leq n - 1, \quad \lambda_n = (-1)^n/(2n)$$

for every $[x, y] \subset [a, b]$.

It is easily verified that for $U_n = \Pi_{n-1}$ and any points $-1 = \tilde{t}_o < \cdots < \tilde{t}_n = 1$, the coefficients in (1.1) are independent of the subintervals of $[a, b]$. But this nice property fails, in general, if we replace Π_{n-1} by any Haar space U_n.

First we give an equivalent formulation of statement (1.10).

Lemma 1.3. *(1.10) is equivalent to the following statement:*

> The space spanned by $\{u_1, \ldots, u_n, f\}$ is a Haar subspace of $C[a, b]$ where $\{u_1, \ldots, u_n\}$ denotes any basis of Π_{n-1}.
> $\qquad$ (1.11)

Proof: Let (1.10) be satisfied and assume that $f + u$ has $n + 1$ zeros in $[a, b]$ where $u \in \Pi_{n-1}$. Then by (1.10), $f + u$ does not vanish identically on a subinterval of $[a, b]$. This implies that $f' + u'$ has at least n sign changes in (a, b). Hence, $f^{(n)}$ changes sign in (a, b), a contradiction. This proves (1.11). Now assume that (1.11) is satisfied. Then it follows from Zielke [8, Theorem 11.3] that span $\{1, f^{(n-1)}\}$ is a Haar space on (a, b). Hence, $f^{(n-1)}$ is strictly monotone in (a, b), which proves (1.10). ∎

In the following, we are interested in generalizing Theorem 1.1 to certain classes of continuous functions without assuming differentiability. Since we need the λ_i in (1.1) to be independent of the subintervals of $[a, b]$, we restrict ourselves to the case when $U_n = \Pi_{n-1}$. To generalize Theorem 1.1, let

$$C(U_n) := \{f \in C[a, b] : \text{ span } \{u_1, \ldots u_n, f\}$$
$$\text{is a weak Chebyshev subspace of } C[a, b]\}, \qquad (1.12)$$

the *convexity cone* generated by the Haar space $U_n = \text{span } \{u_1, \ldots, u_n\}$, i.e., $f \in C(U_n)$ if and only if every $u \in \text{span } \{u_1, \ldots, u_n, f\}$ has at most n sign changes in (a, b). It is well known that this is equivalent to the property that for any $a \le z_o < \cdots < z_n \le b$,

$$\sigma D \begin{pmatrix} u_1 & \cdots & u_n & f \\ z_o & \cdots & & z_n \end{pmatrix} \ge 0,$$

where $\sigma \in \{-1, 1\}$ (see e.g. [8]).

Moreover, we need the subclass $\tilde{C}(U_n)$ of $C(U_n)$ defined by

$$\tilde{C}(U_n) := \{f \in C[a, b] : \text{ span } \{u_1, \ldots u_n, f\}$$
$$\text{is a Haar subspace of } C[a, b]\}. \qquad (1.13)$$

For any set T consisting of points $-1 = \tilde{t}_0 < \cdots < \tilde{t}_n = 1$, let the functional L be defined as in (1.1). For $[x, y] \subset [a, b]$ with $x < y$, let

$$d(x, y) := d_T(x, y) := |L(f, \Pi_{n-1}, [x, y], T)|. \qquad (1.14)$$

Then Theorem 1.1 can be generalized as follows.

Theorem 1.4. *For* $U_n = \Pi_{n-1}$ *and* $f \in C[a,b]$, *the following statements are equivalent:*

$$\text{The function } d_T \text{ in } (1.14) \text{ satisfies } (1.2)\text{–}(1.4) \text{ for every } T. \qquad (1.15)$$

$$f \in C(\Pi_{n-1}). \qquad (1.16)$$

Theorem 1.5. *Under the same hypotheses as in Theorem 1.4, the following statements are equivalent:*

$$\text{The function } d_T \text{ in } (1.14) \text{ satisfies } (1.2)\text{-}(1.4) \text{ and } (1.7) \text{ for every } T. \quad (1.17)$$

$$f \in \tilde{C}(\Pi_{n-1}). \qquad (1.18)$$

Proof of the Theorems. We first show that (1.16) implies (1.15) (resp. (1.18) implies (1.17)). If $n = 1$, then $\lambda_o = -\lambda_1 = 1/2$ and $d_T(x,y) = |f(x) - f(y)|/2$ for $a \leq x < y \leq b$. Since by hypothesis f is monotone (resp. strictly monotone) in $[a,b]$, statement (1.15) (resp. statement (1.17)) is easily verified. Let $n \geq 2$ and $a \leq x < y \leq b$. Moreover, let any $-1 = \tilde{t}_0 < \cdots < \tilde{t}_n = 1$ be given. To define L as in (1.1) for $[x,y]$, set $t_i = ((y-x)\tilde{t}_i + (y+x))/2$, $0 \leq i \leq n$. Define $u_i(t) := t^{i-1}$, $1 \leq i \leq n$, and let $\tilde{u} \in \Pi_{n-1}$ such that $\tilde{u}(t_i) = f(t_i)$, $0 \leq i \leq n-1$. Then we obtain

$$(f - \tilde{u})(t) = cD\begin{pmatrix} u_1 & \cdots & u_n & f \\ t_0 & \cdots & t_{n-1} & t \end{pmatrix},$$

where

$$c = 1/D\begin{pmatrix} u_1 & \cdots & u_n \\ t_0 & \cdots & t_{n-1} \end{pmatrix} > 0.$$

Since $L(\tilde{u}, \Pi_{n-1}, [x,y], T) = 0$, where $T = \{\tilde{t}_0, \ldots, \tilde{t}_n\}$, it follows that

$$L(f, \Pi_{n-1}, [x,y], T) = L(f - \tilde{u}, \Pi_{n-1}, [x,y], T)$$

$$= \sum_{i=0}^{n} \lambda_i(f(t_i) - \tilde{u}(t_i)) = \lambda_n(f(y) - \tilde{u}(y)) \qquad (1.19)$$

$$= \lambda_n\, cD\begin{pmatrix} u_1 & \cdots & u_n & f \\ t_0 & \cdots & t_{n-1} & y \end{pmatrix}.$$

(This equation holds for every $f \in C[a,b]$.) Since $f \in C(\Pi_{n-1})$, we have that

$$\sigma D\begin{pmatrix} u_1 & \cdots & u_n & f \\ t_o & \cdots & t_{n-1} & y \end{pmatrix} \geq 0$$

where $\sigma \in \{-1, 1\}$. (Strict inequality holds if $f \in \tilde{C}(\Pi_{n-1})$.) Hence, since λ_n is constant, the functional L is always nonnegative or always nonpositive for every $y \in (x, b)$.

By [8, Theorem 11.3], for every $g \in \text{span} \{u_1, \ldots, u_n, f\}$ and $t \in (a, b)$, there exists

$$D_+ g(t) := \min_{z \to t+} \frac{g(z) - g(t)}{z - t}. \tag{1.20}$$

Moreover, span $\{D_+ u_2, \ldots, D_+ u_n, D_+ f\}$ is a weak Chebyshev subspace of $C(a, b)$. Obviously, $D_+ u_i(t) = (i - 1)t^{i-2}, 2 \leq i \leq n$. (If $f \in \tilde{C}(\Pi_{n-1})$, then we even obtain a Haar subspace of $C(a, b)$.) Let x be fixed. Since $L(u, \Pi_{n-1}, [x, y], T) = 0$ for all $u \in \Pi_{n-1}$ and all $y \in (x, b)$, it follows that

$$D_+ L(u, \Pi_{n-1}, [x, t], T)|_{t=y} = 0$$

for all $u \in \Pi_{n-1}$ and all $y \in (x, b)$. Moreover, since λ_i is independent of y, $0 \leq i \leq n$, it is easily verified that

$$D_+ L(u, \Pi_{n-1}, [x, t], T)|_{t=y} = \sum_{i=1}^{n} \lambda_i(\tilde{t}_i + 1)/2 \frac{d}{dt} u(t)|_{t=t_i}$$

and

$$D_+ L(f, \Pi_{n-1}, [x, t], T)|_{t=y} = \sum_{i=1}^{n} \lambda_i(\tilde{t}_i + 1)/2 D_+ f(t)|_{t=t_i}$$

for all $y \in (x, b)$. Let $\hat{u} \in \text{span} \{D_+ u_2, \ldots, D_+ u_n\}$ satisfying

$$\hat{u}(t_i) = D_+ f(t)|_{t=t_i}, \quad 1 \leq i \leq n - 1,$$

and let $\bar{u} \in \Pi_{n-1}$ such that $D_+ \bar{u} = \hat{u}$. Then by the above arguments,

$$D_+ L(f, \Pi_{n-1}, [x, t], T)|_{t=y} = D_+ L(f - \bar{u}, \Pi_{n-1}, [x, t], T)|_{t=y}$$

$$= \sum_{i=1}^{n} \lambda_i(\tilde{t}_i + 1)/2(D_+ f(t) - D_+ \bar{u}(t))|_{t=t_i} \tag{1.21}$$

$$= \lambda_n(D_+ f(y) - \hat{u}(y)),$$

since $\tilde{t}_o = -1$ and $\tilde{t}_n = 1$. Then arguing as in (1.19), we obtain

$$D_+ L(f, \Pi_{n-1}, [x, t], T)|_{t=y} = \lambda_n \tilde{c} D\left(\begin{matrix} D_+ u_2 & \cdots & D_+ u_n & D_+ f \\ t_1 & \cdots & t_{n-1} & y \end{matrix}\right), \tag{1.22}$$

where $\tilde{c} = 1/D\left(\begin{smallmatrix} D_+u_2 & \cdots & D_+u_n \\ t_1 & \cdots & t_{n-1} \end{smallmatrix}\right) > 0$. This implies that D_+L is always nonnegative or always nonpositive for every $y \in (x, b)$. (Strict inequality holds if $f \in \tilde{C}(\Pi_{n-1})$.)

To simplify the notation, set $L(f, t) := L(f, \Pi_{n-1}, [x, t], T)$, if $t \in [x, b)$. We show that $d_T(x, y) \leq d_T(x, \tilde{y})$ (resp. $d_T(x, y) < d_T(x, \tilde{y})$, if $f \in \tilde{C}(\Pi_{n-1})$), if $a \leq x \leq y < \tilde{y} < b$. Without loss of generality we may assume that $L(f, t) \geq 0$ for every $t \in (x, b)$. Assume that for some $y, \tilde{y} \in (x, b)$, with $y < \tilde{y}$,

$$L(f, y) > L(f, \tilde{y}) \quad (\text{resp. } L(f, y) \geq L(f, \tilde{y}), \text{ if } f \in \tilde{C}(\Pi_{n-1})).$$

Let $\bar{y} \in [y, \tilde{y})$ such that $L(f, \bar{y}) \geq L(f, t)$ for every $t \in [y, \tilde{y})$. Then, since

$$D_+L(f, t) = \lim_{s \to t+} ((L(f, s) - L(f, t))/(s - t),$$

$D_+L(f, \bar{y}) \leq 0$. Hence by the above arguments, $D_+L(f, t) \leq 0$ for every $t \in (x, b)$ (resp. $D_+L(f, t) < 0$, if $f \in \tilde{C}(\Pi_{n-1})$). Since L is continuous at t, if $t \in [x, b)$, and $L(f, x) = 0$, there exists $\bar{x} \in (x, \bar{y})$ such that $0 \leq L(f, \bar{x}) < L(f, \bar{y})$. Define

$$\varphi(t) := L(f, t) - (L(f, \bar{y}) - L(f, \bar{x}))(t - \bar{x})/(\bar{y} - \bar{x}), \quad t \in [\bar{x}, \bar{y}].$$

Then $\varphi(\bar{x}) = \varphi(\bar{y}) = L(f, \bar{x})$ and $D_+\varphi(t) = D_+L(f, t) - (L(f, \bar{y}) - L(f, \bar{x}))/(\bar{y} - \bar{x}) < 0$ for every $t \in [\bar{x}, \bar{y}]$. Let $\tilde{x} \in [\bar{x}, \bar{y})$ such that $\varphi(\tilde{x}) \leq \varphi(t)$ for every $t \in [\bar{x}, \bar{y})$. It follows from $D_+\varphi(\tilde{x}) = \lim_{s \to \tilde{x}+} (\varphi(s) - \varphi(\tilde{x}))/(s - \tilde{x})$ that $D_+\varphi(\tilde{x}) \geq 0$, a contradiction. (If D_+L is continuous at t with $t \in (x, b)$, then the monotonicity of d easily follows by using the mean value theorem. But by [8, Lemma 12.2] D_+L is only continuous from the right in general.) Thus, we have shown that

$$d_T(x, y) \leq d_T(x, \tilde{y}) \quad (\text{resp. } d_T(x, y) < d_T(x, \tilde{y}),$$

$$\text{if } f \in \tilde{C}(\Pi_{n-1})), \quad \text{if } a \leq x \leq y < \tilde{y} < b.$$

Analogously if follows that

$$d_T(\tilde{x}, y) \leq d_T(x, y) \quad (\text{resp. } d_T(\tilde{x}, y) < d_T(x, y),$$

$$\text{if } f \in \tilde{C}(\Pi_{n-1})), \quad \text{if } a < x < \tilde{x} \leq y \leq b.$$

This implies (1.4), and, if $f \in \tilde{C}(\Pi_{n-1})$, also (1.7). Obviously, d_T also satisfies (1.2) and (1.3). Thus we have shown that (1.16) implies (1.15) (resp. (1.18) implies (1.17)).

Now we show that (1.17) implies (1.18). Assume that $f \notin \tilde{C}(\Pi_{n-1})$. Then there exist points $a \leq z_0 < \cdots < z_n \leq b$ such that $f(z_i) = 0$, $0 \leq i \leq n$. Let $\hat{T} = \{\hat{t}_0, \ldots, \hat{t}_n\}$ such that $z_i = \hat{t}_i(z_n - z_0)/2 + (z_n + z_0)/2$, $0 \leq i \leq n$. Then $-1 = \hat{t}_0 < \cdots < \hat{t}_n = 1$ and by (1.1),

$$L(f, \Pi_{n-1}, [z_0, z_n], \hat{T}) = 0.$$

Hence by (1.14), $d_{\hat{T}}(z_0, z_n) = 0$ which contradicts (1.7) and therefore (1.17). Thus we have shown that (1.17) implies (1.18).

Now we show that (1.15) implies (1.16). Assume that $f \notin C(\Pi_{n-1})$. Then there exist $u \in \Pi_{n-1}$ and points $a < y_0 < \cdots < y_{n+1} < b$ such that

$$(f - u)(y_i)(f - u)(y_{i+1}) < 0, \ 0 \leq i \leq n.$$

Without loss of generality, we may assume that $u \equiv 0$ and sgn $f(y_i) = (-1)^i$, $0 \leq i \leq n+1$. Let

$$z_i = \max \{x \in (y_i, y_{i+1}) : f(x) = 0\},$$

$0 \leq i \leq n$. This implies that $f(z_i) = 0$, $0 \leq i \leq n$. As above, set $\hat{T} = \{\hat{t}_0, \ldots, \hat{t}_n\}$ such that $-1 = \hat{t}_0 < \cdots < \hat{t}_n = 1$ and $z_i = \hat{t}_i(z_n - z_0)/2 + (z_n + z_0)/2$, $0 \leq i \leq n$. Since $(-1)^{i+1} f(t) > 0$ for every $t \in (z_i, y_{i+1})$, $0 \leq i \leq n$, there exists $\epsilon > 0$ such that $(-1)^{i+1} f(t) > 0$ for every $t \in (z_i, z_i+\epsilon]$, $0 \leq i \leq n$. Define

$$\hat{z}_i = \hat{t}_i(z_n - z_0 - \epsilon)/2 + (z_n + z_0 + \epsilon)/2,$$

$0 \leq i \leq n$. Then $z_o + \epsilon = \hat{z}_0 < \hat{z}_1 < \cdots < \hat{z}_n = z_n$ and $z_i < \hat{z}_i \leq z_i + \epsilon$, $0 \leq i \leq n-1$. Since $f(z_i) = 0$, $0 \leq i \leq n$, we obtain from (1.1) that

$$d_{\hat{T}}(z_0, z_n) = |L(f, \Pi_{n-1}, [z_0, z_n], \hat{T})| = 0.$$

Moreover, it follows from (1.4) that, since $[\hat{z}_0, \hat{z}_n] \subset [z_0, z_n]$,

$$d_{\hat{T}}(\hat{z}_0, \hat{z}_n) = |L(f, \Pi_{n-1}, [\hat{z}_0, \hat{z}_n], \hat{T})| = 0.$$

Then by (1.19) (which is true for every $f \in C[a, b]$),

$$D\left(\begin{matrix} u_1 & \cdots & u_n & f \\ \hat{z}_0 & \cdots & & \hat{z}_n \end{matrix}\right) = 0,$$

which implies that for some $u \in \Pi_{n-1}$,

$$u(\hat{z}_i) = f(\hat{z}_i),$$

$0 \le i \le n$. Hence $(-1)^{i+1}u(\hat{z}_i) > 0$, $0 \le i \le n-1$, and $u(\hat{z}_n) = 0$ which implies that the nonzero function u has at least n zeros, a contradiction. Thus we have shown that (1.15) implies (1.16). This completes the proof of Theorems 1.4 and 1.5. ∎

Remark 1.6. (i) The proof of (1.15) $\Rightarrow$ (1.16) (resp. (1.17) $\Rightarrow$ (1.18)) uses only the fact that U_n is a Haar subspace of $C[a,b]$. Hence these statements hold for every Haar space U_n. But the converse (1.16) $\Rightarrow$ (1.15) is not satisfied in general.

(ii) The above statements show that for every T, the function d_T defined in (1.14) has similar properties to the minimal deviation $d(f, \Pi_{n-1}, \cdot)$, if $f \in C(\Pi_{n-1})$. Hence, by the arguments in [4], the leveled sets of knots w.r.t. d_T yield nearly best uniform piecewise polynomial approximations of f, if $f \in C(\Pi_{n-1})$.

(iii) In order to get (1.21), it is important that the λ_i are independent of the subintervals $[x, y]$ of $[a, b]$. They depend only on the partition T of $[a, b]$. Hence we conjecture that if a Haar subspace U_n has this property for every T, then $U_n = \Pi_{n-1}$. In fact, this is true (see the Appendix).

§2. L_1-**Approximation.**

Now we consider segment approximation of continuous functions with respect to best L_1-approximation. To do this, assume that μ is a non-atomic positive finite measure defined on $[a, b]$ such that $\mu(A) > 0$ for every open subset A of $[a, b]$. Then

$$\|f\|_{[a,b]} := \int_a^b |f| \, d\mu, \qquad f \in C[a, b],$$

defines a norm on $C[a, b]$, the $L_1(\mu)$-*norm*. We start with a well-known result due to Hobby and Rice (for details see Pinkus [6, p. 208]).

Theorem 2.1. *Let U_n be an n-dimensional subspace of $C[a,b]$. Then there exist points $a = t_0 < \cdots < t_m < t_{m+1} = b$ such that $m \le n$, and*

$$\sum_{j=1}^{m+1} (-1)^j \int_{t_{j-1}}^{t_j} u\,d\mu = 0 \tag{2.1}$$

for all $u \in U_n$ ($\{t_1, \ldots, t_n\}$ is called a μ-canonical set for U_n).

If U_n is a weak Chebyshev subspace of $C[a, b]$, then every μ-canonical set satisfies $m = n$. If U_n is actually a Haar subspace of $C[a, b]$, then there exists a unique set of points $a = t_o < \cdots < t_n < t_{n+1} = b$ satisfying (2.1). There is a complete characterization of those n-dimensional spaces which have unique μ-canonical sets consisting of n points for all measures $d\mu = wd\lambda$ where λ denotes the Lebesgue measure and $w \in C[a, b]$ with $w > 0$ on $[a, b]$ (see Kroó, Schmidt and Sommer [1] and the references therein).

Now assume that U_n is a weak Chebyshev subspace of $C[a, b]$ such that $U_n|_{[x,y]}$ has a unique μ-canonical set on any subinterval $[x, y]$ of $[a, b]$. By a result of the author (see [6, p. 202]), $U_n|_{[x,y]}$ is a weak Chebyshev subspace of $C[x, y]$ with dimension $m \leq n$. Hence, this unique set contains precisely m points. Let $f \in C[a, b]$. We define a functional L by

$$L(f, U_n, [x, y]) := \sum_{j+1}^{m+1} (-1)^j \int_{t_{j-1}}^{t_j} f \, d\mu, \tag{2.2}$$

where $x = t_0 < \cdots < t_m < t_{m+1} = y$ denote the μ-canonical points for U_n on $[x, y]$, $[x, y] \subset [a, b]$. Then it follows from (2.1) that

$$L(u, U_n, [x, y]) = 0 \tag{2.3}$$

for all $u \in U_n$. We set

$$d(x, y) := |L(f, U_n, [x, y])|, \tag{2.4}$$

where $[x, y] \subset [a, b]$, $x \leq y$. If $\tilde{u} \in U_n$ is a best approximation of f from U_n on $[x, y]$ with respect to the $L_1(\mu)$-norm, then

$$d(x, y) = |L(f - \tilde{u}, U_n, [x, y])| \leq ||f - \tilde{u}||_{[x,y]}$$
$$= \min_{u \in U_n} ||f - u||_{[x,y]} =: d(f, U_n, [x, y]).$$

Let the subclass $C(U_n)$ of $C[a, b]$ be analogously defined as in (1.12). Then, if $f \in C(U_n)$, span $(f \cup U_n)$ is a weak Chebyshev subspace of $C[a, b]$ and by the above mentioned result in [6, p. 202], span $(f \cup U_n)|_{[x,y]}$ is a weak Chebyshev subspace of $C[x, y]$ for any $[x, y] \subset [a, b]$. Using this property and the statements in [6, p. 210] we obtain that the functional L even determines the minimal deviation from U_n for every $f \in C(U_n)$.

Theorem 2.2. ([6, p. 210]). *Let $f \in C(U_n)$. Then $d(x, y) = d(f, U_n, [x, y])$ for all $a \leq x \leq y \leq b$. Moreover, there exists a best $L_1(\mu)$-approximation $\tilde{u}$ of f from U_n on $[x, y]$ such that $\tilde{u}(t_j) = f(t_j)$, $1 \leq j \leq m$, where $\{t_1, \ldots, t_m\}$ denotes the μ-canonical set for U_n on $[x, y]$, and $m = \dim U_n|_{[x,y]}$.*

Hence the function d in (2.4) can be taken as a simple approximation of the minimal deviation. Then following the algorithm given in [4], in phase one we determine a leveled set of knots for piecewise approximation of an element f in $C[a,b]$ by U_n with respect to the function d in (2.4). Starting with this set of knots, in phase two of the algorithm we compute an optimal set of knots for best piecewise approximation of f by U_n with respect to the $L_1(\mu)$-norm. In order to obtain convergence in phase one, the function d in (2.4) must satisfy (1.2)-(1.4). Obviously, (1.3) is satisfied for every $f \in C[a,b]$. Moreover, it follows from the uniqueness of the μ-canonical set for $U_n|_{[x,y]}$ on any $[x,y] \subset [a,b]$ and the proof of (2.1) (see [6, p. 208]) that (1.2) is also satisfied for every $f \in C[a,b]$. We now give sufficient conditions for (1.4) and (1.7), respectively.

Theorem 2.3. (i) *If $f \in C(U_n)$, then d satisfies (1.4).*

(ii) *If $f \in C(U_n)$ and $\mu(Z(f - u)) = 0$ for every $u \in U_n$, where $Z(f - u) :=$ $\{x \in [a,b] : f(x) = u(x)\}$, then d satisfies (1.4) and (1.7).*

(iii) *If U_n is a Haar subspace of $C[a,b]$ and $f \in \tilde{C}(U_n)$ (defined in (1.13)), then d satisfies (1.4) and (1.7).*

Proof: (i) Let $f \in C(U_n)$ and $a \le x < y < \tilde{y} \le b$. Assume that $u, \tilde{u} \in U_n$ such that

$$||f - u||_{[x,y]} = d(f, U_n, [x,y]), \quad ||f - \tilde{u}||_{[x,\tilde{y}]} = d(f, U_n, [x,\tilde{y}]).$$

Then by Theorem 2.2,

$$d(x,y) = d(f, U_n, [x,y]) \le d(f, U_n, [x,\tilde{y}]) = d(x,\tilde{y})$$

which implies (1.4).

(ii) Under the same hypotheses as in (i) we have that $||f - u||_{[x,y]} \le ||f - \tilde{u}||_{[x,\tilde{y}]}$. Since $\mu(Z(f - \tilde{u})) = 0$, we obtain $\int_{[y,\tilde{y}]} |f - \tilde{u}| \, d\mu > 0$ and therefore,

$$d(x,\tilde{y}) = ||f - \tilde{u}||_{[x,\tilde{y}]} > ||f - \tilde{u}||_{[x,y]} \ge ||f - u||_{[x,y]} = d(x,y)$$

which implies (1.7).

(iii) If U_n is a Haar space, then $f \in \tilde{C}(U_n)$ satisfies the hypotheses in (ii), and therefore the statement follows. ∎

We conjecture that the converse of each statement of Theorem 2.3 is also true, but we can only prove it for $U_n = \Pi_{n-1}$ and a subclass of $C(\Pi_{n-1})$.

Theorem 2.4. *Let $U_n = \Pi_{n-1}$ and $f \in C[a,b] \cap C^n(a,b)$. Then the following statements hold:*

(i) *If d satisfies (1.4), then $f \in C(U_n)$.*

(ii) *If d satisfies (1.4) and (1.7), then $f \in \tilde{C}(U_n)$.*

Proof: (i) We show that $\sigma f^n(x) \geq 0$ for all $x \in (a,b)$, where $\sigma \in \{-1,1\}$. Then by the proof of Lemma 1.3, $f \in C(U_n)$. Let $[x,y] \subset (a,b)$, $x < y$ and $x = t_o < \cdots < t_n < t_{n+1} = y$ be the (unique) μ-canonical points for Π_{n-1} on $[x,y]$. Assume that $u \in \Pi_{n-1}$ such that $u(t_i) = f(t_i)$, $1 \leq i \leq n$. Then by a well-known interpolation result,

$$
\begin{aligned}
d(x,y) =& |L(f, U_n, [x,y])| = |L(f - u, U_n, [x,y])| \\
=& \left| \sum_{j=1}^{n+1} (-1)^j \int_{t_{j-1}}^{t_j} (f(t) - u(t)) d\mu(t) \right| \\
=& \left| \sum_{j=1}^{n+1} (-1)^j \int_{t_{j-1}}^{t_j} (t - t_1) \cdot \ldots \cdot (t - t_n) f^{(n)}(z(t))/n! \, d\mu(t) \right|,
\end{aligned}
\tag{2.5}
$$

where $z(t) \in [x,y]$ for all $t \in [x,y]$. Now assume that $f^{(n)} < 0$ in $[x,y] \subset (a,b)$, and $f^{(n)} > 0$ in $[\tilde{x}, \tilde{y}] \subset (a,b)$. Then, since sgn $((t - t_1) \cdots (t - t_n)) = (-1)^{n+1-j}$, if $t \in (t_{j-1}, t_j)$, $1 \leq j \leq n+1$, it follows from the above arguments that

$$
\begin{aligned}
& \text{sgn } L(f, U_n, [x,y]) \\
&= \text{sgn } \left[\sum_{j=1}^{n+1} \int_{t_{j-1}}^{t_j} (-1)^j (t - t_1) \cdot \ldots \cdot (t - t_n) f^n(z(t))/n! d\mu(t) \right] \\
&= (-1)^n,
\end{aligned}
$$

and analogously,

$$
\text{sgn } L(f, U_n, [\tilde{x}, \tilde{y}]) = (-1)^{n+1}.
$$

Without loss of generality, we may assume that $y < \tilde{x}$. Since d satisfies (1.4), we have

$$
|L(f, U_n, [x,y])| \leq |L(f, U_n, [x, \hat{y}])|
$$

for all $y \leq \hat{y} \leq \tilde{y}$. This implies that sgn $L(f, U_n, [x, \tilde{y}]) = (-1)^n$, because otherwise by continuity of L, $L(f, U_n, [x, \bar{y}]) = 0$ for some $\bar{y} \in (y, \tilde{y}]$. Then $d(x,y) > d(x, \bar{y}) = 0$, a contradiction of (1.4), since $\bar{y} > y$. Thus we have shown that sgn $L(f, U_n, [x, \tilde{y}]) = (-1)^n$. Also, sgn $L(f, U_n, [\tilde{x}, \tilde{y}]) = (-1)^{n+1}$.

Now arguing analogously as above we arrive again at a contradiction. This implies that $\sigma f^{(n)} \geq 0$ on (a, b), where $\sigma \in \{-1, 1\}$.

(ii) It follows from (i) that $\sigma f^{(n)} \geq 0$ on (a, b), where $\sigma \in \{-1, 1\}$. In addition to this we show that $f^{(n)}$ does not vanish identically on some $[x, y] \subset (a, b)$, $x < y$. Then by Lemma 1.3, $f \in \tilde{C}(\Pi_{n-1})$. Assume that $f^{(n)}(t) = 0$ for all $t \in [x, y] \subset (a, b)$, $x < y$. Hence by (2.5), $d(\tilde{x}, \tilde{y}) = 0$ for all $x \leq \tilde{x} < \tilde{y} \leq y$, a contradiction of hypothesis (1.7). ∎

Remark 2.5. (i) If $f \in C(U_n)$, then by Theorem 2.2 the minimal deviation of f from U_n on any interval $[x, y]$ is given by the function d in (2.4). Hence in this case the algorithm in [5] easily computes an optimal set of knots for best segment $L_1(\mu)$-approximation by using the simple function d. Thus the results in Theorem 2.3 and 2.4 seem to be of more theoretical interest.

But in numerical examples it turns out that also in the case when $f \notin C(U_n)$, the two-phase method of the algorithm developed in [5] works very well, where in phase one a leveled set of knots is computed w.r.t. the function d in (2.4). In phase two the minimal deviation of f from U_n on each knot interval is determined by an algorithm given in Watson [7].

(ii) If $U_n = \Pi_{n-1}$ and $\mu = \lambda$, the Lebesgue measure, then by a well-known result of Bernstein, $t_j = \cos(j\pi/(n+1))$, $1 \leq j \leq n$, are the λ-canonical points for Π_{n-1} on $[-1, 1]$.

Appendix

In Section 1 we made the conjecture that if $L(u, U_n, [a, b], T) = 0$ for every $u \in U_n$, where the λ_i are independent of $[a, b]$, and for every T, then $U_n = \Pi_{n-1}$. Recently Professor A. Pinkus and his colleague Professor V. Lin (Haifa) made a stronger conjecture and were able to prove it. I would like to thank them for allowing me to give their proof.

Theorem (Pinkus, Lin). *Let any partition $-1 = \tilde{t}_0 < \cdots < \tilde{t}_n = 1$ be given. Moreover, for any $-1 \leq a < b \leq 1$, set*

$$t_i = ((b - a)\tilde{t}_i + (b + a))/2, \quad 0 \leq i \leq n.$$

Assume there exist $\lambda_0, \ldots, \lambda_n$ such that $\sum_{i=0}^{n} |\lambda_i| \neq 0$ and

$$\sum_{i=0}^{n} \lambda_i u(t_i) = 0$$

for some $u \in C[-1,1]$ and for every $-1 \leq a < b \leq 1$. Then $u \in \Pi_{n-1}$.

Proof: Set $s = (b-a)/2$, $r = (b+a)/2$, $-1 \leq a < b \leq 1$. Then the domain of (s,r) is $0 < s \leq 1$, $|r| \leq 1 - s$. Let $0 < \epsilon < 1$. Then by hypothesis,

$$0 = \sum_{i=0}^{n} \lambda_i u(s\tilde{t}_i + r),$$

$0 < s \leq \epsilon$, $|r| \leq 1 - \epsilon$. Let $\varphi \in C^{\infty}(\mathbb{R})$ with support in $(-1+\epsilon, 1-\epsilon)$. Then

$$0 = \sum_{i=0}^{n} \lambda_i u(s\tilde{t}_i + r)\varphi(r).$$

Integrating (and setting $u = 0$ outside $[-1,1]$) we obtain

$$0 = \sum_{i=0}^{n} \lambda_i \int_{-\infty}^{\infty} u(s\tilde{t}_i + r)\varphi(r)dr,$$

$0 < s \leq \epsilon$. By a change of variable,

$$0 = \sum_{i=0}^{n} \lambda_i \int_{-\infty}^{\infty} u(x)\varphi(x - s\tilde{t}_i)dx,$$

$0 < s \leq \epsilon$. Differentiating k times ($k \in \mathbb{N}$) with respect to s, and setting $s = 0$, we obtain

$$0 = \left(\sum_{i=0}^{n} \lambda_i \tilde{t}_i^k\right) \int_{-\infty}^{\infty} u(x)\varphi^{(k)}(x)dx.$$

Since the λ_i are not all zero, there exists a $k \in \{0, 1, \ldots, n\}$ such that

$$\sum_{i=0}^{n} \lambda_i \, \tilde{t}_i^k \neq 0.$$

Hence for this k,

$$\int_{-\infty}^{\infty} u(x)\varphi^{(k)}(x)dx = 0.$$

This holds for all $\varphi \in C^{\infty}(\mathbb{R})$ with compact support S_{φ} in $(-1+\epsilon, 1-\epsilon)$. For every $m \geq 2$ ($m \in \mathbb{N}$), let $\varphi_m \in C^{\infty}(\mathbb{R})$ such that $S_{\varphi_m} = [-1/m, 1/m]$,

$\varphi_m \geq 0$ on S_{φ_m}, and $\int_{-\infty}^{\infty} \varphi_m(x)dx = 1$ (for existence see e.g. [9, p. 1640]). Define for $z \in T_m := [-1 + 2/m, 1 - 2/m]$,

$$u_m(t) := \int_{-\infty}^{\infty} u(x)\varphi_m(t - x)dx.$$

Then $\lim_{m\to\infty} u_m = u$ on every subinterval $[\alpha, \beta]$ of $(-1, 1)$. Moreover, it follows from the above arguments that

$$u_m^{(k)}(t) = \int_{-\infty}^{\infty} u(x)\frac{\partial^k}{\partial t^k}\varphi_m(t - x)dx = 0$$

for every $t \in T_m$, if $1/m > \epsilon$. Therefore, $u_m \in \Pi_{k-1}$ on T_m. Since $\epsilon > 0$ can be chosen sufficiently small, $u_m \in \Pi_{k-1}$ on T_m for every $m \geq 2$. Hence $u \in \Pi_{k-1}$ on every $[\alpha, \beta] \subset (-1, 1)$. Then $u \in C[-1, 1]$ implies that $u \in \Pi_{k-1}$ on $[-1, 1]$. ∎

References

1. Kroó, A., D. Schmidt, and M. Sommer, On A-spaces and their relation to the Hobby-Rice Theorem, Constr. Approx. **7** (1991), 329–339.

2. Lawson, C. L., Characteristic properties of the segmented rational minimax approximation problem, Numer. Math. **6** (1964), 293–301.

3. Meinardus, G., *Approximation of Functions: Theory and Numerical Methods*, Springer-Verlag, Berlin, 1967.

4. Meinardus, G., G. Nürnberger, M. Sommer, and H. Strauss, Algorithms for piecewise polynomials and splines with free knots, Math. of Computat. **53** (1989), 235–247.

5. Nürnberger, G., M. Sommer, and H. Strauss, An algorithm for segment approximation, Numer. Math. **48** (1986), 463–477.

6. Pinkus, A., *On L^1-Approximation*, Cambridge University Press, 1989.

7. Watson, G. A., An algorithm for linear L_1-approximation of continuous functions, IMA J. Numer. Anal. **1** (1981), 157–167.

8. Zielke, R., *Discontinuous Čebyšev Systems*, Lecture Notes in Mathematics 707, Springer-Verlag, Berlin, 1979.

9. Dunford, N. and J. T. Schwartz, *Linear Operators, Part II*, Interscience Publishers, New York, 1963.

10. Pinkus A., Private communication.

Numerical Methods of Approximation Theory, Vol. 9

Dietrich Braess and Larry L. Schumaker (eds.), pp. 347–357.

International Series of Numerical Mathematics, Vol. 105

Construction of Monotone Extensions to Boundary Functions

Cornelis Traas

Dedicated to the memory of Lothar Collatz

Abstract. Suppose we are given monotonically increasing, smooth, univariate functions along the edges of the unit square. The problem is to construct an extension $F(x, y)$ to the whole square which is monotone and of class C^1. A nonlinear method is presented which defines F in terms of a set of level lines, each of which is represented as a cubic Bézier curve. As the level changes, the corresponding control points shift along trajectories which contain appropriate kinks.

§1. Introduction

Let monotonically increasing smooth univariate functions be given along the edges of the unit square such that the combination of the four functions is continuous around the square. We suppose that the global minimum and the global maximum are assumed at the points $(0, 0)$ and $(1, 1)$, respectively. The problem is to construct a C^1 extension $F(x, y)$ of these boundary functions to the whole square which is monotone in the sense that $\partial F/\partial x > 0$, $\partial F/\partial y > 0$ for every point in the square. Of course, the given boundary functions must be compatible with such F.

The problem has been investigated in [1], where it is shown that, in general, there does not exist a linear method for constructing F. They present a nonlinear method in which F is defined implicitly in terms of a set of straight level lines. However, this method only gives a continuous F. Moreover, since the level lines which intersect the corners $(0,1)$ and $(1,0)$ are, in general, not compatible with the given gradients of F at those points, the discontinuity in grad F can be quite severe. In addition, the authors give a (non-constructive) proof of the existence of a monotone extension F which possesses the same order of smoothness as the boundary functions.

In the present article, a nonlinear method is presented for constructing a monotone extension F of class C^1. In this method, F is defined implicitly in terms of a set of level lines which are represented as cubic Bézier curves. The control points $\boldsymbol{b}_0$ and $\boldsymbol{b}_3$ of these curves are located on the edges of the square, while $\boldsymbol{b}_1$ and $\boldsymbol{b}_2$ are prescribed to lie on certain chosen curves *inside* the square. The control polygons should be non-increasing in order to ensure the monotonicity of F.

The chosen curves inside the square contain kinks in order to ensure the C^1 property of F. These kinks compensate for the corners $(1,0)$ and $(0,1)$ in the boundary along which the control points $\boldsymbol{b}_0$ and $\boldsymbol{b}_3$ move.

§2. Constructing a C^1 Extension

We use the notation $F(x,y)$ for the extension which is to be constructed. The components of the gradient of F are written as

$$p := \frac{\partial F}{\partial x} \quad \text{and} \quad q := \frac{\partial F}{\partial y}.$$

The boundary functions are f_1, f_2, f_3 and f_4, along the four edges, respectively, of the unit square. Here $F(x,0) = f_1(x)$, $F(1,y) = f_2(y)$, $F(x,1) = f_3(x)$ and $F(0,y) = f_4(y)$.

Furthermore $p_1 = df_1/dx$, $q_2 = df_2/dy$, etc. Obviously, p_1, q_2, p_3 and q_4 are known functions. Once the functions $q_1 := (\partial F/\partial y)(\cdot,0), p_2, q_3$ and p_4 have been defined, consistent with their known boundary values ($q_1(0) = q_4(0)$, $q_1(1) = q_2(0)$, etc.), the directions of the level lines are known everywhere on the boundary of the square.

We consider the situation in which $F(0,1) > F(1,0)$. The case $F(1,0) > F(0,1)$ can be treated symmetrically, and the treatment of the special case $F(0,1) = F(1,0)$ follows directly from the first case.

On the y-axis there exists a point $0 < y^* < 1$ such that $f_4(y^*) = F(1,0)$, and on the line $x = 1$ there exists a point y^{**} with $0 < y^{**} < 1$ such that $f_2(y^{**}) = F(0,1)$.

In the present construction, which defines $F(x, y)$ implicitly in terms of level lines, the level lines are represented as cubic Bézier curves with respect to some parameter. The control points $\boldsymbol{b}_1$ and $\boldsymbol{b}_2$ are prescribed to lie on certain chosen curves, $k_1(x, y) = 0$ and $k_2(x, y) = 0$, respectively, inside the square. Each of these chosen curves contains two kinks. The first kink lies on the control polygon of the level line passing through the point $(1, 0)$, and the second kink on the control polygon of the level line passing through the point $(0, 1)$. The control points $\boldsymbol{b}_0$ and $\boldsymbol{b}_3$ always lie on the edges of the square; $\boldsymbol{b}_0$ on the lower and right edges. We introduce a parameter u along the x-axis, defined by $u = x$ for every point $(x, 0)$. (This may seem redundant. However, the symbol x has a meaning everywhere in the square; the symbol u, by definition, has a meaning on the x-axis only). Similarly, we define a parameter v along the y-axis, with $v = y$, and a parameter w along the line $x = 1$, with $w = y$. Also, $v^* := y^*$ and $w^* := y^{**}$.

2.1. Defining the Kinks

We write the components of the control point $\boldsymbol{b}_i$ as $b_{i,x}$ and $b_{i,y}$, respectively. Assume $b_{0,x} < 1$. Then the control point $\boldsymbol{b}_1$ can be found as a solution to the system:

$$\left.\begin{aligned} p_1(u)b_{1,x} + q_1(u)b_{1,y} &= up_1(u) \\ k_1(b_{1,x}, b_{1,y}) &= 0, \end{aligned}\right\} \tag{1}$$

where $u = b_{0,x}$ is the point of intersection of the level line with the x-axis. The point of intersection with the y-axis is $v = b_{3,y}$. In (1) the function q_1 is assumed to have been defined.

Differentiating (1) with respect to u we find $d\boldsymbol{b}_1/du$, expressed in terms of $q_1'(u)$, $\partial_1 k_1$ and $\partial_2 k_1$, among others. The control point $\boldsymbol{b}_2$ can be found as a solution to the system

$$\left.\begin{aligned} p_4(v)b_{2,x} + q_4(v)b_{2,y} &= vq_4(v) \\ k_2(b_{2,x}, b_{2,y}) &= 0. \end{aligned}\right\} \tag{2}$$

Here the function p_4 is assumed to have been defined.

Differentiating (2) with respect to v, we find $d\boldsymbol{b}_2/dv$, expressed in terms of $p_4'(v)$, $\partial_1 k_2$ and $\partial_2 k_2$. The relation between v and u is $f_4(v) = f_1(u)$, and thus

$$\frac{d\boldsymbol{b}_2}{du} = \frac{p_1(u)}{q_4(v)} \frac{d\boldsymbol{b}_2}{dv}.$$

All level lines which are labelled with u belong to one family of lines, and thus define a smooth surface (the other functions which are relevant in this context are assumed smooth). We also consider an "extended" level line, i.e., a level line for which $1 < u < \bar{u}$, for some fixed $\bar{u}$, $\bar{u} - 1 << 1$. This one also belongs to the above family of lines. The idea is to use the intersection point of the extended level line with the line $x = 1$ as a subdivision point, generating new control points inside the square. These new control points define the kinks in the lines $k_1 = 0$ and $k_2 = 0$ at the location corresponding to $u = 1$. The second parts of k_1 and k_2 can now be defined freely, taking into account the new directions found in the kinks.

2.2. The Subdivision Process

Let the level line

$$\boldsymbol{x}(t) = \sum_{i=0}^{3} \boldsymbol{b}_i B_i^3(t)$$

intersect the x-axis at the point $(u, 0)$, $1 < u < \bar{u}$, $\bar{u} - 1 << 1$. The B_i^3 are the cubic Bernstein polynomials and $t \in [0, 1]$ is some parameter. This level line intersects the line $x = 1$ at

$$y = \frac{u - 1}{1 - b_{1,x}(1)} b_{1,y}(1) + \mathcal{O}(u - 1)^2,$$

where the parameter value t is

$$t = \frac{u - 1}{3(1 - b_{1,x}(1))} + \mathcal{O}(u - 1)^2.$$

The direction of the tangent in $(1, y)$ is defined by

$$D_t \boldsymbol{x}(t) = 3\left(\boldsymbol{b}_1(u) - \boldsymbol{b}_0(u)\right) + 6t\left(\boldsymbol{b}_2(u) - 2\boldsymbol{b}_1(u) + \boldsymbol{b}_0(u)\right) + \mathcal{O}(t^2)$$

where $\boldsymbol{b}_0(u) = \binom{u}{0}$. Expanding the right hand member into powers of $u - 1$ with respect to the point $(1, 0)$, gives

$$D_t \boldsymbol{x}(t) = 3\left(\boldsymbol{b}_1(1) - \binom{1}{0}\right) + 3(u - 1)\left(\frac{d\boldsymbol{b}_1}{du}(1) - \binom{1}{0}\right)$$
$$+ 6t\left(\boldsymbol{b}_2(1) - 2\boldsymbol{b}_1(1) + \binom{1}{0}\right) + \mathcal{O}(t^2).$$

Since the direction of this tangent is also fixed by

$$\binom{-q_2(y)}{p_2(y)} = \binom{-q_2(0)}{p_2(0)} + y\binom{-dq_2(0)/dy}{dp_2(0)/dy} + \mathcal{O}(y^2),$$

in view of consistency of directions, we can derive relation (3). From the first order equality,

$$(u - 1)\left(\frac{d\boldsymbol{b}_1}{du}(1) - \begin{pmatrix} 1 \\ 0 \end{pmatrix}\right) + 2t\left(\boldsymbol{b}_2(1) - 2\boldsymbol{b}_1(1) + \begin{pmatrix} 1 \\ 0 \end{pmatrix}\right)$$

$$= \frac{yb_{1,y}(1)}{p_2(0)}\begin{bmatrix} -dq_2(0)/dy \\ dp_2(0)/dy \end{bmatrix}, \qquad \forall u \in (1, \bar{u}), \quad 0 < \bar{u} - 1 << 1,$$

it follows that

$$\frac{d\boldsymbol{b}_1}{du}(1) - \begin{pmatrix} 1 \\ 0 \end{pmatrix} + \frac{2}{3(1 - b_{1,x}(1))}\left(\boldsymbol{b}_2(1) - 2\boldsymbol{b}_1(1) + \begin{pmatrix} 1 \\ 0 \end{pmatrix}\right)$$

$$= \frac{b_{1,y}^2(1)}{p_2(0)(1 - b_{1,x}(1))}\begin{pmatrix} -dq_2(0)/dy \\ dp_2(0)/dy \end{pmatrix} \tag{3}$$

From the first component of this vector relation, we compute $db_{1,x}(1)/du$, and then

$$\frac{db_{1,y}}{du}(1) = -\frac{\partial_1 k_1(\boldsymbol{b}_1(1))}{\partial_2 k_1(\boldsymbol{b}_1(1))} \cdot \frac{db_{1,x}}{du}(1),$$

assuming the slope of $k_1 = 0$ has been chosen to be finite and positive.

Since $d\boldsymbol{b}_1/du$ is related to dq_1/du (see the result of differentiating the first equation in (1)), we can compute $dq_1(1)/du = (\partial^2 F/\partial x \partial y)(1, 0)$.

On the other hand, from the second component of (3) we can compute $dp_2(0)/dy = (\partial^2 F/\partial y \partial x)(1, 0)$. In general we will get two different values for the twist at the corner point $(1, 0)$: F_{xy} and F_{yx}. We can influence the difference by modifying the slope $\partial_1 k_1(\boldsymbol{b}_1(1))/\partial_2 k_1(\boldsymbol{b}_1(1))$. The sum of both, i.e., $q_1'(1) + p_2'(0)$, is independent of this slope. (Later on, the two twist values will serve as boundary conditions in the actual definitions of the functions q_1 and p_2, respectively).

Subdivision of the level line with respect to the point $(1, y)$ gives new control points

$$\boldsymbol{b}_0^*(u) = \begin{pmatrix} 1 \\ y \end{pmatrix}$$

$$\boldsymbol{b}_1^*(u) = \boldsymbol{b}_1(u) + 2t(\boldsymbol{b}_2(u) - \boldsymbol{b}_1(u)) + \mathcal{O}(t^2)$$

$$= \boldsymbol{b}_1(1) + (u - 1)\frac{d\boldsymbol{b}_1(1)}{du} + 2t(\boldsymbol{b}_2(1) - \boldsymbol{b}_1(1)) + \mathcal{O}(t^2) \tag{4}$$

$$= \boldsymbol{b}_1(1) + (u - 1)\left\{\frac{d\boldsymbol{b}_1}{du}(1) + \frac{2}{3}\frac{\boldsymbol{b}_2(1) - \boldsymbol{b}_1(1)}{1 - b_{1,x}(1)}\right\} + \mathcal{O}(u - 1)^2$$

$$
\begin{aligned}
\boldsymbol{b}_2^*(u) &= \boldsymbol{b}_2(u) + t(\boldsymbol{b}_3(u) - \boldsymbol{b}_2(u)) + \mathcal{O}(t^2) \\
&= \boldsymbol{b}_2(1) + (u-1)\frac{d\boldsymbol{b}_2}{du}(1) + t(\boldsymbol{b}_3(1) - \boldsymbol{b}_2(1)) + \mathcal{O}(t^2) \\
&= \boldsymbol{b}_2(1) + (u-1)\left\{ \frac{d\boldsymbol{b}_2}{du}(1) + \frac{1}{3}\frac{\boldsymbol{b}_3(1) - \boldsymbol{b}_2(1)}{1 - b_{1,x}(1)} \right\} + \mathcal{O}(u-1)^2.
\end{aligned}
\tag{5}
$$

The expressions between the curly brackets in (4) and (5) define the directions of the second part of $k_1 = 0$ at the point $\boldsymbol{b}_1(1)$, and of the second part of $k_2 = 0$ at the point $\boldsymbol{b}_2(1)$, respectively.

We next consider the level line passing through the corner $(0,1)$ and intersecting the line $x = 1$ at the point $(1, w^*)$. We also consider a neighbouring "extended" level line intersecting the y-axis at $(0, v)$, $1 < v < \bar{v}$, for some fixed $\bar{v}$, $\bar{v} - 1 \ll 1$. The associated control point $\boldsymbol{b}_2$ is found as a solution to the system (2), where k_2 now pertains to the second part of this curve. By differentiating we find $d\boldsymbol{b}_2/dv$ expressed in terms of $p_4'(v)$, $\partial_1 k_2$ and $\partial_2 k_2$. The relation between v and w is $f_4(v) = f_2(w)$, and thus

$$
\frac{d\boldsymbol{b}_2}{dw} = \frac{q_2(w)}{q_4(v)}\frac{d\boldsymbol{b}_2}{dv}.
$$

The control point $\boldsymbol{b}_1$ is found as a solution to the system

$$
\left.
\begin{aligned}
p_2(w)b_{1,x} + q_2(w)b_{1,y} &= p_2(w) + wq_2(w) \\
k_1(b_{1,x}, b_{1,y}) &= 0.
\end{aligned}
\right\}
\tag{6}
$$

By differentiating we find $d\boldsymbol{b}_1/dw$ expressed in terms of $p_2'(w)$, $\partial_1 k_1$ and $\partial_2 k_1$. The level line intersects the line $y = 1$ at

$$
x = \frac{v-1}{1 - b_{2,y}(1)} + \mathcal{O}(v-1)^2,
$$

where the parameter value t is

$$
t = 1 - \frac{v-1}{3(1 - b_{2,y}(1))} + \mathcal{O}(v-1)^2.
$$

The direction of the tangent at $(x, 1)$ is defined by

$$
D_t\boldsymbol{x}(t) = 3(\boldsymbol{b}_3(v) - \boldsymbol{b}_2(v) - 2(1-t)(\boldsymbol{b}_3(v) - 2\boldsymbol{b}_2(v) + \boldsymbol{b}_1(v))) + \mathcal{O}(1-t)^2,
$$

where $\boldsymbol{b}_3(v) = \binom{0}{v}$.

Expanding this into powers of $v - 1$ with respect to the point $(0, 1)$, and noticing that this direction should correspond with the direction defined by $(-q_3(x) \ p_3(x))^T$, we can derive the following relation in view of consistency of directions,

$$\begin{pmatrix} 0 \\ 1 \end{pmatrix} - \frac{d\boldsymbol{b}_2}{dv}(1) - \frac{2}{3(1 - b_{2,y}(1))} \left[\begin{pmatrix} 0 \\ 1 \end{pmatrix} - 2\boldsymbol{b}_2(1) + \boldsymbol{b}_1(1) \right]$$

$$= \frac{b_{2,x}^2(1)}{q_3(0)(1 - b_{2,y}(1))} \begin{pmatrix} -q_3'(0) \\ p_3'(0) \end{pmatrix}. \tag{7}$$

From the second component of (7) we compute $db_{2,y}(1)/dv$, and then

$$\frac{db_{2,x}}{dv}(1) = -\frac{\partial_2 k_2(\boldsymbol{b}_2(1))}{\partial_1 k_2(\boldsymbol{b}_2(1))} \frac{db_{2,y}}{dv}(1).$$

Since $d\boldsymbol{b}_2/dv$ is related to dp_4/dv, we can compute $dp_4(1)/dv = (\partial^2 F/\partial y \partial x)(0, 1)$. From the first component of (7) we can compute $dq_3(0)/dx = (\partial^2 F/\partial x \partial y)(0, 1)$.

Subdivision of the level line with respect to the point $(x, 1)$ gives the new control points

$$\boldsymbol{b}_3^*(v) = \begin{pmatrix} x \\ 1 \end{pmatrix}$$

$$\boldsymbol{b}_2^*(v) = \boldsymbol{b}_2(1) + (v - 1)\left\{ \frac{d\boldsymbol{b}_2}{dv}(1) + \frac{2}{3} \frac{\boldsymbol{b}_1(1) - \boldsymbol{b}_2(1)}{1 - b_{2,y}(1)} \right\} + \mathcal{O}(v - 1)^2 \tag{8}$$

$$\boldsymbol{b}_1^*(v) = \boldsymbol{b}_1(1) + (v - 1)\left\{ \frac{d\boldsymbol{b}_1}{dv}(1) + \frac{1}{3} \frac{\boldsymbol{b}_0(1) - \boldsymbol{b}_1(1)}{1 - b_{2,y}(1)} \right\} + \mathcal{O}(v - 1)^2, \tag{9}$$

where $\boldsymbol{b}_0(1) = \begin{pmatrix} 1 \\ w_* \end{pmatrix}$ (v is here taken as argument for all $\boldsymbol{b}_i$).

The expressions between the curly brackets in (8) and (9) define the directions of the third part of $k_2 = 0$ at the point $\boldsymbol{b}_2(1)$, and of the third part of $k_1 = 0$ at the point $\boldsymbol{b}_1(1)$, respectively.

2.3. Nonintersection of Successive Level Lines

In general, it may occur that successive level lines mutually intersect. In this section we derive sufficient conditions for nonintersection. A sufficient condition is that the curves $k_1 = 0$ and $k_2 = 0$ are monotonically increasing functions. Considering (4), we conclude that the following should hold:

$$\frac{db_{1,x}}{du}(1) + \frac{2}{3}\frac{b_{2,x}(1) - b_{1,x}(1)}{1 - b_{1,x}(1)} > 0.$$

From (3) it follows that

$$\frac{db_{1,x}}{du}(1) + \frac{2}{3}\frac{b_{2,x}(1) - b_{1,x}(1)}{1 - b_{1,x}(1)} = \frac{1}{3} - \frac{b_{1,y}^2(1)}{p_2(0)(1 - b_{1,x}(1))}\frac{dq_2}{dy}(0).$$

For $dq_2(0)/dy > 0$, the inequality thus reads

$$\frac{3b_{1,y}^2(1)}{1 - b_{1,x}(1)} < \frac{p_2(0)}{dq_2(0)/dy}.$$

Since

$$\frac{b_{1,y}(1)}{1 - b_{1,x}(1)} = \frac{p_2(0)}{q_2(0)},$$

we can rewrite the inequality as

$$b_{1,y}(1) < \frac{1}{3}\frac{q_2(0)}{dq_2(0)/dy}. \tag{10}$$

For $dq_2(0)/dy \le 0$, no constraint is put on $b_{1,y}(1)$ within the context of this analysis. A suitable $b_{1,y}(1)$ can thus always be chosen.

Considering (5), we conclude that the following should hold:

$$\frac{db_{2,x}}{du}(1) + \frac{1}{3}\frac{b_{3,x}(1) - b_{2,x}(1)}{1 - b_{1,x}(1)} > 0, \tag{11}$$

where $b_{3,x}(1) = 0$. From differentiating (2) with respect to u, we find

$$\frac{db_{2,x}}{du}(1) = \frac{\partial_2 k_2(\boldsymbol{b}_2(1))}{p_4(v^*) \cdot \partial_2 k_2(\boldsymbol{b}_2(1)) - q_4(v^*) \cdot \partial_1 k_2(\boldsymbol{b}_2(1))}\frac{dv}{du}(1).$$

$$\cdot \left[q_4(v^*) + v^*\frac{dq_4}{dv}(v^*) - \frac{dp_4}{dv}(v^*)b_{2,x}(1) - \frac{dq_4}{dv}(v^*)b_{2,y}(1) \right]. \tag{12}$$

Since $\frac{dp_4}{dv}(v^*)$ and the slope $\partial_1 k_2/\partial_2 k_2$ can be chosen freely, it is always possible to satisfy the inequality, in view of the character of (12) and as a consequence of earlier choices.

With regard to the direction of $k_1 = 0$ after the second kink, we consider (9). The following should hold:

$$\frac{db_{1,y}}{dv}(1) + \frac{1}{3}\frac{w^* - b_{1,y}(1)}{1 - b_{2,y}(1)} > 0. \tag{13}$$

The control point $\boldsymbol{b}_1$ is obtained as a solution to the system (6). Differentiation with respect to v gives

$$\frac{db_{1,y}}{dv}(1) = -\partial_1 k_1(\boldsymbol{b}_1(1)) \cdot [p_2'(w^*) + q_2(w^*) + w^* q_2'(w^*)$$

$$-p_2'(w^*)b_{1,x}(1) - q_2'(w^*)b_{1,y}(1)]\,\frac{q_4(1)}{q_2(w^*)}\Big/$$

$$\{p_2(w^*) \cdot \partial_2 k_1(\boldsymbol{b}_1(1)) - q_2(w^*) \cdot \partial_1 k_1(\boldsymbol{b}_1(1))\}. \tag{14}$$

Since $\frac{dp_2}{dw}(w^*)$ and the slope $\partial_1 k_1/\partial_2 k_1$ can be chosen freely, it is always possible to satisfy the inequality.

With regard to $k_2 = 0$, the following should hold (see(8)):

$$\frac{db_{2,y}}{dv}(1) + \frac{2}{3}\frac{b_{1,y}(1) - b_{2,y}(1)}{1 - b_{2,y}(1)} > 0.$$

From (7) follows

$$\frac{db_{2,y}}{dv}(1) + \frac{2}{3}\frac{b_{1,y}(1) - b_{2,y}(1)}{1 - b_{2,y}(1)} = \frac{1}{3} - \frac{b_{2,x}^2(1)}{q_3(0)(1 - b_{2,y}(1))}\frac{dp_3}{dx}(0).$$

For $dp_3(0)/dx > 0$ the inequality thus reads

$$\frac{3b_{2,x}^2(1)}{1 - b_{2,y}(1)} < \frac{q_3(0)}{dp_3(0)/dx}.$$

Since $\frac{b_{2,x}(1)}{1 - b_{2,y}(1)} = \frac{q_3(0)}{p_3(0)}$, we can rewrite the inequality as

$$b_{2,x}(1) < \frac{1}{3}\frac{p_3(0)}{dp_3(0)/dx}. \tag{15}$$

For $dp_3(0)/dx \leq 0$ no constraint is put on $b_{2,x}(1)$. A suitable $b_{2,x}(1)$ can thus always be found.

§2.4 Implementing the Construction.

As stated above, the functions f_1, f_2, f_3 and f_4 are known, and the gradient components q_1, p_2, q_3 and p_4 have to be chosen as positive functions along the edges of the square. These choices are subject to certain constraints as described above.

The first step towards an actual construction is the determination of the special points v^* (on the y-axis) and w^* (on the line $x = 1$) from the equations

$$f_4(v^*) = F(1,0)$$
$$f_2(w^*) = F(0,1).$$

Next there are three main steps:

1) Compute everything which is associated with the level line passing through v^* and $(1,0)$.

2) Do the same for the level line passing through $(0,1)$ and w^*.

3) Define all other level lines.

In the first main step, values are chosen for $b_{1,y}(u = 1)$, $b_{2,y}(u = 1)$ or $b_{2,x}(u = 1)$, $p_4(v^*)$, the slopes of the first parts of $k_1 = 0$ and $k_2 = 0$ at their end points, and $p_4'(v^*)$, subject to certain constraints as described above. All other quantities, in particular the directions of the second parts of $k_1 = 0$ and $k_2 = 0$ at their initial points, and the two twist values at $(1,0)$, can then be computed.

In the second main step, values are chosen for $b_{2,x}(v = 1)$, $b_{1,x}(v = 1)$, $p_2(w^*)$, the slopes of the second parts of $k_1 = 0$ and $k_2 = 0$ at their end points, and $p_2'(w^*)$. All other quantities, in particular the directions of the third parts of $k_1 = 0$ and $k_2 = 0$ at their initial points, and the two twist values in $(0,1)$, can be computed.

In the third main step, define $k_1 = 0$ and $k_2 = 0$ as monotone increasing functions, taking into account the discontinuous slopes in the kinks. Define next the functions $q_1(x)$, $p_2(y)$, $q_3(x)$ and $p_4(y)$ as positive functions using the information about these functions defined and computed above. If the directions of the level lines at the lower and right-hand edges of the square are converging towards the interior of the square, then check whether the function $k_1 = 0$ is located entirely between the envelope of these directions and these edges. A similar remark applies for the upper and left-hand edges of the square.

A few experiments have been performed in order to test the above construction numerically. The analysis of the numerical results seems to confirm the monotonicity of the obtained $F(x,y)$ and its C^1 property.

§3. Conclusions

A method has been presented which allows actual construction of a monotone C^1 surface, given smooth monotone boundary functions. The available degrees of freedom allow a further control of the monotone surface, if desired. The method has been described on the basis of a unit square domain, but it is certainly not restricted to such a domain.

References

1. W. Dahmen, R.A. DeVore, C.A. Micchelli, On Monotone Extensions of Boundary Data, Numer. Math. 60 (1992) 477–492.

Department of Mathematics
University of Twente
7500 AE Enschede
The Netherlands
traas@math.utwente.nl

ISNM

A series with a long-standing reputation

Since its foundation in 1963 more than 100 volumes have been published by Birkhäuser Verlag in the **International Series of Numerical Mathematics**.

John Todd's *Introduction to the Constructive Theory of Functions*, published as Volume 1, was a remarkable start. Proceedings volumes and further monographs such as Fenyö/Frey, *Moderne mathematische Methoden in der Technik*, Ghizzetti/Ossicini, *Quadrature Formulae*, Todd, *Basic Numerical Mathematics* (two volumes) and Heinrich, *Finite Difference Methods on Irregular Networks* followed, always presenting the state of the art in exposition and research.

Originally the Editorial Board consisted of Ch. Blanc, A. Ghizzetti, A. Ostrowski, J. Todd, H. Unger, A. van Wijngaarden. Despite a number of changes, it has shown long years of continuity; Prof. Ostrowski and Prof. Henrici, for instance, had been members of the Board all their life.

At present the series is being edited by
Karl-Heinz Hoffmann, München, **Hans D. Mittelmann**, Tempe, **John Todd**, Pasadena.

As in the past, we do not intend to restrict the series a priori to certain subjects. The series is open to all aspects of numerical mathematics. At the same time, we wish to include practical applications in science and engineering, with emphasis on mathematical content.

Some of the topics of particular interest to the series are:
Free boundary value problems for differential equations, phase transitions, problems of optimal control and optimization, other nonlinear phenomena in analysis;
nonlinear partial differential equations, efficient solution methods, bifurcation problems;
approximation theory.

If possible, the topic of each volume should be discussed from three different angles, namely those of Mathematical Modelling, Mathematical Analysis, Numerical Case Studies.

The editors particularly welcome research monographs; furthermore, the series is to contain advanced graduate texts, dealing with areas of current research interest, as well as selected and carefully refereed proceedings of major conferences or workshops sponsored by various research centers. Historical material in these areas would also be considered.

We encourage preparation of manuscripts in LaTeX or AMSTeX for delivery in camera-ready copy which enables a rapid publication, or in electronic form for interfacing with laser printers or typesetters.